Internal Combustion Engines 2017

Birmingham, United Kingdom
6 - 7 November 2017

ISBN: 978-1-5108-7388-9

Printed from e-media with permission by:

Curran Associates, Inc.
57 Morehouse Lane
Red Hook, NY 12571

Some format issues inherent in the e-media version may also appear in this print version.

Printed by Curran Associates, Inc. (2019)

For permission requests, please contact Institution of Mechanical Engineers (IMechE) at the address below.

Institution of Mechanical Engineers (IMechE)
One Birdcage Walk
London, SW1H 9JJ
United Kingdom

library@Imeche.org
www.imeche.org

Additional copies of this publication are available from:

Curran Associates, Inc.
57 Morehouse Lane
Red Hook, NY 12571 USA
Phone: 845-758-0400
Fax: 845-758-2633
Email: curran@proceedings.com
Web: www.proceedings.com

CONTENTS

DIESEL ENGINES AND THEIR FUELS

POWERTRAIN DESIGN

GASOLINE ENGINES

VEHICLE AND SIMULATION

ORGANISING COMMITTEE

Powertrain Systems and Fuels Group,
The Institution of Mechanical Engineers

MEMBER CREDITS

Hua Zhao (Chair)	Brunel University London
Rik Alewijnse	AVL Powertrain UK
Frank Atzler	Continental Automotive
Choongsik Bae	Korea Advanced Institute of Science and Technology
Hugh Blaxill	MAHLE Powertrain
Brian Cooper	Jaguar Land Rover
Roger Cracknell	Shell
Colin Garner	Loughborough University
Roy Horrocks	Consultant
Yasuo Moriyoshi	Chiba University
Richard Osborne	Ricardo UK
Michael Richardson	Jaguar Land Rover
Paul Shayler	University of Nottingham
Khizer Tufail	Ford Motor Company
Steve Whelan	Eminox

DIESEL ENGINES AND THEIR FUELS

Effect of fuel quality on emissions over RDE and a comparison of RDE and CD test emissions from Euro 6 diesel cars

R Williams[1], H Hamje[2], J Andersson[3], P Ziman[1], D Rickeard[2], L Pellegrini[4], C Fittavolini[4]

[1] Shell Global Solutions, UK
[2] Concawe, Belgium
[3] Ricardo, UK
[4] ENI S.p.A., Italy

ABSTRACT

Real driving pollutant emissions from passenger cars have come under close scrutiny in recent times due to a perceived divergence between stated certification performance and measured on-road performance. This has been most pointed in the case of NOx emissions from diesel cars. These observations have fuelled the development and implementation of new regulatory test procedures to include the Chassis Dynamometer (CD) based World Harmonised Light Duty Test Procedure (WLTP) and the Real Driving Emissions (RDE) procedure. To build understanding of the emissions performance of latest standard diesel passenger cars, Concawe has conducted a study of the CD and RDE performance of three vehicle types with different exhaust gas after-treatment configurations. For each vehicle, three tests were conducted over each of the RDE, WLTC and NEDC cycles and three further RDE tests were run on a second fuel to build understanding of fuel effects in the new test. The data generated provide insights into the on-road emissions performance of latest diesel passenger cars and how this compares to regulated emissions limits and results from CD tests run under comparable road load conditions.

GLOSSARY

ASC	Ammonia Slip Catalyst
CD	Chassis Dynamometer
CF	Conformity Factor
CLEAR	Power binning correction tool for RDE data
CLD	Chemi-Luminescence Detector
CVS	Constant Volume Sampling
DCT	Dual Clutch Transmission
DEF	Diesel Exhaust Fluid: aqueous urea solution used as a reductant to reduce NOx emissions via Selective Catalytic Reduction
DI	Direct Injection
DPF	Diesel Particulate Filter
(HP & LP) EGR	High & Low Pressure Exhaust Gas Recirculation
EMROAD	Moving average CO2 correction tool for RDE
EPEFE	European Programme on Emissions, Fuels and Engine technologies
F1, F2	Fuel 1, Fuel 2
FAME	Fatty Acid Methyl Ester
FBP	Final Boiling Point
FID	Flame Ionisation Detector
IBP	Initial Boiling Point
IDI	Indirect Injection
LD	Light-Duty
LNT	Lean NOx Trap
MAW	Moving Average Window
NCV	Net Calorific Value
NDIR	Non-Dispersive Infra Red
NEDC	New European Drive Cycle

NTE	Not To Exceed pollutant limit applicable for RDE testing	PN	Particle Number
OBD	On-Board Diagnostics	RDE	Real Driving Emissions on road emissions test procedure
PAH	Polycyclic Aromatic Hydrocarbons	SCR(F)	Selective Catalytic Reduction (on Filter)
PEMS	Portable Emissions Measurement System	WLTC	World harmonised Light duty Test Cycle
PM	Particulate Matter	WLTP	World harmonised Light duty Test Procedure
PMP	Particulate Measurement Programme		

1 INTRODUCTION

Emissions have been the focus of worldwide legislation for more than twenty-five years. European emissions legislation has set limits for particulate matter (PM) and oxides of nitrogen (NOx) emissions from diesel vehicles since the early 1990's along with hydrocarbons (HC), which were initially included with NOx until 2000, and carbon monoxide (CO). Limits for PM have become progressively tighter, reaching 4.5 mg/km for the Euro 6 (2014) standards. PM is determined by gravimetric means, and is suitable for pass-fail evaluations, but the variability is such that small differences in emissions are difficult to detect. A non-volatile particle number (PN) counting method was developed based on the European Particulate Measurement Programme (PMP, 1-3) to increase accuracy and resolution, and after first introduction at Euro 5, a limit of $6 \cdot 10^{11}$ particles/km in now in place for both diesel and gasoline vehicles.

NOx emissions limits have also reduced steadily and for diesel Euro 6 vehicles the limit has reached 80mg/km. Similarly stringent standards have been introduced in other parts of the world. The introduction of these limits along with the introduction of clean fuels and advanced vehicle and after-treatment technologies has resulted in a substantial reduction in automotive particulate mass (PM) emissions (4, 5) with a corresponding improvement in air quality. However, ambient NOx levels have not reduced to the same degree as the Euro standards (6-8). Part of the reason is that emissions regulations for passenger vehicles have traditionally been based on the New European Driving Cycle (NEDC). Amid concerns that this test cycle does not represent closely enough real road driving in terms of CO_2 and other emissions levels including NOx, two new test procedures are under development - World harmonized Light duty Test Procedure (WLTP) featuring the WLTC cycle, for use on the chassis dynamometer and, for on-road use, the Real Driving Emissions (RDE) procedure. The WLTC cycle is longer than the NEDC and is expected to be more severe as it covers a more realistic range of driving conditions, though the impact of cold start is diminished. A further difference is that vehicle emissions are measured as NOx (total oxides of nitrogen), whereas air quality standards are expressed as ambient NO_2 concentration, applicable when the NOx has reacted in the atmosphere to form NO_2.

One of the enablers to being able to carry out testing of more real world driving on the road is the use of Portable Equipment Measurement Systems (PEMS) which are able to measure gaseous and PN emissions under real driving conditions. The RDE test protocol has been adopted in 2016 together with the Not-to-exceed limit (NTE) for NOx, published in the first two packages (9, 10). The NTE limit is expressed as a Conformity Factor (CF), defined as the permitted emission on the RDE test divided by the certification limit on the CD test cycle. Two extra Euro 6 vehicle stages will be introduced as a consequence, Euro 6d-Temp as of September 2017 with a NOx CF of 2.1 and Euro 6d as of January 2020 with a NOx CF of 1.5.

The RDE test protocol defines limits for the environmental and driving dynamic boundary conditions. PEMS measurement data is post-processed with normalization tools (EMROAD and CLEAR) before the CF is calculated. Two more packages are expected including a PN CF of 1.5, cold-start provisions, hybrids, and in-service conformity testing among other things. These tests will be used to certify vehicles going forward, so there is much interest in how they will compare with the current NEDC certification test and future WLTP test and if new Euro 6b+ vehicles will meet the limits for the new procedures.

Although there have been many studies focused on the effect of diesel fuel properties on emissions and fuel economy many of them are heavy-duty vehicle studies and of the light-duty vehicles studies many of the earlier ones were focused on indirect injection vehicles which were more common in the 1990's. One of the first studies which addressed the effects of fuel properties on emissions and fuel economy from a direct injection light-duty diesel engine was carried out by Tritthart et. al. (11). While there were differences between fuels tested, density, aromatics and cetane number were highly correlated and the effects were not prescribed to any one property. The same study concluded that back-end volatility in the form of T90 did not affect any of the emissions. The EPEFE (European Programme on Emissions, Fuels and Engine technologies) programme, which was carried out in the 1990's on diesel vehicles meeting Euro 2 emissions standards (12) noted that vehicles with electronic direct injection (DI) systems were generally more sensitive to fuel property changes than those with mechanically controlled systems, generally indirect injection (IDI). The same study found that fuel density affected the fuel pump setting, injection timing and EGR operation on a European DI diesel, but the effect of density on particulates could be wholly or partially removed by adjusting engine settings and operation (13). Similar effects were demonstrated with other fuel properties (14). A study by Concawe (15) tested two Euro 3 passenger cars, one fitted with an oxidation catalyst, the other with a diesel particulate filter (DPF). Wide variations in fuel properties (EN590 diesel to Fischer-Tropsch diesel or Swedish Class 1) reduced NOx emissions, but only by a small amount and the differences between the vehicles was much larger. The same fuel changes also reduced PM emissions, but the most dramatic impact was that of the DPF which produced very low emissions. More recently Concawe has conducted a study to look at fuel effects on modern diesel vehicles including the effects of density, cetane number, poly-aromatics and FAME content. (16). This study concluded that fuel effects appear to be getting smaller as after-treatment systems become more sophisticated although this data was generated on the NEDC and a hot WLTC cycle.

This study was conducted to build understanding of the emissions performance of Euro 6 diesel passenger cars, particularly on the road. Concawe has conducted a study of the CD and RDE performance of 3 vehicle types. For each vehicle, three tests were conducted over RDE, WLTC and NEDC cycles and additional RDE tests were run on a second fuel to determine the scope for detecting fuel effects over the new regulatory cycle. Testing on the road and in the laboratory may entail differences in boundary conditions (17) such as road load, vehicle mass, ambient conditions (temperature, pressure and humidity), battery charge level, tyre pressure, etc., therefore particular attention has been awarded to keep these boundary conditions under control so as to enable the test results to be fairly benchmarked. The data generated are intended to provide insights into the on-road emissions performance of latest diesel passenger cars and how this compares to regulated emissions limits and results from CD tests run under comparable road load conditions.

2 EXPERIMENTAL

Three Euro 6 diesel passenger car vehicles were installed on a chassis dynamometer and tested over the NEDC and WLTC drive cycles. In addition on-road RDE tests were run. The following sections describe the vehicles, fuels and testing procedures.

2.1 Description of vehicles

The test vehicles were purchased second-hand from the German market and the chosen vehicles were thought to have state of the art emissions performance and represented a range of exhaust after-treatment technologies and manufacturers.

Table 1: Technical details of test vehicles

	Car A	Car B	Car C
Emissions class	M1	M1	M1
Size category	D	E	C
Emissions certification	Euro 6b	Euro 6c	Euro 6b
Certification combined cycle CO2 (g/km)	119	112	109
Year of registration	2015	2016	2016
Cylinders / valves	4 / 16	4 / 16	3 / 12
Compression ratio	16.2	15.5	16.5
Displacement (l)	1.97	1.95	1.50
Output (kW)	140	143	85
Emissions controls	HP&LP EGR, urea-SCRF, ASC	HP&LP EGR, urea-SCRF, SCR/ASC	HP EGR, LNT, DPF, SCR
Transmission	DCT6	DCT9	M6
SOT mileage (km)	5969	10025	6514
Mass as tested (kg)	1897	2067	1724

2.2 Fuels

Two fuels were tested over the RDE to determine the scope for detecting fuel effects over the new regulatory cycle. Fuels were chosen which differed substantially in terms of density, one being near the EN590 density minimum and one being near the EN590 density maximum. The fuels also differed substantially in terms of Polycyclic Aromatic Hydrocarbon (PAH) content and distillation properties. The low density fuel number 2 was fully EN590 compliant whereas Fuel 1 was EN590 compliant except for T95 which was 10°C over the EN590 limit. Apart from shakedown tests, only Fuel 2 was tested over the CD cycles given that the key objective for those tests was to compare CD and RDE emissions.

Table 2: Test fuel properties

Property	Method	EN590 min	EN590 max	Fuel 1	Fuel 2
CN	EN ISO 5165	51	-	52.2	53
Density kg/m^3	EN ISO 12185	820	845	844	822
Sulfur mg/kg	EN ISO 20846	-	10	7.1	7.9
Viscosity at 40°C mm^2/s	ASTM D445	2	4.5	2.1	2.28
FAME v/v%	EN 14078	-	7	4.0	4.1
PAH %m/m	IP 391 mod	-	8.0	7.5	1.2
Total aromatics %m/m	IP 391 mod	-	-	27.0	11.5
C %m/m	ASTM D3343	-	-	86.32	85.78
H %m/m	ASTM D3343	-	-	13.25	13.84
O %m/m	EN 14078	-	-	0.43	0.44
NCV MJ/kg	ASTM D3338	-	-	42.82	43.12
IBP °C	ASTM D86	-	-	176	160
T50 °C	ASTM D86	-	-	274	255
T95 °C	ASTM D86	-	360	370	344
FBP °C	ASTM D86	-	-	382	355

2.3 Test detail

All CD and RDE testing was conducted in and around Ricardo's Shoreham-by-sea facility in the UK.

2.3.1 Road load models

As a key objective of the study was to compare emissions on CD and on the road, realistic road load models were derived for the CD tests from track based coast-downs with the vehicles ballasted to represent the mass of the PEMS equipment (165kg) fitted during RDE testing. All regulatory requirements for the generation of coast-down terms, including the statistical repeatability of the track times, were achieved.

2.3.2 Testing overview

Initial NEDC and WLTC check tests were performed on the vehicles in duplicate to ensure that there were no obvious problems and checks were done to ensure there were no faults recorded in the OBD. All tests were executed in accordance with the latest relevant protocols including pre-conditioning.

The test sequence for each car was designed to facilitate a valid statistical comparison between the fuels F1 & F2 in the on road RDE test. Thus, the RDE test sequence was comprised of an alternating sequence of F1 & F2. In addition, to facilitate a direct comparison of the emissions from the RDE, NEDC and WLTC on F2, the CD cycles were carried out each time the car was running on F2 with the order of the three cycles varying each time. The designed sequence is shown in Table 3.

Table 3: showing the designed test sequence. This was departed from in some instances as some tests, in particular RDE tests, were deemed non-compliant and were repeated out of sequence due to logistical requirements

Fuel change to			
F1	RDE # 1		
F2	RDE # 1	NEDC # 1	WLTC # 1
F1	RDE # 2		
F2	NEDC # 2	WLTC # 2	RDE # 2
F1	RDE # 3		
F2	WLTC # 3	RDE # 3	NEDC # 3

Preconditioning consisted of a full RDE cycle to help manage battery state of charge and DPF loading. The repeat NEDC and WLTC CD cycles were therefore carried out some days apart and interspersed with fuel changes and several RDE cycles (test and preconditioning). Thus, the variability of these NEDC and WLTC tests was expected to be somewhat higher than that typically achieved in back-to-back repeats on the same test procedure and fuel combination. Vehicle fuel systems were drained and flushed to effect fuel changes. In practice, the test order varied from the designed programme, because some of the road tests had to be repeated to generate results which were fully validated in the RDE procedure. The actual test order and results can be found in the Appendix.

2.3.3 Emissions measurement

On road emissions were measured using a Horiba PEMS OBS-ONE unit with a pitot flow tube for exhaust flow determination and CD emissions measured using Horiba MEXA ONE analysers. Given that a key objective was comparison of CD and on-road emissions, focus is given to those pollutants available from both the CD and PEMS testing platforms, i.e. CO_2, CO, PN and NOx. Dilute bagged emissions were measured via CVS on the CD. HC is not required for LD RDE testing and is not usually measured because the heated FID (Flame Ionisation Detector) requires substantial electrical power and consequently additional batteries and space.

Table 4: Emissions measurement on CD and on road

	MEXA ONE (CD)	OBS-ONE (PEMS)
CO	NDIR (Non-dispersive Infra Red)	NDIR
CO_2	NDIR	NDIR
NOx	CLD (Chemi-Luminescence Detector)	CLD
PN	MEXA 2000SPCS	OBS-ONE PN
	condensation particle counter (CPC) with primary diluter, volatile particle removal and secondary dilution. 23nm d50 cut-off as per PMP (Particulate Measurement Programme) protocol	

2.3.4 RDE

The RDE route used is illustrated in Figure 1. The route commences with urban operation wholly in 20 and 30 mile/hr zones. Rural and motorway phases are conducted on major roads to the west and east of Ricardo's site respectively. The requirements of the test are achieved without introducing artificial stop periods and urban severity is achieved through moderate hill climbs and multiple T-junctions. Hill climbs and descents are also present in both rural and motorway sections. Total test time is around 105 minutes and cold start emissions were included in the analysis.

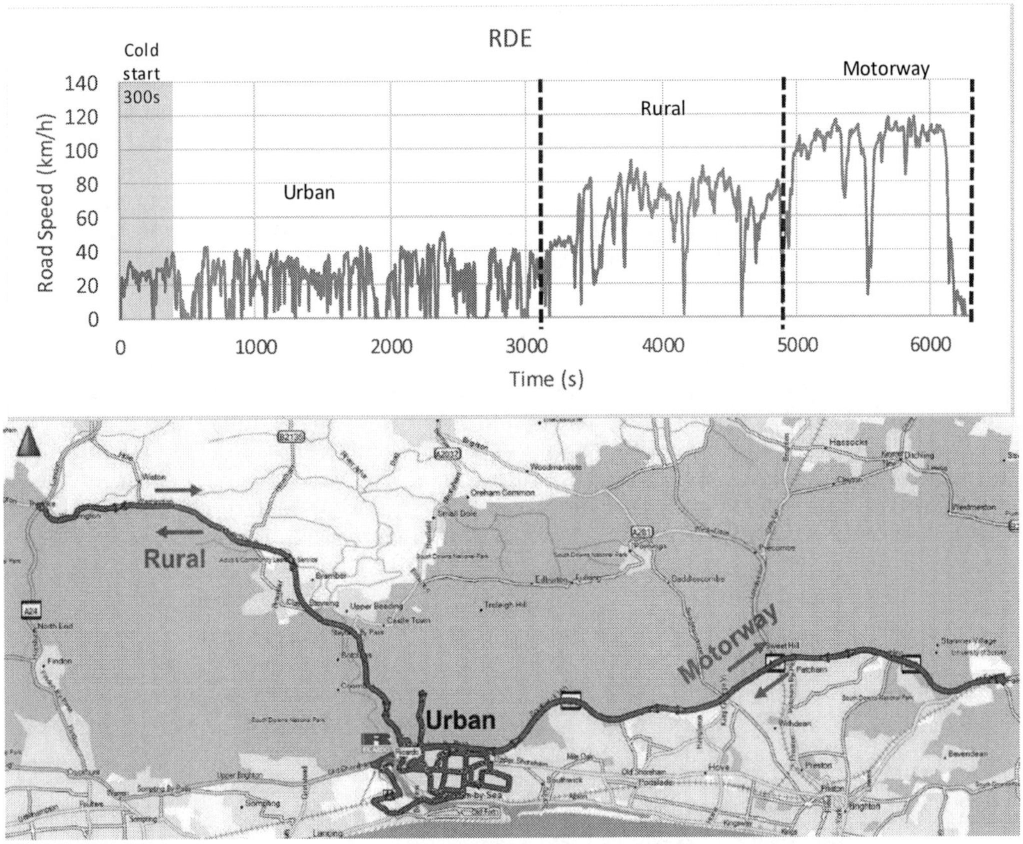

Figure 1: Speed/time and map plot of Ricardo RDE route

Normalisation of raw emissions data in EMROAD has a significant reductive effect on emissions levels reported. One major reason for this normalisation effect is the impact on emissions that have substantial cold start contributions. The first MAW (Moving Average Window) is calculated from t=0 following the release of CO_2 equivalent to that generated over the entire WLTC. This initial MAW then 'averages down' the cold start contribution by combining it with around 1500 subsequent data-points. Later MAW are calculated based upon the same quantity of CO_2 but starting at t=1, then 2 ... n. This approach means that the cold start contributes to fewer MAW in the urban phase, and overall RDE, than does a datum at 1000s, for example. Since the results calculated by EMROAD give all MAW equal weighting, but cold start contributes to relatively few MAW, the influence of the cold start is diminished relative to a direct determination of emissions from raw. The other major mechanism impacting emissions is via the exclusion of MAW data that does not conform to required boundary conditions such as instantaneous speed, temperature and other parameters. The MAW that are excluded because CO_2 was higher or lower than the permitted bands relative to reference WLTC levels may contain high emitting events that would otherwise increase levels in the final emissions calculation. The treatment of cold start, and normalisation, via EMROAD is still subject to discussion within the RDE working groups and may be subject to future regulatory change.

2.3.5 Diesel Exhaust Fluid (DEF) consumption

The DEF consumption of the two urea-SCR equipped vehicles (Car A and Car B) was determined volumetrically by the amount required to top up the tank after a known mileage comprising road and CD running. The measured distance for Car A was 708km and for Car B 554km. The tank range and top up interval was estimated based on the consumption rate and tank size.

3 PEMS AND CD MEASUREMENT COMPARISON

A critical aspect of achieving acceptable emissions measurements outside the test laboratory is the reliable correlation with PEMS-based analysers. In the lab, where environmental conditions are carefully controlled and managed, analysers are expected to have exceptional repeatability and reproducibility, as well as defined accuracy and precision. Analysers for use on the road, where operational temperatures vary and other influences such as vibrations are encountered, must be more robust, compactly sized to aid installation and still provide similar performance to the lab-based systems.

Within the RDE regulation, the required similarity between Lab and PEMS is based upon final cycle "emission"/km figures generated when employing both systems in parallel during a test on the chassis dynamometer. Consequently, the correlation takes into account both flow and measurement system variation. For validation of the PEMS, its emissions are compared with those of a validation space around the values of the lab-based analysers, and a valid correlation test must be performed either before or after an RDE test. The validation space is defined in two ways:

- Lab value ± fixed %
- Lab value ± absolute value

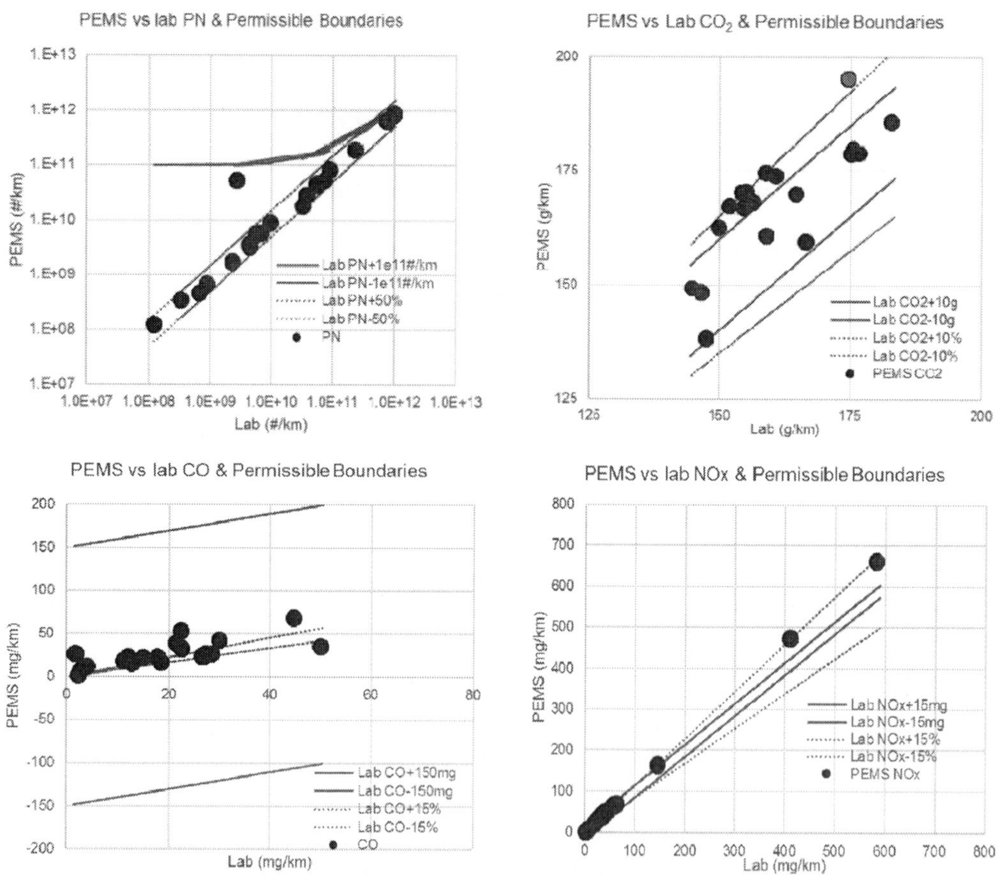

Figure 2: Correlations between PEMS and CD laboratory emissions showing that all measurements subject to the correlation criteria were within the specified limits

Both possibilities are calculated, and the one that creates the largest validation space is used to verify acceptable correlation between PEMS and lab. In the study correlation data were obtained from NEDC and WLTC tests on all three vehicles, with at least 2 repeats on each cycle. In total over 20 correlation tests were performed. The results of these correlations and validations are shown graphically in Figure 2. For carbon dioxide, the PEMS system reported on average ~5% higher than the lab analyser, within the ± 10% boundaries necessary for validation. All tests, with one exception (shown as a red dot), met the validation criteria. The 'invalid' test (~11% above the lab value) was positioned in the test sequence prior to another chassis dyno test and not an RDE, so there was no impact on validity of results in the project.

4 DATA TREATMENT AND STATISTICAL ANALYSIS

4.1 Data Quality

CD tests were considered valid if they were conducted to the relevant regulation and did not include a DPF regeneration. Initially tests in which the start/stop did not engage at idle were rejected and repeated, however these tests were re-admitted for Car B because the start/stop application was found to be erratic despite the consistent pre-conditioning protocol and on-road conditioning to ensure consistent battery state of charge. RDE tests were considered to be valid for statistical analysis if they met the RDE criteria, did not contain DPF regenerations and also validated in EMROAD. Occasionally, some tests considered invalid were repeated out of sequence driven by logistical requirements.

Car C had repeated failures to validate in EMROAD and it was not possible to obtain the designed set of valid RDE tests on this car. To carry out any statistical analysis, four RDE tests that failed because of invalid EMROAD were replaced with a valid preconditioning RDE from the previous day. The only difference between these preconditioning tests and the 'formal' tests was that they were not, themselves, preceded by an RDE. This at least provided a set of RDE tests on each fuel for the purposes of comparing emissions.

4.2 Statistical analysis of RDE tests

The three cars were each analysed separately to examine differences between fuels F1 & F2. Three valid tests were available for analysis on each car-fuel combination except for Car C which only had two tests on F1. Such a small number of tests per fuel did not allow sufficient statistical power to utilise formal outlier tests and all valid tests were included in the analysis.

CO_2, CO and NOx emissions were analysed on the raw scale so that the average emissions for each fuel is the arithmetic mean. A weighted analysis where each measurement is weighted by $1/(\text{fitted value})^2$ was used to account for the fact that higher emissions are generally more variable. In line with standard practice, PN were analysed on the log-scale and are summarised by their geometric means and plotted on a logarithmic scale. In each case a simple t-test was used to assess the significance of any difference between the fuels.

Results are plotted as the mean emissions for each fuel with error bars equivalent to half the least significant difference at 90% confidence between the fuels. Thus, overlapping error bars indicate there is no significant difference between fuels at the 90% confidence level.

The error bars are constructed as follows:

For CO_2, CO & NOx:

Error bars = $\pm 0.5 t_{0.1,df} SE_D$

where $t_{0.1,df}$ is the 10% point of the statistical t-distribution on the appropriate degrees of freedom (*df* = 4 or 3 in the case of Car C emissions) and SE_D is the standard error of the difference in the means.

For PN the error bar is first constructed as ± half the least significant difference, L, on the on the $\log_e$ scale i.e. $L = 0.5 t_{0.1,df} SE_{DL}$ where $t_{0.1,df}$ is as above and SE_{DL} is the standard error of the difference in the means on the $\log_e$ scale. Transforming back to the raw scale, the error bars are given by:

Lower error bar = $e^{\bar{y}} - e^{(\bar{y}-L)}$

Upper error bar = $e^{(\bar{y}+L)} - e^{\bar{y}}$

where $\bar{y}$ is the mean on the $\log_e$ scale.

4.3 Statistical analysis for comparison of emissions across test methods

Arithmetic means are given for CO_2, CO and NOx emissions and geometric means for PN. Error bars on these charts correspond to ±1SD (Standard Deviation) and their purpose is to illustrate the observed level of variability between the repeats within each method.

5 RESULTS AND DISCUSSION

5.1 Summary of RDE results

A summary of the RDE (overall, raw) results for CO, NOx, PN and CO_2 is presented in this section. Results for individual tests including those containing DPF regenerations are shown in chronological order.

5.1.1 CO

The results for CO showed that in all but one test containing a DPF regeneration CO results were at or below 50mg/km, which is an order of magnitude below the Euro 6 limit of 0.5g/km. These data suggest that CO emissions from modern diesel cars are not a focus for concern even in on road driving. There are no obvious correlations between CO and fuel. (See Appendix for data).

5.1.2 NOx

The results in Figure 3 show that all of the overall RDE NOx results for cars A and B fall within the Euro 6 NOx limit applicable to the NEDC (CF1) except for one test on car A containing a DPF regeneration. These results are substantially lower than some published in other studies of on road emissions featuring Euro 6 diesel cars (6-8). Car C however produces much higher levels of NOx, but as a Euro 6b certified car would have been required to pass Euro 6 NOx limit over NEDC and not over RDE. It is outside the scope of this study to determine the root cause of the difference in RDE NOx levels between cars A and B and car C, which could be within the hardware, including the after-treatment; for example higher NOx conversion efficiencies of urea-SCR than LNT especially over high duty conditions; or the calibration: such as use of different EGR rates. NOx tends to be higher on Fuel 1 in car A and on Fuel 2 in car B, however neither of these effects are statistically significant even at the 90% confidence level, (see Figure 6).

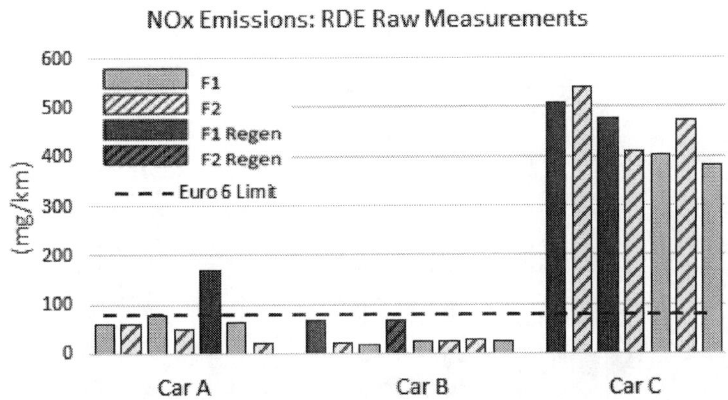

Figure 3: Raw overall RDE results for NOx for the three test vehicles

5.1.3 PN

The results in Figure 4 show that the overall results for all cars and all tests without DPF regenerations fall within the Euro 6 PN limit applicable to the NEDC (CF1). Whereas some tests in which DPF regenerations occur exceed the limit. Overall it appears that contemporary DPF technology is capable of meeting the PN targets for near future legislative requirements. No fuel effects are apparent.

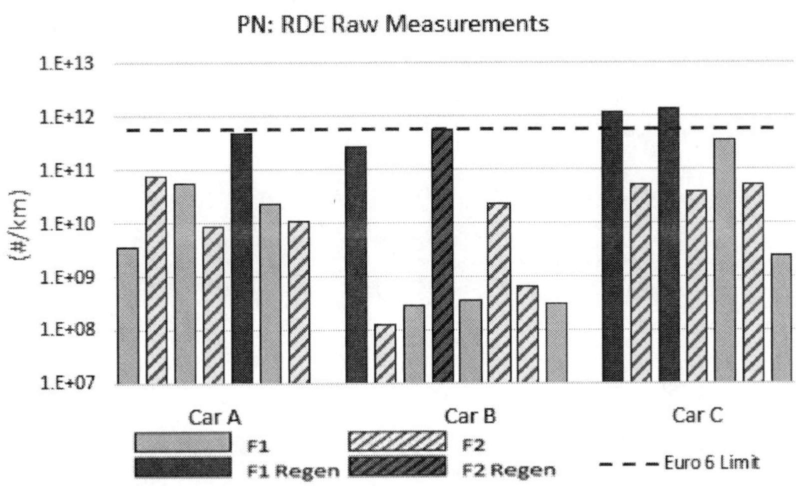

Figure 4: Raw overall RDE results for NOx for the three test vehicles

5.1.4 CO_2

The results in Figure 5 show that, as expected, tests containing DPF regenerations have higher CO_2 emissions. There is evidence of trends or unrepeatability in the data which are to be expected from the uncontrolled variables associated with on-road testing such as weather and traffic conditions. No fuel effects are apparent.

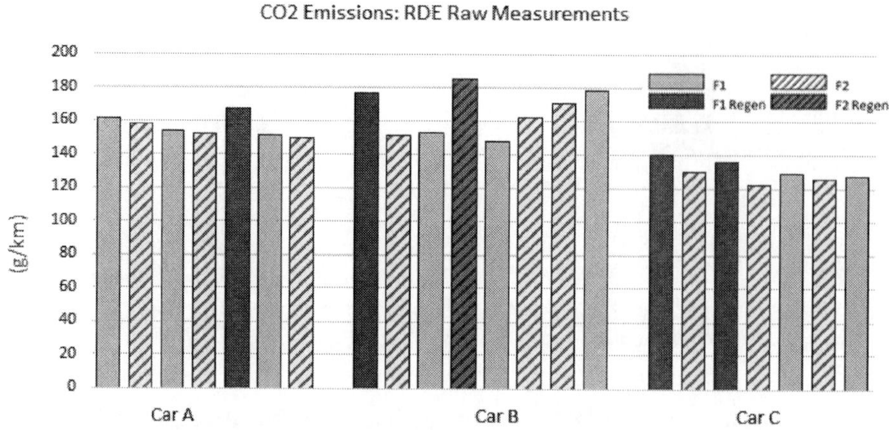

Figure 5: RDE CO_2 results for the three test vehicles

5.2 Fuel Effects on RDE Results

A comparison of mean RDE emissions on the two test fuels is given in this section. The error bars correspond to ±half the Least Significant Difference at the 90% confidence level. Given that there are no cases where the error bars for mean values on fuels 1 and 2 do not overlap, there are no statistically significant fuel differences even at this lower confidence level. Measures to increase the likelihood of measuring significant fuel differences could be to increase the number of repeat tests and testing fuels that differ in quality beyond the extremes of the EN590 specification. Where the EMROAD Moving Average Windows (MAW) has a substantial impact on the emissions, this usually reduces the result, in particular for CO (see Appendix for data).

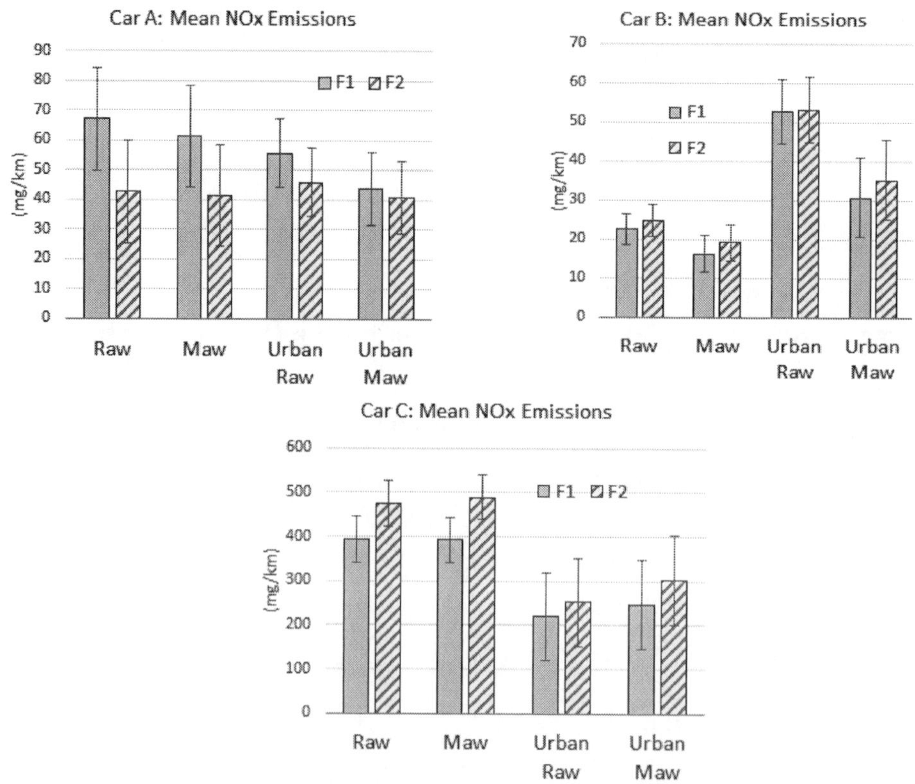

Figure 6: Fuel effects on RDE NOx. The fuel difference in the car C MAW data is not significant at the 90% confidence level

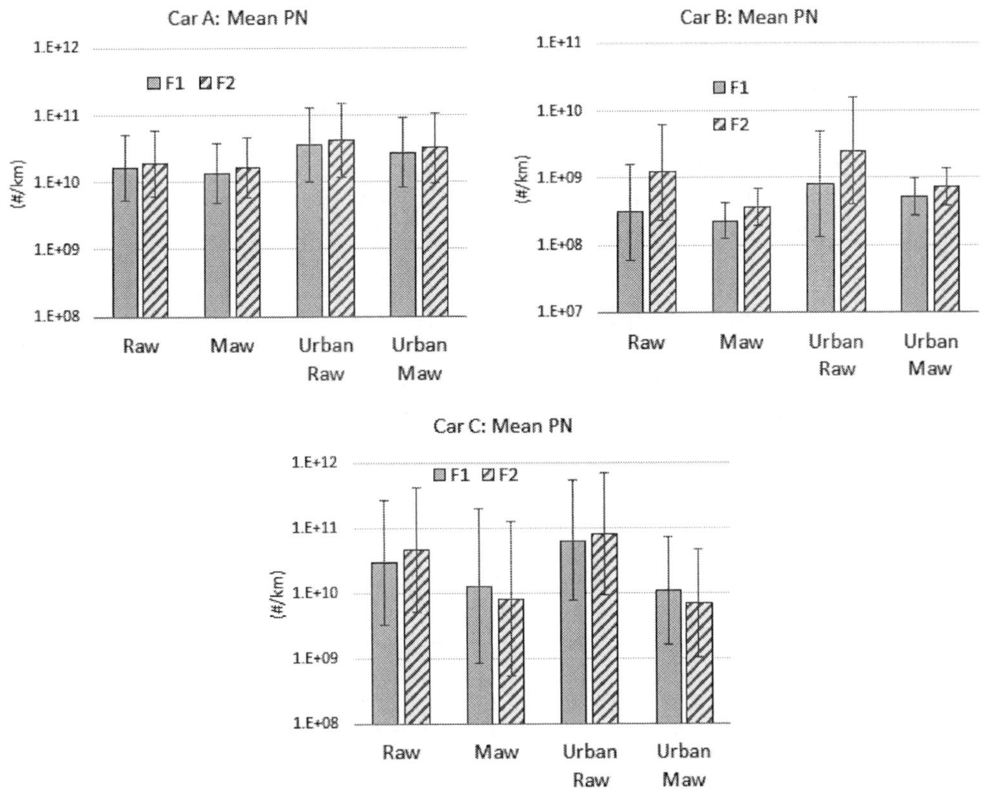

Figure 7: Fuel effects on RDE PN

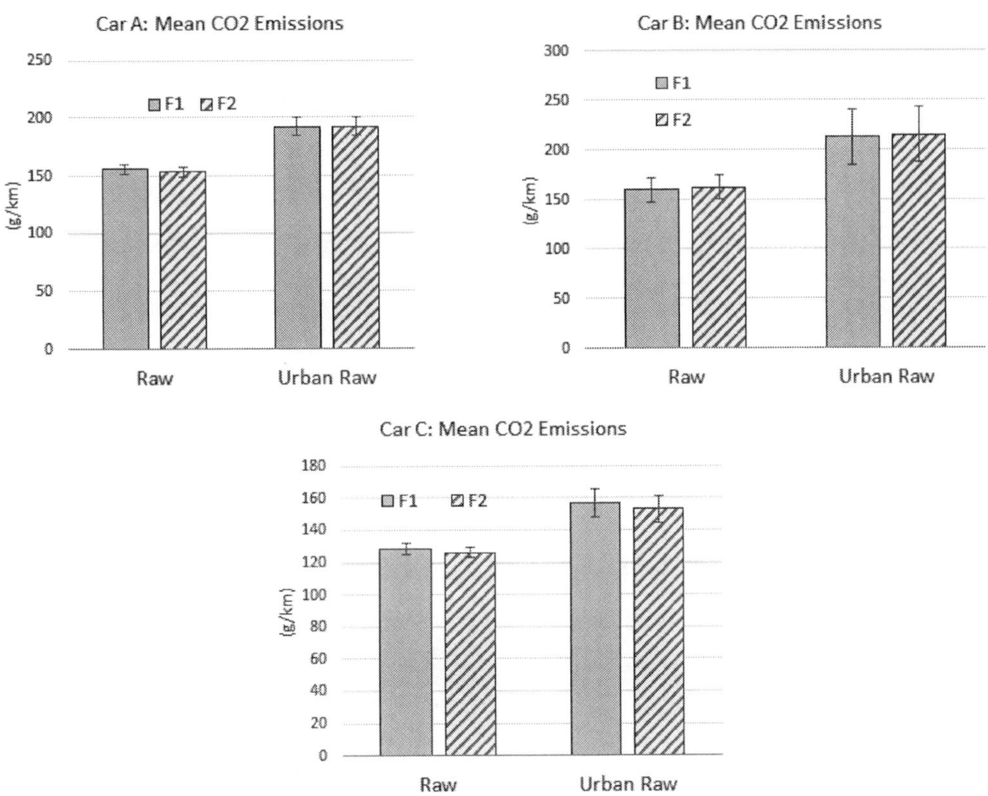

Figure 8: Fuel effects on RDE CO_2

5.3 Comparison of emissions across test scenarios

A comparison of mean emissions for Fuel 2 across the RDE, NEDC and WLTC test cycles can be found in Figure 9. A minimum of three and maximum of four repeat measurements deemed to be valid for inclusion in the analysis were carried out for each test and vehicle combination. Arithmetic means are given for CO_2, CO and NOx emissions and geometric means for PN. Error bars on these charts correspond to ±1SD and their purpose is to illustrate the observed level of variability between the repeats. In general:

- Neither test method was consistently more repeatable than the others. This was somewhat surprising given that the RDE is subject to a lot more variables than the CD tests. Due to logistical considerations, tests on the different methods were interspersed which is sub-optimal for achieving the best repeatability in any single test method.
- Using representative road load models derived from track based coast-downs all three vehicles were compliant with Euro 6 emissions limits over NEDC as certified.
- CO was similar across all cycles and cars and was at least one order of magnitude lower than the Euro 6 limit.
- NOx was similar across cycles for the urea-SCR equipped cars and below the Euro 6 limit in all cases. NOx from the non-urea-SCR car scaled with cycle duty and exceeded the Euro 6 limit over the WLTC and RDE. It is of note that the non-urea-SCR car produces around half of the NOx in the urban portion of the RDE than it does in the RDE (Figure 6) whereas Car A produces similar NOx over urban and whole RDE and Car B produces higher NOx over the urban section of the RDE.
- PN was similar across cycles for the three cars and fell within the Euro 6 limit.
- CO_2 was similar across cycles for the urea-SCR cars and was substantially lower over RDE for the non-urea-SCR car which is lighter and has a smaller engine. It should also be considered that CO_2 measured via PEMS was found to be ~5% higher than measured by the CD system in comparative tests and subtracting 5% from the RDE CO_2 results would give substantially lower CO_2 over RDE for all three cars. The measured CO_2 cannot be directly compared with the stated value for each vehicle from its Certificate of Conformity, which would have been generated over the NEDC following the relevant protocols and flexibilities. In these tests, the real road load applicable in the RDE testing, i.e. including 165kg additional mass to account for the PEMS equipment was used. It should be noted that there is a large range of severity allowed in the RDE and this will influence emission levels.

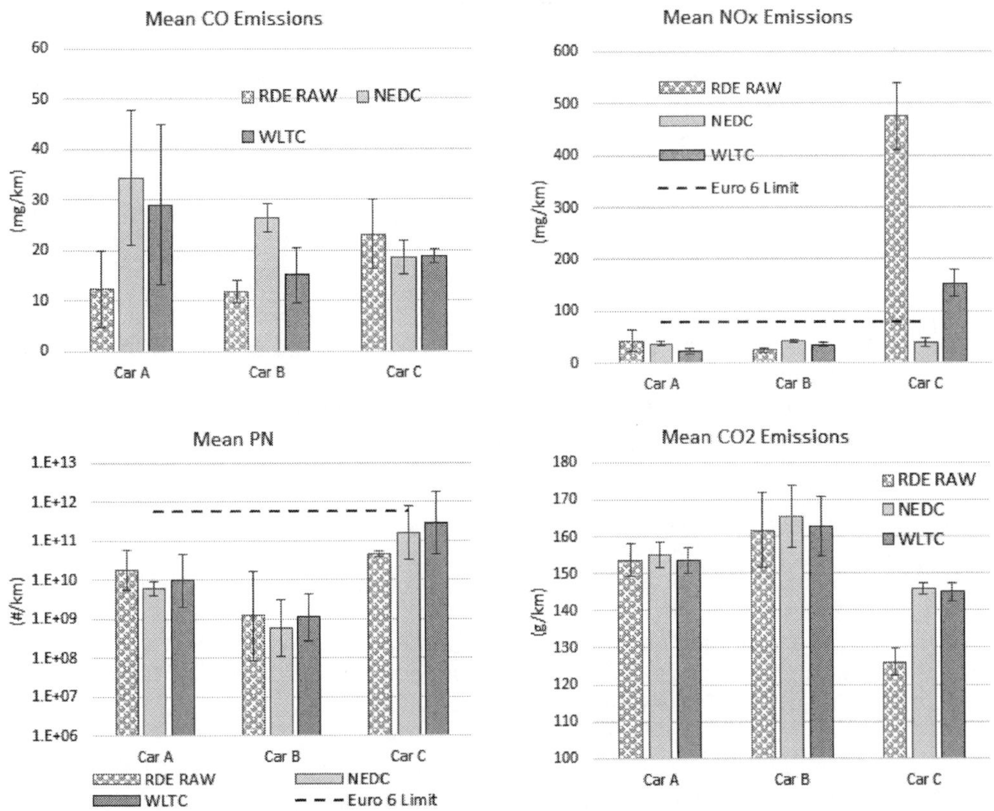

Figure 9: Comparison of mean CO, NOx, PN and CO2 emissions from the RDE, NEDC and WLTC tests for all vehicles

5.4 Diesel Exhaust Fluid (DEF) consumption

Diesel exhaust fluid consumption over the test period (708km for Car A and 554km for Car B) is shown in Figure 10. Based on these data it is apparent that one or more DEF tank replenishments would be required between services.

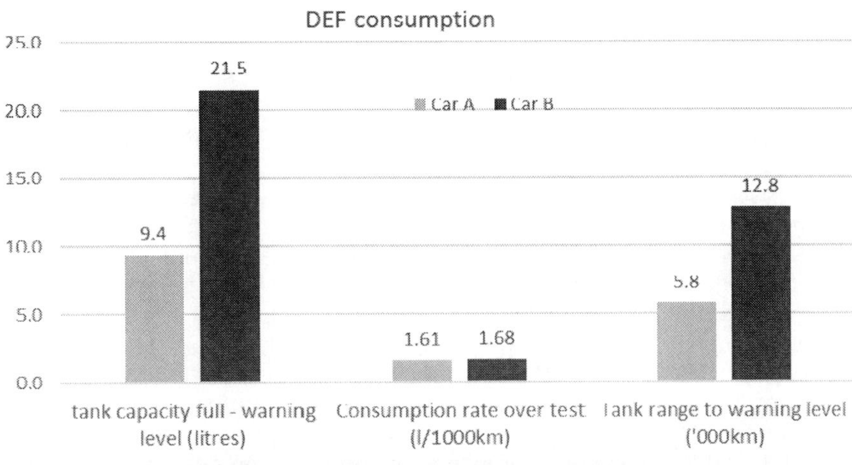

Figure 10: DEF consumption rate and tank range based on tank capacity between full and initial warning level

6 CONCLUSIONS

A set of three Euro 6 diesel passenger cars with differing exhaust after-treatment technology have been tested over WLTC and NEDC CD test cycles using realistic road load models and over RDE with two substantially different fuels. The test results show that:

- A good correlation is achievable between PEMS and CVS emissions measurement.
- Where the EMROAD Moving Average Windows (MAW) has a substantial impact on the emissions, this usually reduces the result.
- CO emissions from modern diesel cars are not a cause for concern even in on road driving.
- State of the art diesel passenger cars are capable of meeting near future NOx emissions requirements of RDE testing commensurate with Euro 6d though this may be difficult without urea-SCR NOx after-treatment systems.
- Modern diesel cars using contemporary DPF technology are capable of meeting the PN requirements for Euro 6d even on road.
- CO_2 measured over RDE was similar to, or lower than that measured over NEDC or WLTC.
- DPF regenerations led to higher CO_2 and PN in some RDE tests.
- Diesel fuel effects on emissions are difficult to detect in RDE testing where the fuel quality lies within the range of EN590.
- Neither test method was consistently more repeatable than the others. Better repeatability over the CD test methods could be expected if the tests were run successively.
- Diesel Exhaust Fluid consumption for these urea-SCR cars complying with RDE emissions requirements was in the range of 1.6-1.7 litres/1000km. Given the vehicles' DEF tank sizes, this would mean that one or more top ups was needed between services.

REFERENCES

(1) Andersson, J. and Wedekind, B. (2001), "DETR/SMMT/Concawe Particulate research programme 1998-2001 - Summary report". Report No. DP01/0515, Shoreham-by-Sea: Ricardo Consulting Engineers Ltd.

(2) Andersson, J. et al (2007) "Particle Measurement Programme (PMP) Light-duty interlaboratory correlation exercise (ILCE_LD)". European Commission, JRC, Institute for Environment and Sustainability, Final Report No. EUR 22775 EN. Ispra.

(3) "Comparison of particle emissions from advanced vehicles using DG TREN and PMP measurement protocols", Report No. 2/09. Brussels: Concawe (2009).

(4) EU (1998) "Directive 98/69/EC of the European Parliament and of the Council of 13 October 1998 relating to measures to be taken against air pollution by emissions from motor vehicles and amending Council Directive 70/220/EEC". Official Journal of the European Communities No. L350, 28.12.1998.

(5) EU (2000) "Directive 1999/96/EC of the European Parliament and of the Council of 13 December 1999 on the approximation of the laws of the Member States relating to measures to be taken against the emission of gaseous and particulate pollutants from compression ignition engines for use in vehicles, and the emission of gaseous pollutants from positive ignition engines fuelled with natural gas or liquefied petroleum gas for use in vehicles and amending Council Directive 88/77/EEC". Official Journal of the European Communities No. L44, 16.02.2000.

(6) Carslaw, D.C. et al. Recent evidence concerning higher NOx emissions from passenger cars and light duty vehicles, Atmos. Environ. 45, (2011), 7053-7065.
(7) Chen, Y. & Borken-Kleefeld, J. Real driving emissions from cars and light commercial vehicle - Results from 13 years remote sensing at Zurich/CH, Atmos. Environ, 88 (2014) 157-164.
(8) Weiss et al., Will Euro 6 reduce the NOx emissions of new diesel cars? - Insights from on-road tests with Portable Emissions Measurement systems (PEMS), Atmos. Environ. 62 (2012), 657-665.
(9) Commission Regulation (EU) 2016/427 of 10 March 2016 amending Regulation (EC) No 692/2008 as regards emissions from light passenger and commercial vehicles (Euro 6).
(10) Commission Regulation (EU) 2016/646 of 20 April 2016 amending Regulation (EC) No 692/2008 as regards emissions from light passenger and commercial vehicles (Euro 6).
(11) Tritthart, P., Chichocko, R. and Cartellieri, W., "Fuel Effects on Emissions in Various Test Cycles in Advanced Passenger Car Diesel Vehicles", SAE Paper 932684, 1993.
(12) Hublin, M., Gadd, P.G., Hall, D.E and Schindler, K.P. "European Programmes on Emissions, Fuels and Engine Technologies (EPEFE) Light Duty Diesel Study", SAE Paper 961073, 1995.
(13) Mann, N., Kvinge, F. and Wilson, G. "Diesel Fuel Effects on Emissions - Towards a Better Understanding", SAE Paper 982486, 1998.
(14) Kwon, Y., Mann, N., Rickeard, D.J., Haugland, R., et al., "Fuel Effects on Diesel Emissions - A New Understanding", SAE Paper 2001-01-3522, 2001.
(15) Concawe Report 2/05, Fuel effects on emissions from advanced diesel engines and vehicles, www.concawe.org, 2005.
(16) Williams R. et al Effect of Diesel Properties on Emissions and Fuel Consumption from Euro 4, 5 and 6 European Passenger Car, SAE paper no. 2016 - 01 - 2246.
(17) Chappell, E., Burke, R., Lu, P., Gee, M. et al., "Analysis of a Diesel Passenger Car Behavior On-Road and over Certification Duty Cycles," SAE Int. J. Engines 9(4):2201-2214, 2016, doi:10.4271/2016-01-2328.

APPENDIX – TEST ORDER AND DATA

Car A							
Date	**Fuel**	**Cycle**	**CO**	**CO2**	**NOX**	**PN**	**Comment**
			[mg/km]	**[g/km]**	**[mg/km]**	**[#/km]**	
11-Apr-17	F1	RDE	8.038	161.5	58.95	3.594E+09	
13-Apr-17	F2	RDE	5.714	158.5	59.98	7.310E+10	
19-Apr-17	F2	NEDC	49.871	154.9	39.30	9.558E+09	
25-Apr-17	F1	RDE	8.232	154.4	78.40	5.500E+10	
28-Apr-17	F2	NEDC	26.427	158.8	41.01	4.337E+09	
03-May-17	F2	WLTC	29.796	156.2	28.10	5.386E+10	
04-May-17	F2	RDE	10.326	152.1	47.69	8.449E+09	
09-May-17	F1	RDE	11.160	167.7	170.02	4.769E+11	Regen during test
10-May-17	F1	RDE	18.796	151.4	63.84	2.360E+10	
16-May-17	F2	WLTC	44.563	154.5	21.55	6.682E+09	
17-May-17	F2	RDE	20.569	150.1	20.30	1.089E+10	
19-May-17	F2	NEDC	26.762	151.8	33.17	5.495E+09	
23-May-17	F2	WLTC	12.658	149.8	18.50	2.666E+09	
Car B							
Date	**Fuel**	**Cycle**	**CO**	**CO2**	**NOX**	**PN**	**Comment**
			[mg/km]	**[g/km]**	**[mg/km]**	**[#/km]**	
01-Jun-17	F1	RDE	13.448	177.1	65.19	2.725E+11	Regen during test
05-Jun-17	F2	RDE	14.021	151.1	19.61	1.226E+08	
06-Jun-17	F2	NEDC	22.290	174.6	40.83	1.220E+08	Stop/Start failed
08-Jun-17	F2	WLTC	11.903	155.8	31.56	3.238E+08	
15-Jun-17	F1	RDE	14.339	153.0	17.50	2.826E+08	
20-Jun-17	F2	NEDC	27.064	154.1	42.76	4.507E+09	
22-Jun-17	F2	WLTC	11.903	160.6	32.46	8.711E+08	
23-Jun-17	F2	RDE	16.072	185.4	65.81	5.758E+11	Regen during test
27-Jun-17	F1	RDE	14.000	148.0	25.68	3.471E+08	
05-Jul-17	F2	NEDC	28.574	165.2	44.59	1.978E+08	Stop/Start failed
06-Jul-17	F2	RDE	9.497	162.5	25.96	2.247E+10	
07-Jul-17	F2	NEDC	27.461	167.6	38.88	1.189E+09	Stop/Start failed
10-Jul-17	F2	WLTC	21.369	171.4	41.14	4.915E+09	Stop/Start failed
12-Jul-17	F2	RDE	11.280	171.4	28.88	6.423E+08	No fuel change
14-Jul-17	F1	RDE	9.873	178.9	24.69	3.093E+08	
Car C							
Date	**Fuel**	**Cycle**	**CO**	**CO2**	**NOX**	**PN**	**Comment**
			[mg/km]	**[g/km]**	**[mg/km]**	**[#/km]**	
28-Jul-17	F1	RDE	28.059	140.1	507.99	1.182E+12	F1 Regen in urban
31-Jul-17	F2	RDE	28.576	129.9	540.59	5.098E+10	replaces 01-Aug test
02-Aug-17	F2	NEDC	14.791	144.6	38.75	7.583E+11	
04-Aug-17	F2	WLTC	17.541	146.5	183.24	1.004E+12	
08-Aug-17	F1	RDE	105.483	135.9	475.96	1.340E+12	F1 Regen in Mway
11-Aug-17	F2	NEDC	21.245	147.4	49.13	3.170E+10	
15-Aug-17	F2	WLTC	18.292	146.3	143.38	3.653E+10	
15-Aug-17	F2	RDE	15.288	122.6	411.19	3.884E+10	replaces 16-Aug test
17-Aug-17	F1	RDE	44.788	129.7	404.31	3.571E+11	replaces 18-Aug test
23-Aug-17	F2	NEDC	19.315	145.5	33.26	1.824E+11	
24-Aug-17	F2	RDE	25.273	125.8	473.72	5.053E+10	F2
31-Aug-17	F1	RDE	28.267	127.3	381.75	2.465E+09	replaces 1-Sep test
07-Sep-17	F2	WLTC	20.448	141.9	132.57	7.325E+11	

Understanding the factors that influence testing precision for modern diesel vehicles

R D Burke, E C Chappell
Powertrain and Vehicle Research Centre, University of Bath, UK

M Gee, R Williams
Shell Global Solutions, UK

ABSTRACT

Modern diesel vehicles employ many technologies to reduce emissions whilst minimising fuel consumption. Many of these technologies have a time-dependent behaviour which results in unexplained test-to-test variation when conducting short tests on chassis dynamometers. This vehicle-borne imprecision increases the effort required to demonstrate small but significant differences between automotive technologies. The aim of this paper is to identify and quantify variability inherent to the vehicle and to propose methodologies that improve testing precision.

Two vehicles with DPF and SCR technology were characterised on-road and on chassis dynamometer. The on-road analysis gave an understanding of the long-term vehicle behaviour and was used to inform the chassis dynamometer tests. The chassis dynamometer tests were conducted in a twin-axle facility, located in a climatic chamber, and driven by a robot driver. Data was recorded through the vehicle OBD port and via dedicated emissions and fuel flow measurement instruments.

The duty cycle was seen to affect the precision with higher speed cycles offering higher precision: the RMSE of the WLTC Low phase was 1.6% compared with 0.6 % for the Very High phase. Interestingly, the WLTC had better repeatability than the NEDC cycle. The vehicle start/stop system affected precision by 0.2-0.6 %-points and the battery smart charging strategy caused a 1 %-point decrease in precision. Understanding and managing each of these factors is key to achieving high precision vehicle testing.

NOMENCLATURE

Abbreviations

CD	Chassis Dynamometer
DOC	Diesel Oxidation Catalyst
DPF	Diesel Particulate Filter
ECU	Engine Control Unit
EGR	Exhaust Gas Recirculation
EM	Emissions Measurement point
FC	Fuel Consumption measurement point
MF	gas Mass Flow measurement point
NEDC	New European Drive Cycle
NOx	Nitrous Oxides emissions
OBD	On Board Diagnostics
RMSE	Root Mean Square Error
SCR	Selective Catalytic Reduction
THC	Total Hydrocarbon emissions
WLTC	World Light-duty Transient Cycle

Notation

c_X	Volumetric concentration of species X	ppm or %
K_{dw}	Dry to wet correction factor	-
m_f	Fuel consumed	g
$\dot{m}_f$	Fuel consumption rate	g/s
$\dot{m}_{EX}$	Exhaust mass flow rate	g/s
N_{eng}	Engine Speed	r/min
n	Number of repeated tests	-
T_f	Fuel temperature	K
t	time	s
$t_{95\%}$	Student's two-tailed t-value at p=0.05	-
u_X	Relative density constant for species X (from BSI)	
$\dot{V}_f$	Volumetric flow rate of fuel	m^3/s
w_c	Carbon content of fuel	-
ρ_f	Fuel density	Kg/m^3
σ	Standard deviation of the sample	-

Subscripts

air	Fresh air flow
CB	Carbon Balance FC method
CO	Carbon monoxide
CO_2	Carbon dioxide
ECU	Engine control unit measurement
EX	Exhaust flow
FM	Flow meter FC method
inj, i	Fuel injection event i during an engine cycle
THC	Total Hydrocarbons

1 INTRODUCTION

The ability to demonstrate cost/benefits is vital to the automotive sector, but this is challenging because of the complexity of vehicle powertrains [1]. In addition, there is increasing emphasis on demonstrating benefits in on-road driving conditions. Demonstrating benefits on-road has the advantage of capturing a wide range of conditions and allows for the use of many vehicles over long distances. The major disadvantage is the need for a large quantity of data to detect small differences because of the inherent variability of on-road driving [2]. In contrast, Chassis Dynamometer testing can achieve high levels of precision over short distances using prescribed cycles [3]. Considering these points mean there is now a need for high precision chassis dynamometer testing, conducted under realistic transient operating conditions. High precision testing on the World-harmonised Light-duty Cycle (WLTC) is the first step to achieving this goal [4].

This paper is part of a series of publications from a project which aims to develop understanding that will improve the current limits of precision on chassis dynamometer facilities to assist with the development and demonstration of future fuels and lubricants. The aim of this paper is to assess the repeatability of two modern diesel vehicles to identify sources of vehicle-borne imprecision that will manifest during NEDC and WLTC testing. The impact of factors inherent to the test vehicle will first be quantified before proposing methods to control or eliminate the resulting variability. The work has been undertaken at the Engineering and Physical Research Council (EPSRC) funded Centre for Low Emissions Vehicle Research (CLEVeR) at the University of Bath. The centre is built around a high precision chassis dynamometer facility [3] to ensure that imprecision from the test cell is minimised.

2 BACKGROUND

Before undertaking the work described in this paper, the behaviour of the two test vehicles was monitored over 10,000 km on the road [5]. This investigation highlighted the differences between on-road driving and the standard chassis dynamometer based tests. These differences can be classified as:

1. Environmental: the ambient conditions (temperature, pressure, humidity) within which the vehicle is driven;
2. Operational: this is split into two aspects:
    a. The speeds and accelerations the vehicle is subjected to, which reflect in the engine speed and torque operating points;
    b. The duration of the duty cycles.

Ultimately, the on-road investigation allows for large amounts of data that can show trends in vehicle behaviour such as the frequency of regeneration of the particulate filter [5, 6]. However, the lack of precisely repeated driving conditions means that detailed analysis of the vehicle behaviour is challenging.

Testing on a chassis dynamometer allows for better control of the testing conditions, however to demonstrate small differences with statistical confidence requires either a high level of test-to-test repeatability or many repeated tests [7]. The latter has obvious cost implications and high precision is therefore critical to this approach. Previous studies have shown that variation in key noise factors can have both large and statistically significant effects on vehicle fuel consumption over legislative chassis dynamometer test cycles [3, 8]. These studies have focused on sources of imprecision that result from the test facility or from the testing procedure. They have shown that factors such as ensuring battery charge equality between tests, managing engine auxiliary loads, driver speed error, drive aggressiveness, test cell climatic conditions, and tyre inflation pressure all need to be well controlled to achieve high precision from chassis dynamometer tests. This is true regardless of the vehicle type that is under test.

Modern diesel vehicles are complex systems with many subsystems, some of which have strong time-varying characteristics such as Diesel Particulate Filters (DPF) [6], Selective Catalytic Reduction (SCR) catalysts [9], Lean NOx Traps (LNT) [10] and smart battery chargers [11, 12]. These components and their controllers have inherent dynamic behaviour that affect the engine performance over time periods typically greater than standard CD tests. For example, the diesel particulate filter (DPF) accumulates soot over time, causing an increase in engine pumping losses and requires periodic regeneration, a process strongly affecting fuel consumption [13, 14]. The SCR system is subject to the rapid transients of engine operation which will impact the composition of the gases entering the catalyst (NO_x, O_2, Urea...). However, the SCR also accumulates ammonia on the substrate of the catalyst, and its capacity is temperature dependent; the dynamics of this aspect of the system are of the order of 1-10mins [9]. The instantaneous performance of the SCR catalyst depends on both these aspects.

The on-road testing showed that the time-based variation of these systems is often much longer than chassis dynamometer tests meaning their behaviour cannot be fully captured during a short drive cycle. A lack of understanding of these systems can lead to unexplained variations in the vehicle controller. There is no universal solution of how to account for these systems in vehicle testing. The simplest solution is to remove these devices from the vehicle under test [15], but the removal of such a vital system component can have significant influences beyond the emissions it is designed to reduce – in the case of a DPF this would significantly reduce engine back pressure and pumping losses.

It would be desirable to be able to capture the long-term vehicle effects over shorter testing campaigns, such as those conducted in CD facilities. A common compromise solution is to test with the DPF in place, but to ignore any tests with active regeneration [16]. However, the most recent EU commission regulations for Real Driving Emissions legislation foresee the inclusion of after-treatment regenerations within the testing procedure [17]. The practicalities of this approach depend on the frequency of regeneration events which vary from technology to technology. For the DPF system, this approach allows the effects of soot accumulation to be observed but omits the impact of regeneration events. New technologies appearing in vehicles, such as lean NOx traps and SCR catalysts, also have long term dynamic characteristics [18]. There is therefore a requirement for new, high precision testing approaches that capture and account for all the long-term variations of the vehicle system within short distance CD tests.

For on-road testing where large amounts of data are collected, the effects of the DPF can be captured over long driving distances and be included in overall statistical analysis. It is important to realise that when conducting chassis dynamometer tests over a short period, these long-term effects remain relatively constant for an individual test cycle. However, when many repeated tests are conducted, each individual test is affected by these transitional behaviours in a different way. This results in vehicle-borne imprecision that affects the test-to-test repeatability during a testing campaign.

3 METHODOLOGY

Two modern diesel vehicles were installed on a high precision chassis dynamometer and repeatedly tested over the NEDC and WLTC drive cycles. The testing repeatability was subsequently quantified and compared for both vehicles before an in-depth study into the vehicle behaviour was conducted to identify the sources of imprecision inherent to the vehicle. The following sections describe the vehicles, their installation and testing procedure. The subsequent results section will show the repeatability and the methods used to improve testing precision.

3.1 Description of vehicles

Two modern diesel vehicles were used in this work, both incorporating after-treatment systems with DPF and SCR systems. Each vehicle was procured with 8,000-16,000 km ensuring that the vehicles had been run-in before testing. The vehicles were also serviced 2000 miles before the chassis dynamometer testing phase to ensure that the engine lubricant was de-greened [19].

The summary of the two vehicle specifications are given in table 1. The major differences between the two vehicles, apart from engine size, are the emissions specification and the layout of the exhaust after-treatment. Vehicle 1 was a 2014 vehicle with EURO 5+ specification incorporating a diesel oxidation catalyst (DOC) and DPF located in close proximity to the turbocharger outlet. The SCR catalyst was located downstream of the DPF, under the main body of the vehicle. Vehicle 2 was a 2016 EURO 6 specification vehicle with the SCR located upstream of the DPF. For this vehicle, both DPF and SCR were located under the vehicle chassis, some 2 metres from the turbocharger outlet.

Table 1 Test Vehicle Specifications

	Vehicle 1	Vehicle 2
Engine displacement	2.0L	1.6L
Max. power (kW)	100	90
Max. Torque (Nm)	320	300
Emissions level	EURO 5+	EURO 6
After-treatment	DOC + DPF + SCR	DOC + SCR + DPF
Unladen mass (kg)	1800	1250
Gearbox	Manual 6 Spd	Manual 6 Spd

3.2 Test setup and instrumentation

Each of the test vehicles was installed in the same way with similar levels of instrumentation onto the CLEVeR chassis dynamometer facility. The vehicle installation is summarised in figure 1. The facility is a twin axle machine such that even though the vehicles in this study were both 2-wheel drive, the rear wheels were motored to avoid vehicle controller errors during testing. The dynamometer is situated within a climatic chamber that controls temperature and humidity. The vehicle was driven by a Strahle Autopilot SAP2000[1] robot. Vehicle cooling was provided by an NEDC specification frontal fan where fan speed was controlled as a function of vehicle speed.

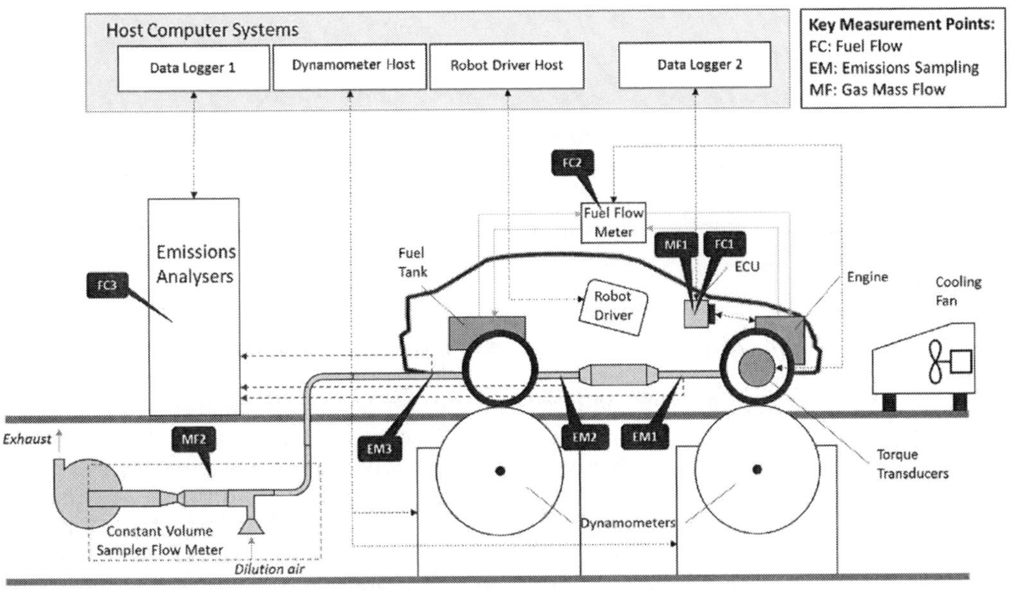

Figure 1: Chassis dynamometer layout illustrating key measurement and data acquisition devices

Data logging was undertaken from four separate computer systems:

1. An Influx Technologies[2] Rebel data logger was connected to the vehicle ECU through the On-board diagnostics (OBD) port and measured approximately 70 different data channels relating to fuel injection, exhaust gas recirculation (EGR), turbocharger, engine temperatures, DPF control and SCR control.

[1] Full specification available from: http://www.stahle.com/en/produkte.php?gr=1&prod=1 (August 2017)
[2] Full specification available from: http://www.influxtechnology.com/ (August 2017)

2. The robot host computer which logged all information relating to the robot pedal positions, pedal actuation forces and internal controllers.
3. An AVL RoadSim[3] dynamometer host computer which logged speeds and forces on the individual axles of the system.
4. A Sierra CP Cadet[4] (test cell control computer which logged all information from emissions analysers.

In total around 150 data channels were recorded across all the data systems and all data was logged at 2 Hz (which was limited by the OBD logger).

The key metric for repeatability in this work was fuel consumption. Consequently, fuel consumption was measured simultaneously using three different measurement principles all highlighted in figure 1:

> **FC1:** Estimation of fuel consumption from fuelling quantities read from the ECU
> **FC2:** Volumetric flow of fuel to the engine
> **FC3:** Carbon balance equating the known carbon content of the fuel to measured exhaust emissions of CO, CO_2 and hydrocarbons.

Each of the three fuel consumption measurements were recorded as a fuel flow rate $\dot{m}_f$ at the 2 Hz logging frequency. The total fuel consumption during each drive cycle or during each phase of the drive cycles was obtained using equation 1.

$$m_f = \oint \dot{m}_f dt \qquad \textbf{1}$$

where m_f is the fuel consumption since the start of the test and $\dot{m}_f$ the instantaneous fuel flow rate in g/s.

Each of the fuel consumption techniques will be detailed in the following sections. The information relating to the instrumentation devices used for each of these techniques is given in table 2.

3.2.1 Fuel consumption estimate using engine control unit

In modern engines, the ECU will contain signals giving an indication of the amount of fuel being injected into the cylinder. It is important to note that this is not a measurement, but rather a calibrated value within the ECU. Nevertheless, this fast signal can be a reasonable estimate of fuel consumption that offers a high resolution during transient events. In a modern engine, the numbers are usually available as a mass of fuel injected for each individual injection, and therefore the total fuel consumption is obtained by summation of the *n* individual injections as shown in equation 2. For the test vehicles in this work, up to 4 individual injections were observed during normal running, increasing to 7 during DPF regeneration.

$$\dot{m}_{f,ECU} = \frac{2N_{eng}}{60} \sum_{i=1}^{n} m_{f,inj\ i} \qquad \textbf{2}$$

where N_{eng} is the engine rotational speed (rev/min) and $m_{f,inj\ i}$ is the fuel injected (g) during an engine cycle for injection number i of a total number of injections n.

[3] Full specification available from: http://www.avl.com/-/avl-roadsim-chassis-dynamometer (August 2017)
[4] Full specification available from: http://www.sierrainstruments.com/autotest/products/automation-systems/cadet_v14 (August 2017)

3.2.2 *Fuel flow meter*

The fuel flow meter directly measures the volumetric flow of fuel to the engine (FC2 in figure 1). Both test vehicles were fitted with high pressure common rail systems that incorporated a return flow to tank. The return flow consists of fuel that leaks within the high-pressure pump and return flow from the injectors that is used both to cool the injectors and core to the hydraulic operation of the fuel injection system [20]. The fuel flow meter is a complete system that splits the fuelling circuit into two as shown in figure 2. With this installation, the pressure regulator ensures the local engine circuit is at the desired delivery pressure from the electric feed pump such that the flowmeter is seamless to the engine. The heat exchanger ensures the fuel circulating in the local engine circuit does not overheat. Finally, the positive displacement flow sensor measures only the fuel that flows between the two circuits that is replacing the fuel burnt in the combustion chambers.

The meter outputs a volumetric flow reading which is converted into a mass flow reading using equation 3. The fuel density is a function of temperature [21]. However, as the fuel temperature at the location of the sensor remains stable throughout the test, the effect of temperature is minimal [22].

$$\dot{m}_{f,FM} = \rho_f(T_f)\dot{V}_f \qquad \textbf{3}$$

where $\rho_f(T_f)$ is the fuel density (kg/m3) which is a function of fuel temperature and $\dot{V}_f$ is the volumetric fuel flow rate (m3/s).

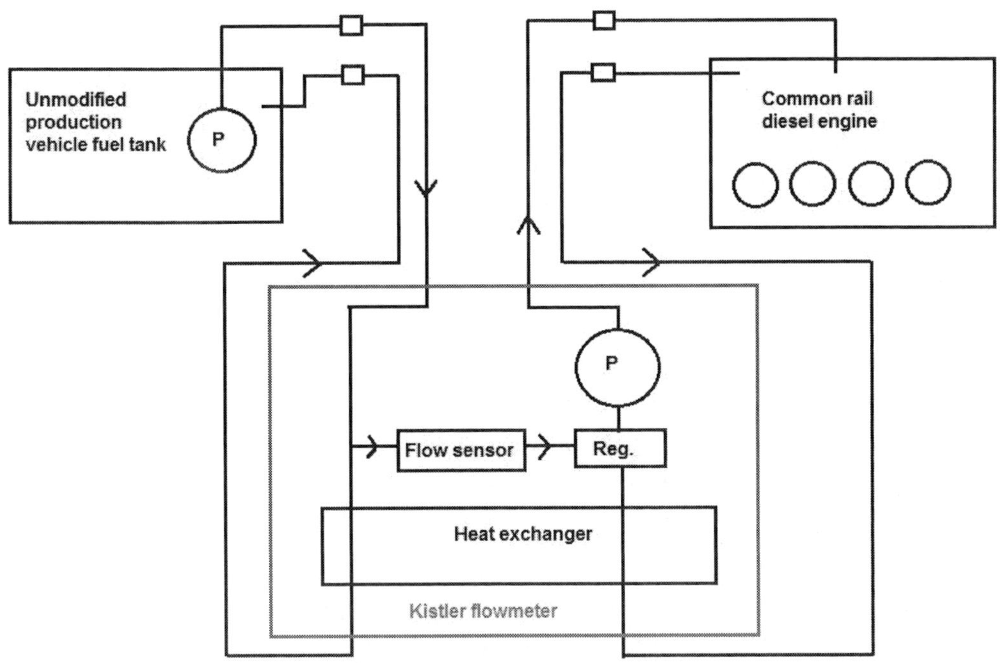

Figure 2: Detail of fuelling circuit illustrating the location of fuel flow meter sensor (P: Electric pump)

Table 2 Instrumentation specification for fuel consumption measurements

Location	Analyser type	Make/Model	Measurement principle
FC1	ECU Fuelling quantity estimate	N/A	None – derived from calibration
FC2	Fuel flow meter	Kistler DFL3x-5bar	Positive displacement volumetric sensor
EM1/EM2	Emissions analyser	AVL SESAM i60 FT	Fourier Transform infra-red spectrometer (FTIR)
EM3	Emissions analyser	Horiba Mexa 7000 series	CO/CO_2: Non-dispersive infra-red (NDIR) HC: Flame ionization Detector (FID)
MF1	ECU mass flow measure	Unknown	Vehicle hot film flow meter
MF2	Exhaust mass flow meter	Horiba CVS 7000	Constant Volume Sampler + dilution air Venturi flow meter

3.2.3 Carbon balance

The carbon balance method equates the measured carbon in the exhaust gases to the carbon concentration of the fuel to determine how much fuel has been burnt and is calculated using equation 4.

$$\dot{m}_{f,CB} = \frac{1}{\frac{w_c}{100}}\left(\frac{w_c}{100}\dot{m}_{THC} + 0.428\dot{m}_{CO} + 0.273\dot{m}_{CO_2}\right) \qquad \mathbf{4}$$

where w_c is the carbon content of the fuel by mass and $\dot{m}_{THC}$, $\dot{m}_{CO}$ and $\dot{m}_{CO2}$ are the exhaust flow rates of unburned hydrocarbons, carbon monoxide and carbon dioxide respectively (calculated using equation 5).

The mass of carbon in the exhaust is determined by the weighted sum of CO, CO_2 and THC emissions, each calculated using equation 5. This equation includes two standard factors which describe the density of the exhaust flow and a compensation for water vapour in the exhaust [23]. The dry to wet factor (K_{dw}) is applied for analysers where the exhaust sample is cooled to remove water vapour which can interfere with certain measurement techniques. This factor should be applied for the measurements of CO and CO_2 using non-dispersive IR (NDIR) analysers, but omitted for THC measurements using Flame Ionisation technique and measurements using the FTIR analysers [22].

$$\dot{m}_X = K_{dw} c_X u_X \dot{m}_{EX} \qquad \mathbf{5}$$

where $\dot{m}_X$ is the mass flow of exhaust species X (g/s), $\dot{m}_{EX}$ is the mass flow of exhaust gases, c_X is the volumetric concentration (in ppm or %, depending on the emission) and K_{dw} and u_X are factors defined in BS ISO 8178-1:2006 [14].

The exhaust flow was measured in two ways in this work highlighted in figure 1:

- **MF1** uses the ECU air flow and ECU fuel flow to calculate fuel consumption using equation 6. It is unconventional to use a fuel flow rate in the calculation of fuel flow rate, however this is used for the calculation of exhaust mass flow and only accounts for a small fraction of total flow. Using the ECU will be sufficient without affecting accuracy thus negating the need to perform iterative calculations.

- **MF2** is a direct measurement using a constant volume sampling system to measure exhaust flow

$$\dot{m}_{EX} = \dot{m}_{air,ECU} + \dot{m}_{f,ECU} \quad \textbf{6}$$

where $\dot{m}_{air,ECU}$ and $\dot{m}_{f,ECU}$ and the air and fuel flow recorded within the engine ECU.

The carbon balance method relies on the measurement of emissions concentrations and exhaust flow rate and can be applied with any combination of the three sampling points (EM1, EM2 and EM3 in figure 1) and two mass flow measurements (MF1 and MF2 in figure 1). This results in six possible estimates of carbon balance fuel consumption. In the interest of conciseness, this paper will focus on the carbon balance estimate using mass flow MF1 and exhaust gas measurements EM3 however comments will be made where significant differences were observed between carbon balance methods.

3.3 Test plan

The vehicles were operated on the chassis dynamometer over both the New European Drive cycle (NEDC) and the World Harmonized Light Duty Cycle (WLTC). The cycles were run from a cold start condition where the vehicle had been soaked overnight at an ambient temperature of 25 °C and 23 °C for NEDC and WLTC respectively. The cycles were also run from a hot starting condition where the vehicle was run for 20 minutes at 70km/h such that the vehicle was thermally soaked at its operating temperature. For these "hot start" tests, the ambient cell temperature was controlled to 25 °C and 23 °C as for the cold start tests. The number of repeated tests is given in table 3.The vehicle speed duty cycles and the engine coolant temperature for vehicle 1 over both the overnight soak and hot start tests are shown in figure 3. Also highlighted are the typical breakdown of these cycles into phases as per the respective regulations defining these cycles. The fuel consumption results presented in this paper will be broken down according to these phases.

Table 3 Total number of repeated experiments for each vehicle and each test type

Vehicle	NEDC		WLTC	
	Overnight soak at 25°C	Hot start	Overnight soak at 23°C	Hot start
1	6	11	7	5
2	7	9	9	8

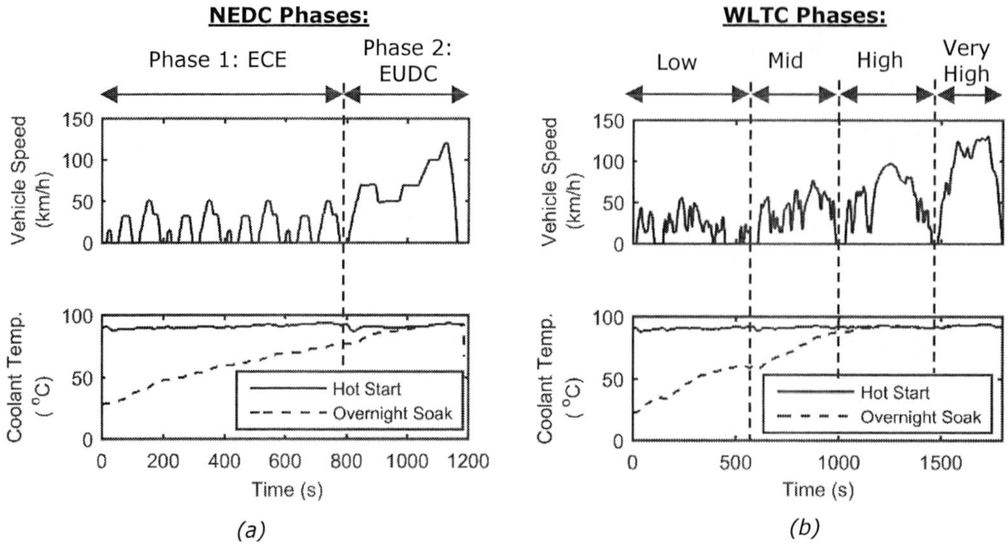

Figure 3: Vehicle speed and engine coolant temperatures for vehicle duty cycles from hot and cold start shows (a) NEDC and (b) WLTC

The gear shifting strategies used varied according to the vehicle and the test cycle. For vehicle 1 running on the NEDC cycle, the standard regulation gear shift points were used [24]. It should be noted that this is different to the gear shifting strategy that would be used in certification testing, where the vehicle shift indicator would overrule the standard regulation. Thus, for vehicle 1 the absolute fuel consumption results from this work will not be comparable with published certification numbers [25, 26]. For vehicle 2, additional effort was made to reproduce the vehicle shift indicator logic within the robot driver such that the gear shifting strategy was closer to that for the certification process. This was achieved by running a series of cycles with a human driver who reacted to the shift indicator lights. The measured gear shift points were used to create a shift pattern for the robot control.

For the WLTC, the prototype gear shifting regulation was used which provides a strategy for calculating the gear shifting points by comparing the theoretically available engine power in each gear to the drive cycle demanded power at each timestamp [27, 28]. The highest possible gear is then chosen to achieve the required cycle power, whilst also applying a selection of additional clauses, such as a minimum in gear time [29]. This regulation was not finalised at the time the experiments were carried out and so a small number of adjustments were made by iteration to avoid vehicle stalls during pull-away manoeuvres and to mitigate transient performance effects until a workable strategy was achieved. This consisted of adjusting the dynamic torque factor within the gear shift algorithm.

4 RESULTS AND DISCUSSION

4.1 Summary of testing precision by section of the drive cycle

Figure 4 presents a summary of the base repeatability of each fuel consumption metric, for both vehicles over the different phase of the hot and cold duty cycles. The repeatability is quantified using the Root Mean Square Error (RMSE), calculated using equation 7. The figure contains much information which will be analysed through a series of comparisons in the following sections. Absolute fuel consumption values for each vehicle are provided in appendix 1.

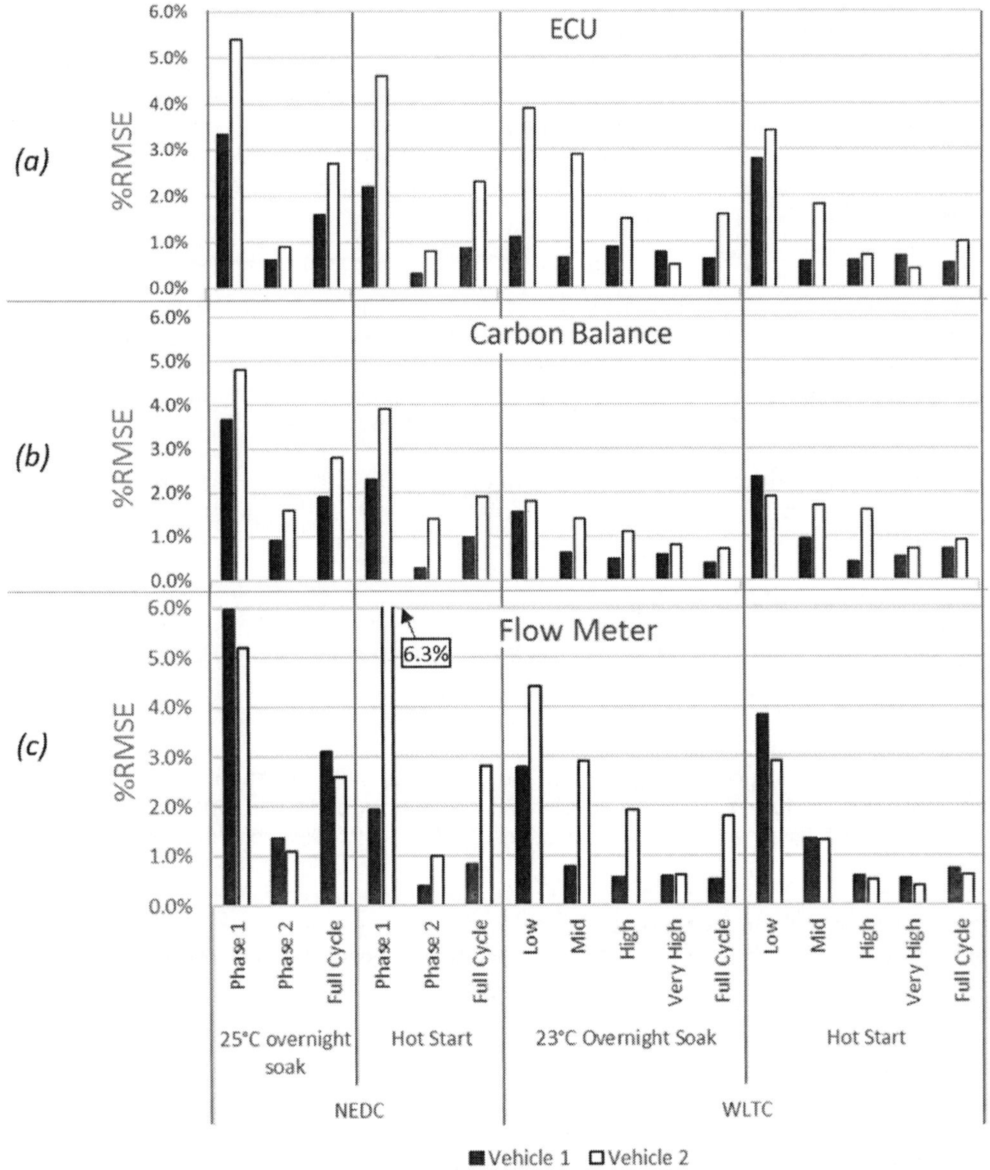

Figure 4: %RMSE of testing by certification cycle and by vehicle for (a) ECU fuel consumption estimate, (b) carbon balance estimate and (c) Fuel flow meter

$$RMSE = \frac{\sqrt{\frac{\sum_{i=1}^{n}\left(\overline{m_f} - m_{f,i}\right)^2}{n}}}{\overline{m_f}} \quad \textbf{7}$$

where i is an individual test and n the total number of repeated tests and:

$$\overline{m_f} = \frac{\sum_{i=1}^{n} m_{f,i}}{n}$$

4.1.1 NEDC cycle vs. WLTC cycle

Overall, the WLTC cycle is more repeatable than the NEDC cycle for both vehicles and for all measurement principles. For example, the RMSE for the fuel flow meter for vehicles 1 and 2 over the cold start NEDC is 2.5-3.1 % whereas for the cold start WLTC it is 0.5-1.8 %. This can seem somewhat surprising as the WLTC is far more transient in nature and therefore would present a harder problem for the robot controllers. Other researchers have also demonstrated similar repeatability of the WLTC and NEDC cycles [30, 31].

Within the NEDC and WLTC cycles, there is a consistent trend for higher power phases to be more repeatable than lower power phases. For example, phase 1 of the NEDC consistently has higher RMSE than phase 2. The low power phase of the WLTC has poorer repeatability than the mid, high, and very high phases. The fact that the NEDC phase 1 has the highest RMSE suggests that this may be due to the number of vehicle stops and pull-aways. The NEDC has 12 vehicle pull-away events (see figure 3); the low phase of the WLTC has 3 whilst all other phases only have a single pull-away. However, it is also conceivable that there are fixed and scaling factors in the total error meaning that overall error and repeatability should be better under high power/high fuel consumption conditions.

4.1.2 Hot vs. cold testing

A comparison of the repeatability of overnight soak tests compared to hot start tests shows that the RMSE for NEDC tests is lower for hot start. The one exception being the flow meter readings for vehicle 2. For both vehicles, most of this difference appears to come from phase 1 on the duty cycle.

For the WLTC, the repeatability for vehicle 1 tended to be slightly worse in the hot start tests for the low phase, but then similar for all subsequent phases. For vehicle 2 the hot start tests appear to have slightly lower RMSE (except for the carbon balance method).

4.1.3 Vehicle 1 vs. Vehicle 2

Some of the vehicle to vehicle differences have already been mentioned in the previous section. However, it is interesting to point out in addition that vehicle 2 generally has worse repeatability than vehicle 1. The main exception to this is the behaviour of vehicle 1 during the overnight soak NEDC cycles which warrant further investigation.

4.1.4 Measurement principle comparison

There are clear differences in the magnitude of RMSE from the different measurement techniques. In particular, the fuel flowmeter has consistently higher RMSE than the other measurement principles over the first phase of the cold start NEDC and WLTC. It is also worth discussing that the carbon balance has lower RMSE than the ECU metric, despite the ECU metric featuring in the calculation for carbon balance method (equation 6). This is because the sensitivity of the carbon balance fuel consumption to the ECU fuel consumption is very small. The ECU metric is only used to calculate exhaust mass flow and around 98% of this mass flow is combusted air, measured via the car's MAF sensor.

4.2 Detailed analysis of the causes of the precision issues

Based on the overall assessment of repeatability, further investigation will be conducted on the following points:

1. Understanding the poor repeatability of vehicle 1 during phase 1 of the NEDC
2. Highlighting the reasons why vehicle 2 has worse repeatability than vehicle 1

4.2.1 *Overnight soak NEDC cycle with vehicle 1*

Figure 5 shows the evolution of fuel consumption for vehicle 1 throughout the 6 repeated overnight-soak cold-start NEDC tests. The plot presents the measured fuel flow rate, the cumulative measured fuel consumed since the start of the test and the difference in cumulative fuel consumed for each test to the mean fuel consumed (calculated using equation 8). This form of representation highlights where divergence occurs in the tests to focus on periods where test-to-test variability is occurring.

$$Cumulative\ Difference\ (test\ i) = m_{f,test\ i}(t) - \frac{\sum_{i=1}^{n} m_{f,i}(t)}{n} \quad \mathbf{8}$$

where i is an individual test and n the total number of repeated tests.

Figure 5 immediately shows that significant divergence between tests occurs between 500-600 seconds. The tests have been grouped into two difference control behaviours. The two tests exhibiting control behaviour 2 have significantly lower fuel consumption during this period compared to the other four tests exhibiting control behaviour 1. This is evidenced by the significant negative gradient during this phase on the cumulative difference plot. Interestingly these tests also have slightly higher fuel consumption between 675-750 seconds (evidenced by the positive gradient for the two tests plotted in dashed lines).

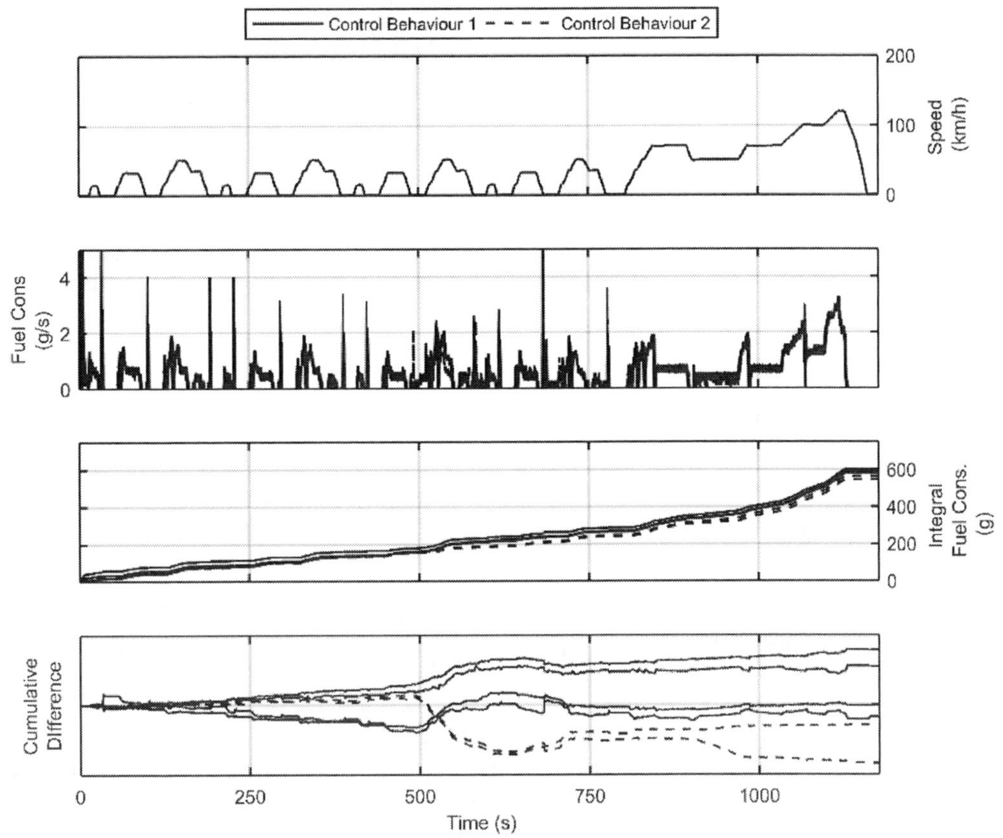

Figure 5: Evolution of fuel flowmeter fuel consumption for vehicle 1 during overnight soak NEDC

Partial least squares regression (PLS) was used to seek the strongest correlations between the measured fuel consumption and the 150 measured data channels. PLS regression is suited to data sets containing many correlated data channels and a small number of repeated tests. This mathematical approach *projects* the individual data channels (*independent variables*) onto *latent variables* which are most correlated with the particular response variable [32, 33]. Each of the latent variables *explains* a certain proportion of the response variable and each independent variable is *loaded* to a different degree onto the latent variables.

The analysis here will seek to analyse the independent variables that have the highest loading on latent variables that have the highest explanation of a single response variable as illustrated in figure 6. The PLS approach was applied to the data set from phase 1 of the NEDC. The response variable was defined as the total fuel consumed over NEDC phase 1, as measured by the Kistler flow meter. The independent variables were a subset of the 150 data channels recorded during the tests, averaged over phase 1 of the cycle. The subset of 63 channels was chosen by eliminating those channels that would automatically correlate with fuel consumption such as CO_2 emissions and other fuel consumption measures. The data preparation is illustrated in figure 7.

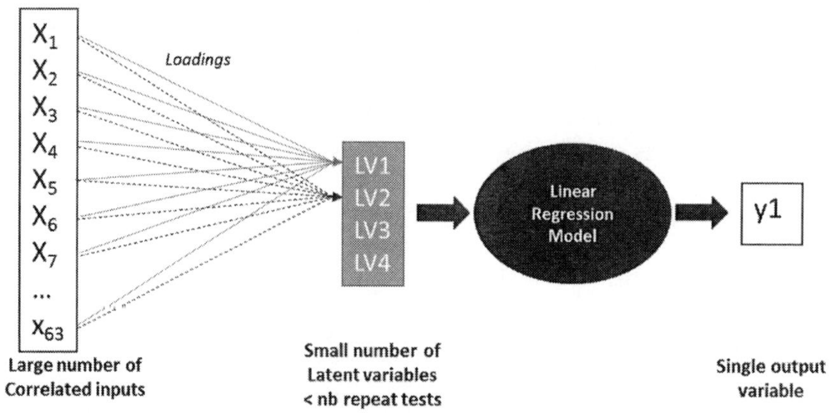

Figure 6: Overview of the PLS modelling approach adapted to seek correlations between measured data channels and a single output variable

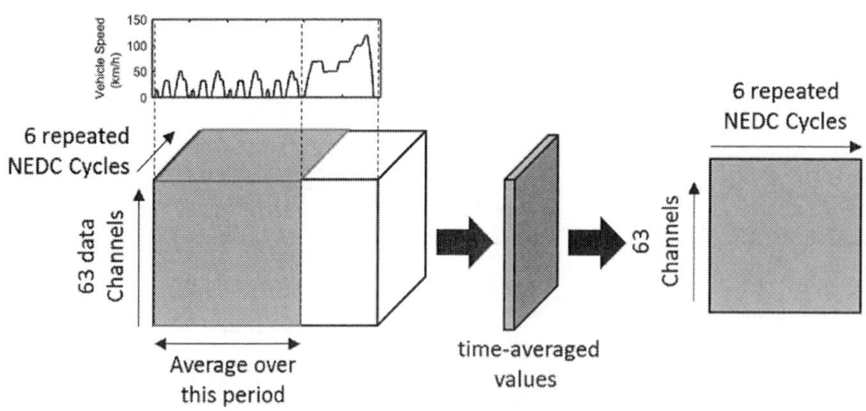

Figure 7: Graphical representation of the data manipulation to create input matrix for PLS regression modelling

The top-ranking variables from the PLS analysis were primarily related to post injection activity, throttle valve and exhaust gas temperature. Figure 8 shows the time-based recording of these variables over the NEDC, again grouped by control behaviour. From these measurements, it is obvious that with control behaviour 1 (solid black lines) at around 490 seconds the intake throttle is closed and post injections are activated. This results in an increase in exhaust gas temperature. For control behaviour 2, a similar behaviour occurs at about 675 seconds. These would directly explain the changes in fuel consumption observed in figure 5. The behaviour of the ECU is comparable to a DPF regeneration; however, previous knowledge of the vehicle confirms that no DPF regeneration is active at this period [5]. This behaviour could be described as an *exhaust heating strategy* that could be implemented to ensure that the exhaust after-treatment devices reach their operating temperature to supress tailpipe emissions. Catalyst light-off remains a critical issue in diesel engines [34].

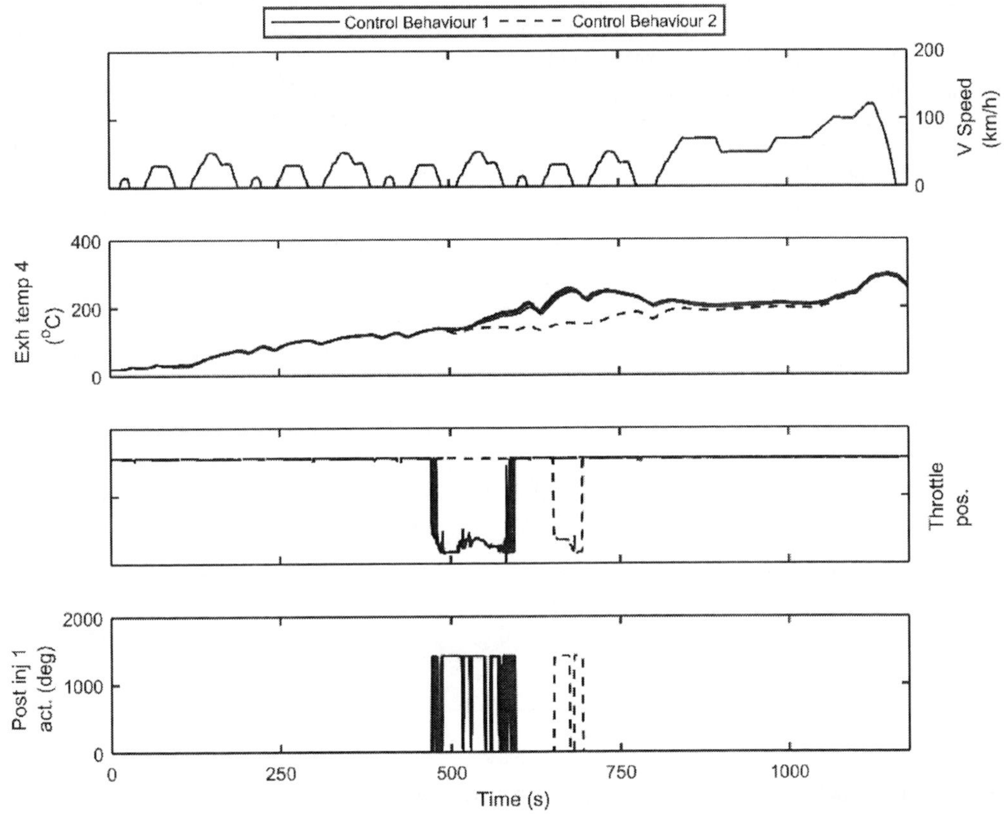

Figure 8: Differences in exhaust gas temperature post-DPF, throttle control and the use of post injections for repeated overnight soak NEDC tests with vehicle 1

To identify the potential reasons for the differences seen in figure 8, PLS regression was used for a second time. In this case, the response variable was the average post injection activity between 450-600 seconds. The independent variables were a subset of 43 variables chosen from the 150 variables, having eliminated those describing post injection activity that would automatically correlate. In addition, these variables have been averaged over the period of the drive cycle 0-450 seconds (i.e. the period before the heating strategy begins). In this way, correlations are sought that may explain the differences in behaviour after 450 seconds.

The highest loaded independent variable is the engine coolant temperature logged from the ECU. This would suggest that the ECU is implementing the differences in strategy based on the warm up rate of the vehicle. This would be logical, as a colder vehicle may not meet emissions targets due to differences in catalyst light-off.

The engine coolant temperature will be a result of the balance between heat rejected to the coolant from the engine and the cooling of the vehicle by the test cell ambient air. Variations did exist in the cell ambient temperature; however the measurements of cell temperature were all between 25.1 °C and 25.7 °C. This cell temperature measurement may not be representative of engine bay temperatures which would also result from variations in the cooling air flow around the vehicle. Previous work has shown similar results where the correlation between engine temperature and engine behaviour is well defined but the correlation between vehicle cooling and vehicle temperature is less repeatable [3].

It was therefore postulated that the heating strategy is sensitive to the warm up rate that can be modulated through ambient temperature. A series of NEDC tests were performed at ambient temperatures of 22 °C, 25 °C and 28 °C. Figure 9 shows the resulting impact on the engine heating strategy. The results reveal that lower ambient temperature tends to advance the start of heating strategy while heating mode was both retarded in the cycle and reduced in magnitude with 28 °C ambient temperature.

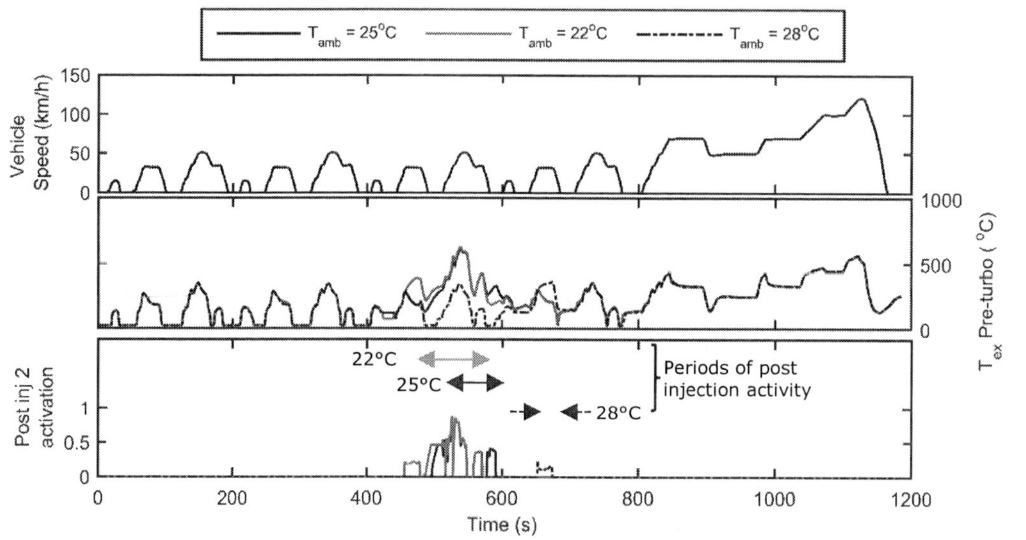

Figure 9: Differences in exhaust gas temperature, throttle control and the use of post injections for repeated overnight soak NEDC tests with vehicle 1

To investigate the repeatability of the heating strategy at a certain ambient temperature, a series of 6 NEDC tests were completed with a soak temperature of 28 °C. Figure 10 shows the mean fuel consumption and 95 % confidence intervals calculated using equation 9. With the higher ambient temperature, the overall fuel consumption over the NEDC is reduced by 15-20 g for flow meter and ECU measurements. Interestingly the carbon balance method is much more sensitive to this change. For all measurement instruments, the RMSE is reduced from 2-3 % to 0.6-0.7 % demonstrating that the vehicle is less sensitive at this higher operating temperature, thus allowing higher levels of test precision. This is illustrated by the smaller confidence intervals for the tests conducted with a 28 °C overnight soak temperature.

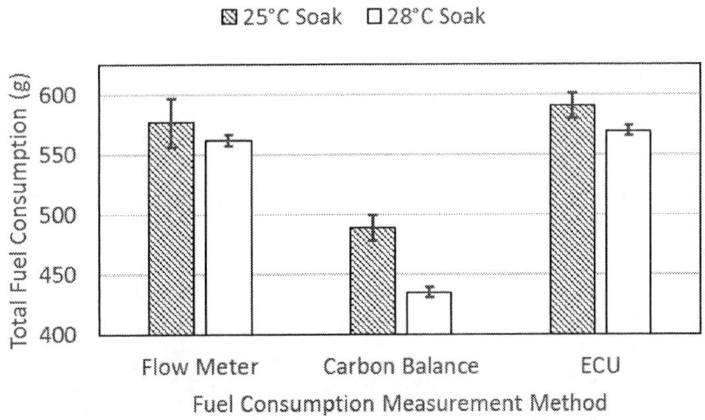

Figure 10: Evolution of fuel flowmeter fuel consumption for vehicle 1 during overnight soak NEDC (error bars showing 95% confidence intervals)

$$95\%CI = \frac{\sigma_j}{\sqrt{n_j}} t_{n_j,95\%} \qquad \mathbf{9}$$

where j represents the 25 °C or 28 °C testing campaign.

4.2.2 Testing precision with vehicle 2

The PLS approach was again used for vehicle 2, seeking correlations between the measured data and the test-to-test variation in fuel consumption. This analysis highlighted two major factors that correlated with the differences in fuel consumption: start stop behaviour and battery smart charger system.

Start stop behaviour

Table 4 compares the start-stop behaviour for the two vehicles in each of the drive cycles. These are presented as the average number of stops over the repeated tests. Also shown in brackets is the standard deviation of the number of start-stop events. For the NEDC, a vehicle that systematically engaged it's start-stop system would have 12 occurrences. As can be seen from the results, neither vehicle systematically engaged its start stop system. However, vehicle 1 has less variation in the number of start stops as evidenced by smaller standard deviations. The same is true of the WLTC cycles where the maximum number of events would be 5. The larger variation for vehicle 2 would result in larger variation in total fuel consumption.

It is worth noting at this point that vehicle 2 had a binary behaviour of the start-stop system: it was either working or not working for the duration of the test cycles. For vehicle 1, the behaviour was more erratic with the start-stop system varying over the course of a particular test.

Table 4: Average and standard deviation (in brackets) of number of times start-stop system engages during repeated tests for each vehicle and each duty cycle

Vehicle	NEDC (max 12)		WLTC (max 5)	
	Overnight Soak	Hot Start	Overnight Soak	Hot Start
1	10.3 (0.5)	11.5 (0.5)	2 (1)	1.4 (1.1)
2	10.3 (4.5)	9.2 (5.2)	3.9 (2.2)	3.6 (2.2)

Battery charging

Vehicle 2 is equipped with a smart charging system that chooses when to recharge the vehicle battery. These systems are used to encourage charging when the vehicle is decelerating, thus making use of kinetic energy recovery. Despite a procedure being in place to leave the battery system on charge between each experiment, significant differences in battery charge behaviour were observed. These different modes are presented in figure 11. The battery behaviour could be categorised in 2 modes:

- Mode 1: The battery is subjected to a constant trickle charge whilst the engine is running and to an increased charge at the point the engine is started. This is illustrated in figure 11 by the large positive spikes at vehicle pull-aways, gradually decaying to a near constant positive charge until the next vehicle stop. When the vehicle is stopped, there is a constant negative current (current draw) from the vehicle electrical systems.
- Mode 2: The vehicle has a persistent current draw on the battery whilst the engine is running and whilst stopped – this is around -10 A. When the vehicle undergoes a deceleration, the smart charger engages resulting in spikes of positive charge to the battery.

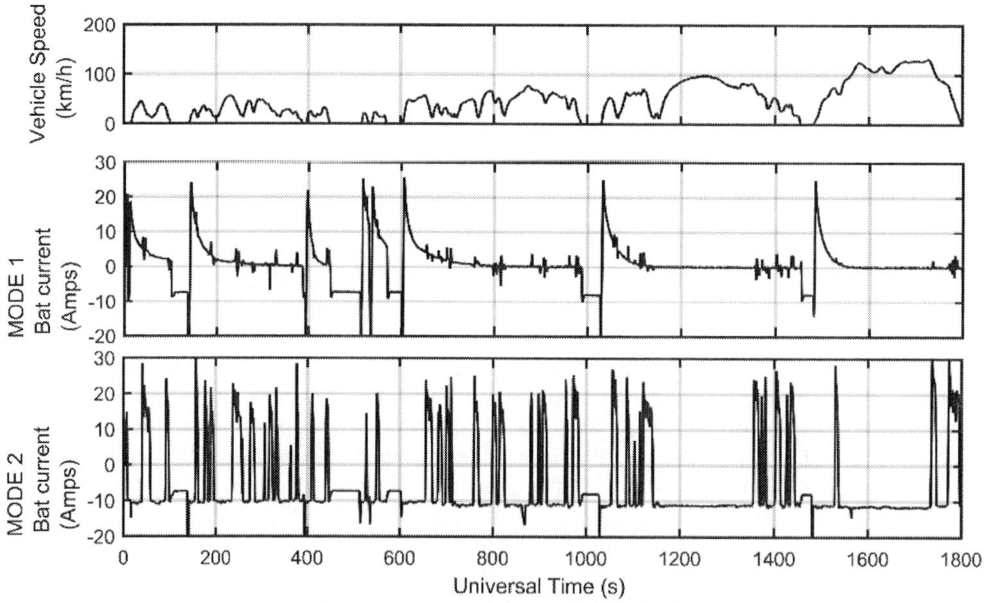

Figure 11: Comparison of different battery charging behaviours with vehicle 2 over WLTC cycles

Switching between these two modes was observed both between and within tests. Mode 2 is designed to reduce the parasitic load on the engine from the alternator and thus reduce fuel consumption. This is clearly illustrated in figure 12. Smart charging systems have been shown to improve fuel consumption by 2-3% [11], therefore it is clear why a switching will compromise the precision for fuel consumption measurement. Although the effect of smart charging behaviour on fuel consumption was obvious, methods to mitigate the variation in the system could not be found within the project timeframe. The proposed solution relies on monitoring the behaviour of the alternator controller to isolate testes with similar charging modes.

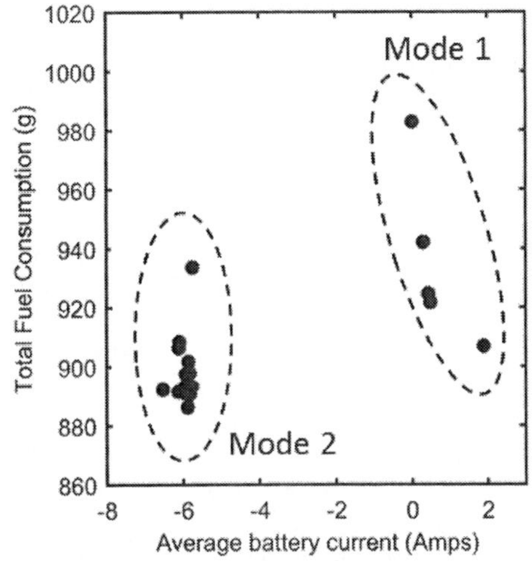

Figure 12: Correlation between fuel consumption and the battery charging mode for vehicle 2 over overnight soak WLTC cycle

4.3 Summary of possible improvements in testing precision

The potential improvements in RMSE through better vehicle control have been captured in this section. These have been collated from:

1. The comparison between different duty cycles (figure 4).
2. The results from the analysis of vehicle 1 behaviour (figure 10).
3. The RMSE for subsets of tests with comparable battery of start-stop behaviour within the vehicle 2 datasets.

Table 5 details these potential benefits and proposes steps to achieve them.

Table 5 Potential improvements to testing precision through the control of various parameters

Factor	Vehicle 1	Vehicle 2	Potential improvement in RMSE (%points)	Possible Steps to improve RMSE
Heating Strategy	X		2.3	Increase ambient temperature (within range) to reduce vehicle sensitivity
Battery Charging		X	1	Monitor Battery behaviour (current/voltage) to ensure consistent operation and aid selection of outliers
Cold vs Hot Testing	X	X	0	No preference
Duty Cycle	X	X	2	Select duty cycles that have higher speed
NEDC vs WLTC	X	X	0.2	Choose WTLC for better repeatability of fuel consumption
Start/Stop	X	X	0.2-0.6	Ensure consistent behaviour of Start/Stop system. System could be disabled. Different measurement systems react differently to Start/Stop

5 CONCLUSIONS

This paper has investigated the precision of testing for two modern diesel vehicles over both the NEDC and the WLTC cycles. The work was conducted with a focus on the imprecision inherent to the vehicle rather than imprecision from the vehicle installation on the test facility.

The duty cycle was shown to have an influence on testing precision, with cycles with higher speed offering greater precision. This was shown by the WLTC offering a reduction in FC RMSE of 0.2 %-points compared to the NEDC. Higher load phases such as the EUDC for the NEDC and the high and extra high load phases offered up to 2 %-points reduction in RMSE compared to the low load cycles. Interestingly, the RMSE was similar for cold and hot start testing.

Some sources of imprecision were specific to each vehicle. One vehicle was seen to have an exhaust heating strategy that was particularly sensitive to ambient temperatures around 25°C on the NEDC cycle. This was shown to cause 2.3 %-points in RMSE of fuel consumption. The second vehicle incorporated a smart charging strategy and test-to-test variation in its behaviour caused 1 %-point increase in FC RMSE.

This study has highlighted some of the challenges when using vehicle testing to demonstrate small differences in fuel economy and some of the vehicle-borne factors that need to be managed.

REFERENCES

(1) Taylor AMKP. Science review of internal combustion engines. Energy Policy. 2008;36(12):4657-67.

(2) Nattrass S.R., Jones W.R., The Application of Telematics to the High-Precision Assessment of Fuel-Borne Fuel Economy Additives. SAE International; 2012, SAE Paper Number 2012-01-1738.

(3) Brace, C. J., Burke, R. and Moffa, J., Increasing accuracy and repeatability of fuel consumption measurement in chassis dynamometer testing. Proceedings of the Institution of Mechanical Engineers, Part D: Journal of Automobile Engineering, 2009. 223 (9): p. 1163-1177.

(4) Tutuianu M, Bonnel P, Ciuffo B, Haniu T, Ichikawa N, Marotta A, Pavlovic, J. and Steven, H., Development of the World-wide harmonized Light duty Test Cycle (WLTC) and a possible pathway for its introduction in the European legislation. Transportation Research Part D: Transport and Environment. 2015;40:61-75.

(5) Chappell, E., Burke, R., Lu, P., Gee, M. and Williams, R., "Analysis of a Diesel Passenger Car Behavior On-Road and over Certification Duty Cycles," SAE Int. J. Engines 9(4):2201-2214, 2016, doi:10.4271/2016-01-2328.

(6) Singh N, Rutland CJ, Foster DE, Narayanaswamy K, He Y. Investigation into Different DPF Regeneration Strategies Based on Fuel Economy Using Integrated System Simulation. SAE International; 2009, SAE Paper number 2009-01-1275.

(7) Burke R. Investigation into the interactions between thermal management, lubrication and control systems of a diesel engine. PhD Thesis, Bath: University of Bath; 2011.

(8) Chappell, E.C., Brace, C.J. and Ritchie, C., The control of chassis dynamometer fuel consumption testing noise factors and the use of response modelling for validation of test repeatability. Proceedings of the Institution of Mechanical Engineers, Part D: Journal of Automobile Engineering, 2013. 227 (6): p. 853-865.

(9) Skaf Z, Aliyev T, Shead L, Steffen T., The State of the Art in Selective Catalytic Reduction Control. SAE International; 2014, SAE Paper Number 2014-01-1533.
(10) Theis JR, Dearth M, McCabe R., LNT+SCR Catalyst Systems Optimized for NOx Conversion on Diesel Applications. SAE International; 2011, SAE Paper number 2011-01-0305.
(11) Montalto, I., Tavella Ing, D., Casavola PhD, A. and De Cristofaro, F., Intelligent Alternator Employment To Reduce Co2 Emission and to Improve Engine Performance, SAE Int. J. Alt. Power. :2012, doi:10.4271/2011-01-2444.
(12) Lakshminarasimhan, V. and Athani, G., An Intelligent Alternator Control Mechanism for Energy Recuperation and Fuel Efficiency Improvement, SAE Int. J. Alt. Power. 2(1):2013, doi:10.4271/2013-01-1750.
(13) Konstandopoulos A.G., Kostoglou M., Skaperdas E., Papaioannou E., Zarvalis D. and Kladopoulou E., Fundamental Studies of Diesel Particulate Filters: Transient Loading, Regeneration and Aging, SAE International; 2000, SAE Paper Number 2000-01-1016.
(14) Stenning, L., Strategies for Achieving pre DPF Regeneration Temperatures using in Cylinder Post Injection on a Common Rail Diesel Engine with EGR, DOC and Intake Throttle, SAE International; 2010, SAE Paper Number 2010-36-0306.
(15) Vojtisek-Lom M., Fenkl M., Dufek M. and Mareš J., Off-cycle, Real-World Emissions of Modern Light Duty Diesel Vehicles; 2009, SAE Paper Number 2009-24-0148.
(16) Andersson, J., May, J., Favre, C., Bosteels, D., de Vries, S., Heaney, M., Keenan, M. and Mansell, J., On-Road and Chassis Dynamometer Evaluations of Emissions from Two Euro 6 Diesel Vehicles. SAE Int. J. Fuels Lubr., 2014. 7 (3): p. 919-934. SAE Technical Paper No: 2014-01-2826.
(17) Commission Regulation (EU) 2017/1154 of 7 June 2017 amending Regulation (EU) 2017/1151 supplementing Regulation (EC) No 715/2007 of the European Parliament and of the Council on type-approval of motor vehicles with respect to emissions from light passenger and commercial vehicles (Euro 5 and Euro 6) and on access to vehicle repair and maintenance information, amending Directive 2007/46/EC of the European Parliament and of the Council, Commission Regulation (EC) No 692/2008 and Commission Regulation (EU) No 1230/2012 and repealing Regulation (EC) No 692/2008 and Directive 2007/46/EC of the European Parliament and of the Council as regards real-driving emissions from light passenger and commercial vehicles (Euro 6) (Text with EEA relevance), Official Journal of the European Union, L175 pp708-732, 7th June 2017.
(18) Majewski, W.A., Selective Catalytic Reduction, Ecopoint Inc., 2015, [online], available from: https://www.dieselnet.com/tech/cat_scr.php [accessed 05/07/2015]
(19) Hawley JG, Bannister CD, Brace CJ, Akehurst S, Pegg I and Avery MR., The effect of engine and transmission oil viscometrics on vehicle fuel consumption. Proceedings of the Institution of Mechanical Engineers, Part D: Journal of Automobile Engineering. 2010;224(9):1213-28.
(20) Guerrassi, N. and Dupraz, P., A common rail injection system for high-speed, direct-injection diesel engines, SAE International Congress and Exposition, SAE International Warrendale Pennsylvania USA: Detroit, Michigan.
(21) Dzida, M. and Prusakiewicz, P. The effect of temperature and pressure on the physicochemical properties of petroleum diesel oil and biodiesel fuel. Fuel, 2008, 87(10-11), 1941-1948.
(22) Burke RD, Brace CJ, Hawley JG., Critical analysis of on-engine fuel consumption measurement. Proc Inst Mech Eng Part D-J Automob Eng. 2011;255(6):829-44.

(23) BSI. BS ISO 8178-1:2006, Reciprocating internal combustion engines. Exhaust emission measurement. Test-bed measurement of gaseous and particulate exhaust emissions. (BSI, 2009)
(24) Anon, ECE 83-05: Uniform provisions concerning the approval of: Vehicles with regard to the emission of pollutants according to engine fuel requirements (online), available from:
http://air.emb.gov.ph/wp-content/uploads/2016/10/DAO2016-23-1.pdf [accessed 01/07/2017]
(25) Amanatidis S, Ntziachristos L, Samaras Z, Kouridis C, Janka K and Tikkanen J., Use of a PPS Sensor in Evaluating the Impact of Fuel Efficiency Improvement Technologies on the Particle Emissions of a Euro 5 Diesel Car. SAE International; 2014, SAE Paper number 2014-01-1601.
(26) Vagg C, Brace C, Hari D, Akehurst S and Ash L., A Driver Advisory Tool to Reduce Fuel Consumption. SAE International; 2013, SAE Paper Number 2012-01-2087.
(27) UN GTR No. 15 - Worldwide harmonized Light vehicles Test Procedure (ECE/TRANS/180/Add.15), 12th March 2014, available from:
https://www.unece.org/trans/main/wp29/wp29wgs/wp29gen/wp29glob_registry.html [accessed 01/08/2017]
(28) Ligterink, N.E., van Mensch, P., Cuelenaere, R.F.A., Hausberger, S., Leitner, D. and Silberholz, G., Correction algorithms for WLTC chassis dynamometer and coast down testing, TNO report number, TNO 2015 R10955, TNO, Delft, NL, 2015 (online) available from:
https://ec.europa.eu/clima/sites/clima/files/transport/vehicles/docs/wltp_correction_algorithms_en.pdf [accessed 01/10/2017]
(29) Anon, Global Technical Regulations (GTRs) of the UNECE, Online, available from:
https://www.unece.org/trans/main/wp29/wp29wgs/wp29gen/wp29glob_registry.html (accessed 27/04/2017)
(30) Bielaczyc P, Woodburn J, Szczotka A. A Comparison of Carbon Dioxide Exhaust Emissions and Fuel Consumption for Vehicles Tested over the NEDC, FTP-75 and WLTC Chassis Dynamometer Test Cycles. SAE International; 2015, SAE Paper Number 2015-01-1065.
(31) Favre C, Bosteels D and May J. Exhaust Emissions from European Market-Available Passenger Cars Evaluated on Various Drive Cycles. SAE International; 2013, SAE Paper Number 2013-24-0154.
(32) Eriksson, L., Johansson, E., Kettaneh-Wold, N., Trygg, J., Wikstrom, C. and Wold, S., 2006, Multi and Megavariate data analysis. Part 1: Basic Principals and Applications, Umetrics Academy, Umea, Sweden.
(33) Wold, S., Geladi, P., Esbensen, K. and Öhman, J. (1987), Multi-way principal components-and PLS-analysis. J. Chemometrics, 1: 41–56. doi:10.1002/cem.1180010107.
(34) Laible T, Pischinger S, Holderbaum B. Internal and External Measures for Catalyst Light-Off Support. SAE International; 2015, SAE Paper Number 2015-24-2501.

APPENDIX 1
Mean fuel consumption data by test and vehicle (see figure 4)

Vehicle	FC Measurment	NEDC						WLTC										
		25°C overnight soak			Host Start			23°C overnight soak					Hot start					
		Phase 1	*Phase 2*	*Cycle Total*	*Phase 1*	*Phase 2*	*Cycle Total*	*Low*	*Mid*	*High*	*V. High*	*Cycle Total*	*Low*	*Mid*	*High*	*V. High*	*Cycle Total*
1	ECU	271	319	590	215	309	524	189	211	309	477	1194	153	199	301	472	1132
	CB	222	266	489	183	258	441	141	168	246	377	937	118	159	240	347	898
	FM	262	315	577	195	310	504	181	207	306	475	1173	140	197	304	473	1119
2	ECU	184	251	435	157	248	405	136	187	255	395	972	116	183	252	394	945
	CB	141	219	360	120	211	331	117	153	212	350	832	98	147	208	347	801
	FM	159	245	404	131	239	370	121	165	239	382	907	97	161	238	381	878

The future diesel engine as part of a hybrid powertrain

H Busch [1], T Koerfer [2], M Koetter [3], J Schaub [3], G Vagnoni [4]
[1] Coventry University, UK
[2] FEV Group Holding GmbH, Germany
[3] FEV Europe GmbH, Germany
[4] Institute for Combustion Engines (VKA), RWTH Aachen University, Germany

ABSTRACT

Future legislative requirements and market drivers challenge pollutant emissions, fuel consumption and costs of Diesel based powertrains. The introduction of hybridization offers multiple new degrees of freedom, not only in terms of the topology of the hybrid system. The hybridization strategy can also affect the configuration of the internal combustion engine, the exhaust aftertreatment and the control strategy of the complete powertrain system.

These additional degrees of freedom require new development methods to maintain short development times. The use of simulation tools allows assessing and validation of different concepts very early in the development cycle.

In the following study a simulation toolchain is used to assess different diesel hybrid powertrain configurations – including variations of engine, aftertreatment, hybridization and control concepts – for application in a medium sized SUV.

ACRONYMS

SUV	-	Sports Utility Vehicle
OEM	-	Original Equipment Manufacturer
DOC	-	Diesel Oxidation Catalst
LNT	-	Lean NOx Trap (also known as NSC – Nox Storage Catalyst)
DPF	-	Diesel Particulate Filter
SDPF	-	SCR on Diesel Particulate Filter
a-ccSDPF	-	active close-coupled SCR on Diesel Particulate Filter
SCR	-	Selective Catalytic Reduction
a-ufSCR	-	active underfloor SCR catalyst
p-ufSCR	-	passive underfloor SCR catalyst
RDE	-	Real Driving Emissions
WLTC	-	Worldwide harmonized Light vehicles Test Cycle
HEV	-	Hybrid Electric Vehicle
MHEV	-	Mild Hybrid Electric Vehicle
P0	-	Mild hybrid concept, the electric machine is integrated into the FEAD
FEAD	-	Front End Accessory Drive
BISG	-	Belt Integrated Starter Generator
P2	-	Full hybrid concept, the electric machine is located on the gearbox side
DCT-7	-	Dual Clutch Transmission with seven forward gear ratios

EGR - Exhaust Gas Recirculation
HP-EGR - High Pressure Exhaust Gas Recirculation
LP-EGR - Low Pressure Exhaust Gas Recirculation
CN6b - Chinese emission standard targeting a NO_x limit of 35 mg/km from 07/2023

1 DRIVERS FOR THE DEVELOPMENT OF ELECTRIFIED DIESEL POWERTRAINS FOR PASSENGER CARS

Passenger cars powered by Diesel engines enjoyed a massive success in the last 25 years in Europe. Excellent fuel economy and good driveability appealed to customers. Improvements of boosting, combustion and aftertreatment systems contributed to the strengths of the Diesel Powertrain. These advantages compared to vehicles powered by spark ignition engines positioned the Diesel engine as cost effective option for the end customer as well as OEM's to meet future CO2 emission regulation. However, tightening pollutant and greenhouse gas emission standards require further developments to maintain the Diesel engine as an attractive powertrain offering. Future legislation in the US targets a reduction of CO2 emissions by 20% from 2020 to 2025, which is in line with discussions to a further reduction of CO2 emissions to a range between 78 and 68 g/km CO2 in Europe[2].

Taking into account the magnitude of the emission reduction and the relatively short timeframe until 2025, meeting these targets mainly by internal combustion engine optimization is neither realistic nor cost efficient, therefore increased levels of powertrain electrification are necessary [1].

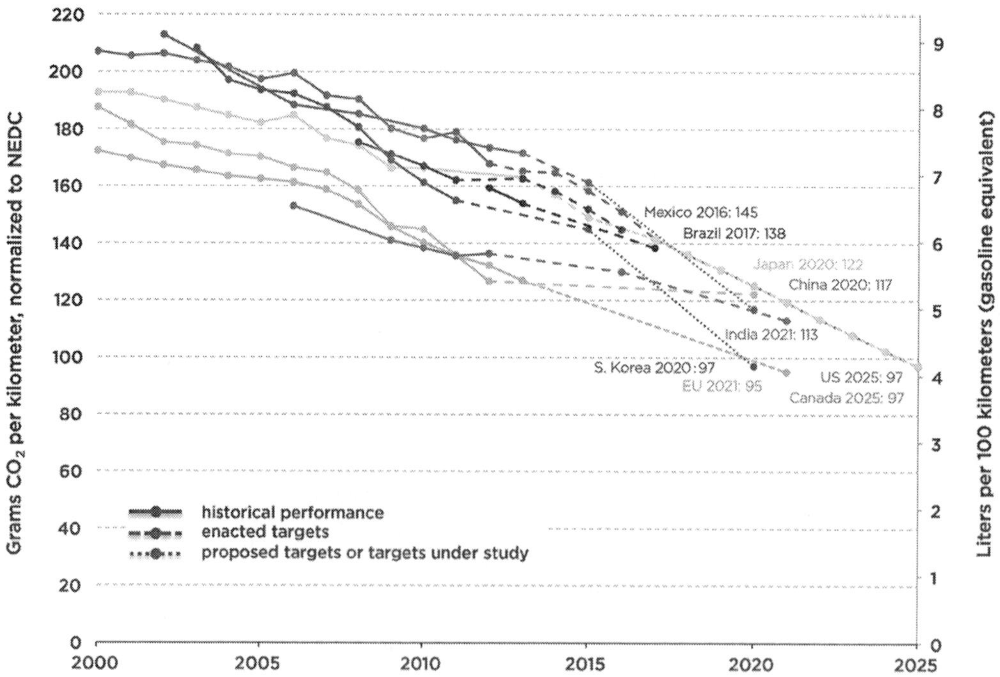

Figure 1 Evolution of CO2-Emission targets [2]

Legislative requirements to reduce vehicle CO2 emissions are accompanied by a further tightening of pollutant emissions in two ways. On the one hand limits for pollutant emissions emitted during legislative cycles measured in a laboratory environment are further tightened, on the other hand meeting pollutant emissions also in real world driving conditions is enforced.

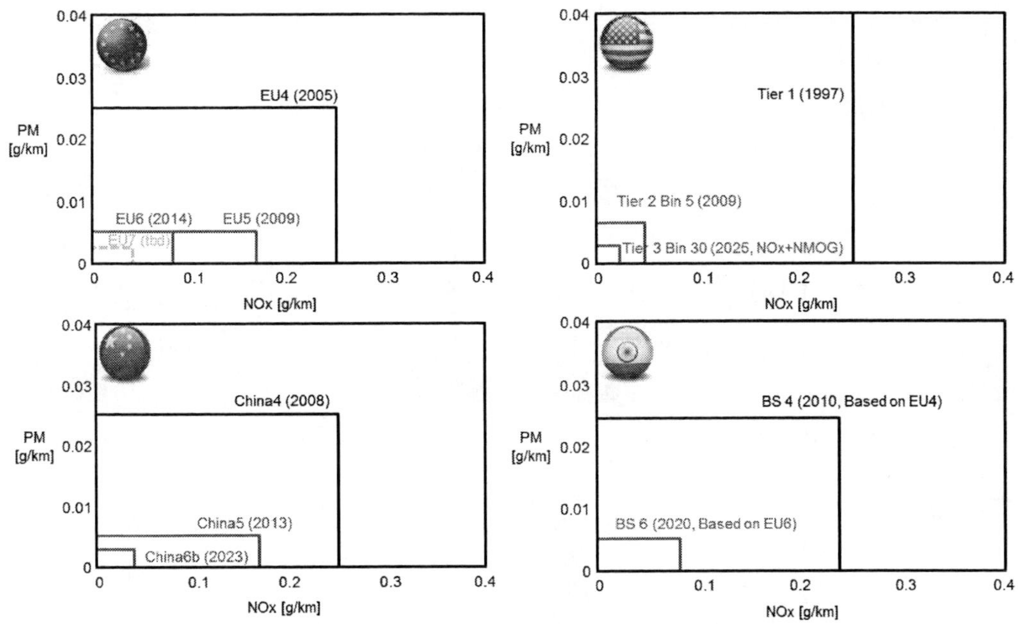

Figure 2 Evolution of emission legislation in world markets

The aim of this study is to identify synergies between hybridisation of a Diesel engine based powertrain and highly efficient exhaust aftertreatment concepts to meet the challenges lying ahead.

2 SELECTION OF HARDWARE CONCEPTS

In this section the selected hardware configurations are highlighted and discussed in two sections. In the first section different hybrid topologies are discussed, in the second section different aftertreatment concepts are shown. All concepts are compared to a baseline configuration, which does not feature hybridisation apart from stop start and relies on a state-of-the art aftertreatment system with a closed coupled diesel oxidation catalyst and a SCR catalyst on DPF.

2.1 Hybrid Concepts

Two different levels of hybridization were investigated, a P0 concept featuring a belt integrated 48 V starter generator being capable of recuperation and boosting up to a level of 12 kW and 100 Nm. This solution requires only moderate changes to the base engine and the avoidance of high voltage systems and large battery capacities limits the system cost increase. This solution enhances the freedom of powertrain management: Torque assist from the electrical machine in highly transient operating conditions reduces the load response requirement from the internal combustion engine and thereby helps to reduce engine-out NOx emissions under these conditions.

Additionally a more powerful P2 high-voltage hybrid concept is investigated. This concept uses a system voltage of 350 V and a stronger 50 kW electric machine is integrated into the transmission. This arrangement allows all electric driving and a more efficient energy recuperation, as these driving modes can be realised without the need to overcome motoring losses of the internal combustion engine. To better exploit this additional potential, the battery size is increased from 1 to 2 kWh compared to the P0 architecture.

A schematic of the investigated hybrid topologies along with a diagram showing the torque capability of the different motors is shown below.

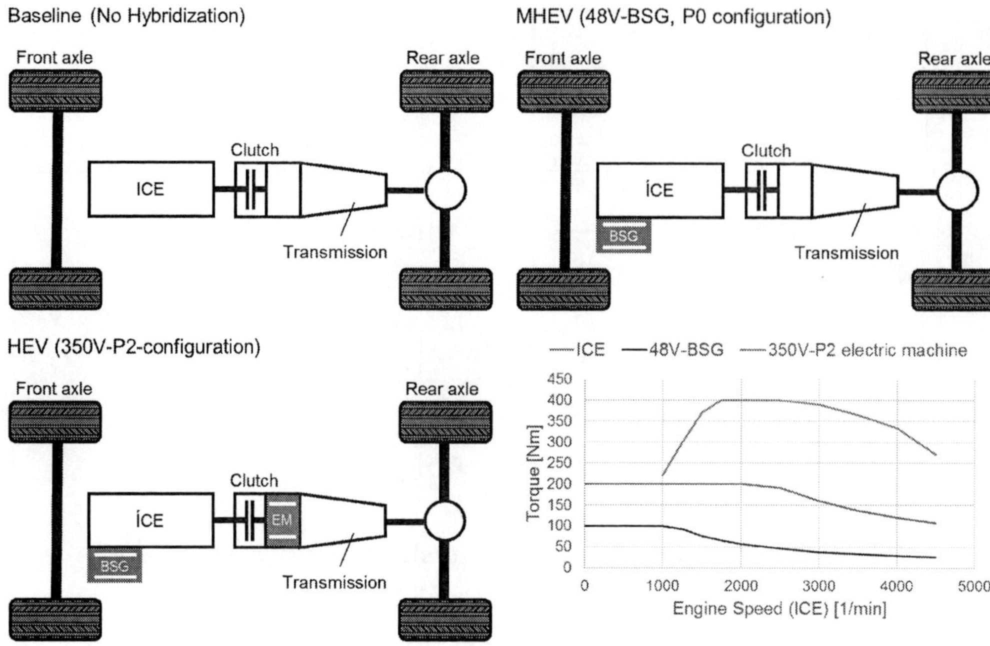

Figure 3 Investigated hybrid topologies

2.2 EAS concepts

Many different exhaust aftertreatment configurations have been introduced for EU6 passenger car Diesel engines. In this study only configurations with a high potential to meet future EU6d emission legislation including RDE requirements were considered.

As a baseline, a system consisting of a close coupled oxidation catalyst with an SDPF and a passive SCR slip catalyst further downstream is investigated. The relatively close coupled SDPF supports an early NOx conversion but is at risk at falling out of its temperature window for high NOx conversion efficiency under more dynamic driving conditions.

An option to enlarge the temperature window for high NOx-Conversion efficiency is the implementation of an LNT instead of the DOC. The LNT can be integrated directly next to the turbine outlet and thereby light off even faster than the previously described SDPF. In order to maintain high NOx conversion efficiency also under highly dynamic driving conditions the SDPF is split into two components - a close coupled DPF and an underfloor SCR catalyst.

The third configuration investigated replaces the DPF with an SDPF and uses a second AddBlue injector to closed-loop control the ammonia storage on the underfloor SCR catalyst.

2.3 Assessment method

The different hybridization levels and aftertreatment configurations were investigated in the simulation environment FEV SimEx. SimEx is a system simulation tool, which relies on 0D-models for engine and 1D-models for aftertreatment performance.

2.3.1 *Engine airpath modelling*

The engine filling is modelled as a mean value model based on the engine volumetric efficiency as a function of engine speed and differential pressure across the engine. The differential pressure is calculated via a map-based turbocharger model. This modelling is accompanied by a 0D-modelling of EGR-flows through the high- and low-pressure EGR-paths. This modelling approach provides a precise estimation of the composition of the in-cylinder charge which is key to simulating engine out emissions in combination with the engine temperature.

2.3.2 *Engine out emission modelling*

The SimEx simulation suite simulates engine out emissions NOx, CO, HC and PM based on the composition of the charge inside the cylinder, the engine operating conditions defined by engine speed and load and the engine temperature. Dependent on the species, different approaches are used. The calibrated NOx-model calculates emissions based on an engine reference mapping, the corresponding oxygen concentration of the fresh charge, the current engine temperature and the current oxygen concentration of the fresh charge. CO-, HC- and PM-Emissions are simulated based on measured emissions at reference conditions and corrected based on the current in-cylinder conditions.

The turbine inlet temperature is modelled based on engine operating conditions, the turbine outlet temperature is simulated based on turbine inlet temperature and turbocharger operating conditions.

2.3.3 *Exhaust aftertreatment modelling*

The exhaust aftertreatment system is modelled with a 0D-/1D-approach. The temperature and pressure along the exhaust system is modelled with a 1D-approach, whereas the conversion in the catalyst bricks are simulated based on maps taking into account the current storage level of NH3 in SCR catalysts and the loading of LNT catalysts with NOx.

2.3.4 *Interaction with Engine control system*

A key feature of SimEx is the combination of the 0D-modelling with a production intent control strategy designed in Simulink, which allows a combined evaluation of engine, aftertreatment, software functionality and system calibration.

2.4 Assessment scope

In order to get a representative characterisation, the system performance was investigated for three different driving conditions – the WLTC to simulate the legislative driving cycle and two RDE cycles from the FEV database. A first RDE cycle named FEV2 represents normal driving conditions, whereas the second cycle named RDE95 represents a very aggressive driving scenario.

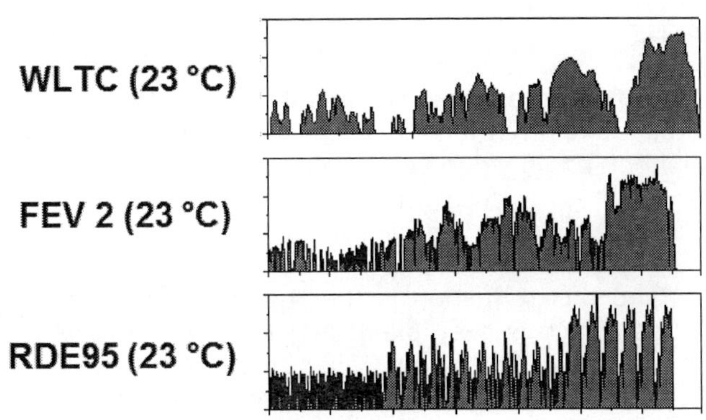

Figure 4 Investigated driving conditions

The assessment of the emission capability is based on a typical mid-size SUV with a modern, state-of-the-art Diesel engine with a displacement of 2.0 Liter. The engine uses a single turbocharger to achieve a specific power output of 70 kW/l and features a 2-way EGR system to best maximise engine efficiency across a wide engine map area.

In total five different configurations of hybridization and exhaust aftertreatment are analysed, see table 1.

1. As baseline, a conventional Diesel Powertrain without hybridization and an exhaust aftertreatment system consisting of DOC, SDPF and a passive SCR catalyst is analysed.
2. A P0 configuration with the same aftertreatment configuration as the baseline. This configuration will be called MHEV.
3. A P0 configuration with a close-coupled LNT and DPF combined with an underfloor SCR catalyst. This concept will be called MHEV+LNT.
4. A P0 configuration with a close coupled LNT, a SDPF and an underfloor SCR which is controlled by a second AdBlue-Injector. This concept will be called MHEV+SCR^2.
5. A P2 configuration with a close coupled LNT, a SDPF and an underfloor SCR which is controlled by a second AdBlue-Injector. This concept will be called HEV.

Table 1 Layout of investigated powertrain configurations

Case			Baseline	MHEV	MHEV+LNT	MHEV+SCR^2	HEV	
Vehicle	Type		C/D-Segment SUV					
	Mass	WLTC	1700					kg
		RDE	2100*					kg
Engine	Displacement		2					L
	Max Power		140					kW
	Max Torque		400					Nm
	EGR		uncooled HP- & cooled LP-EGR					
Hybrid	Topology		-	P0 MHEV (BISG)			P2 HEV	
	e-Motor		-	12			50	kW
	Battery		-	1			2	kWh
	Mass		-	25			100	kg
Transmission	Type		DCT-7					
ATS	Layout		DOC a-ccSDPF p-ufSCR		LNT DPF a-ufSCR	LNT a-ccSDPF a-ufSCR	LNT a-ccSDPF p-ufSCR	
	Volumes		1.7 / 2.0 / 1.5		1.7 / 2.0 / 2.6	1.7 / 2.0 / 2.6	1.7 / 2.0 / 1.5	L

3 EMISSION AND FUEL ECONOMY POTENTIAL OF HYBRIDIZED DIESEL POWERTRAINS

3.1 Emission and fuel economy potential from a 48V-MHEV System

The introduction of a P0 hybridization shows in all investigated scenarios a significant reduction of engine-out NO_x emissions between 20 and 35 %, which underlines the effectiveness even of mild hybrid concepts in minimizing emissions. The benefits increase with more dynamic driving.

The mechanism behind this reduction is the reduction of the load gradient during transient manoeuvres, which is exemplary shown for a comparison between a non-hybridized baseline case and the MHEV case. In the non-hybridized case the load response has to be realized by the internal combustion engine alone, which requires reducing the EGR-Amount to meet the demanded torque. Increased powertrain electrification significantly reduces these occurrences as the load response can be satisfied to a certain degree by the electric machine. As the simulations were all carried out in charge sustaining mode, the availability of this emission benefit will be very high.

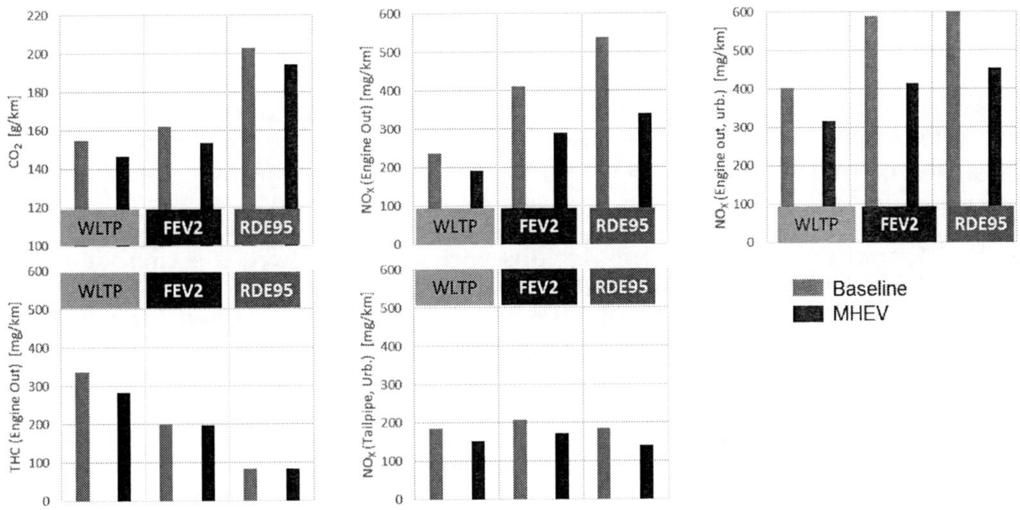

Figure 5 Emission and fuel economy potential with 48V MHEV System

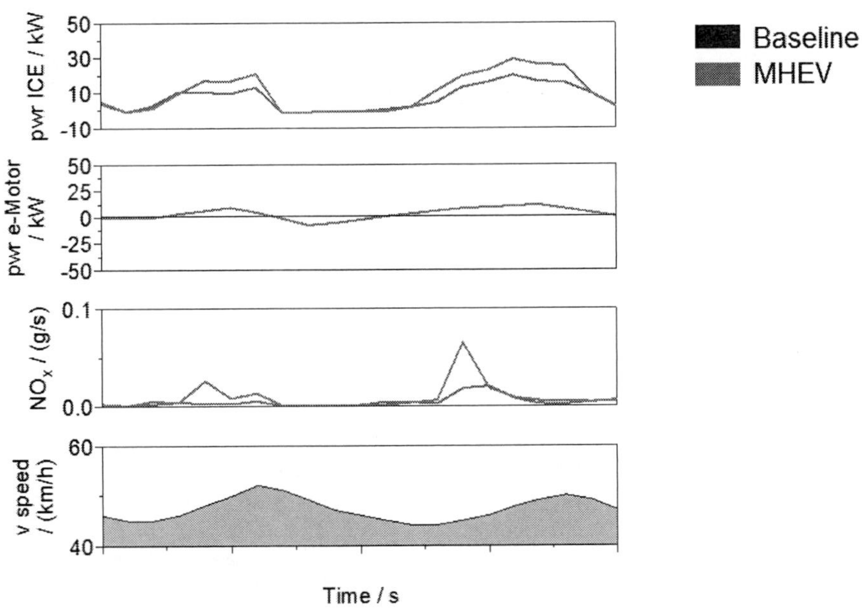

Figure 6 Impact of different hybridization levels on engine out emissions

The reduced engine out emissions with hybridization translate directly into reduced tailpipe emissions of the same magnitude. It is especially noteworthy, that maintaining the relatively simple exhaust aftertreatment system with the SDPF being the only NOx reduction device almost meets EU6d emission legislation even under urban driving conditions, whereas the non-hybridized version emits higher NOx emissions especially under more dynamic driving conditions.

The introduction of the MHEV system also enables an attractive fuel economy reduction between 4.2 and 5.4 % depending on the drive cycle as well as significant reduction of AdBlue consumption of 40%.

Emissions of unburned hydrocarbons can be reduced with the shown hybridization strategy, especially on shorter trips with engine operation more often at low loads and a higher share of engine operation at cold conditions. The ability of the mild hybrid system to increase engine load under these conditions to charge the 48V battery also helps to shift the engine operating conditions to areas with lower engine out hydrocarbon emissions.

However, under lower speed driving conditions, the addition of the hybridization is not sufficient to safely achieve RDE limits, as the temperature in the SDPF is too low, especially after cold start as highlighted by the tailpipe emissions for low speed driving in the WLTC.

3.2 Reduction of pollutant emissions with an MHEV layout – Impact of the exhaust aftertreatment configuration in the WLTC

As configurations with only an SDPF are not capable to robustly demonstrate NOx emission compliance against the RDE legislation, further steps to achieve clean exhaust gases are required.

The addition of an LNT targets an increased NOx conversion efficiency, especially under low load conditions with typically very low exhaust gas temperature. The final addition of a second active underfloor SCR catalyst targets maximum conversion efficiency at higher engine loads.

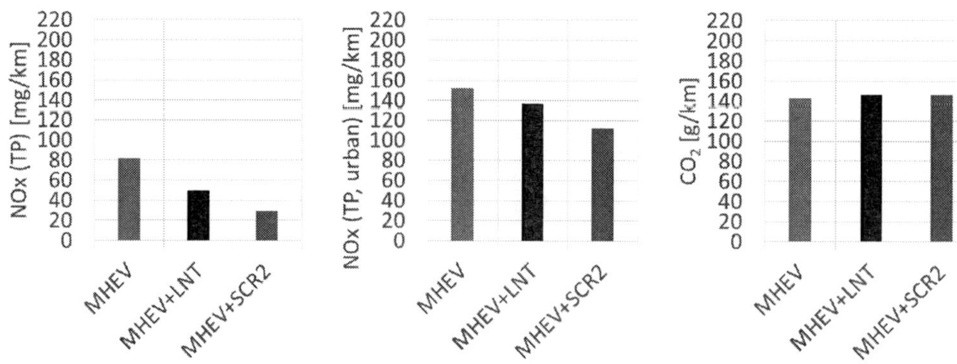

Figure 7 Impact of different exhaust aftertreatment concepts in the WLTC

The simulation results show, that the addition of an LNT has only a small impact on the NOx-tailpipe emissions under low speed driving conditions. The small benefit from the addition of the LNT to the EAS in the first part of the cycle is linked to the low temperature level at the LNT. Although the more close coupled condition of the LNT raises the temperature level by 10 – 15 °C compared to the DOC+SDPF configuration, this is not sufficient for a significant increase of NOx conversion as the conversion in the underfloor SCR is reduced compared to the conversion in the more close coupled SDPF.

The combination of an LNT with an SDPF reduces the overall tailpipe NOx emissions in the low speed part of the WLTC by 10 % compared to the previously discussed configurations, due to a higher temperature level in the SDPF compared to an underfloor SCR enabling improved conversion efficiencies.

All three aftertreatment configurations show a somewhat similar behaviour after 800 s until the end of the WLTC. The exhaust aftertreatment components safely achieve operating temperature and tailpipe emissions are minimal beyond this point. Especially noteworthy is the fact, that the most complex aftertreatment system consisting of an LNT and two active SCR elements, achieves almost 100 % NOx-conversion efficiency after 800 s, which confirms that the passenger car Diesel engine can make a step towards pollutant free mobility with an optimized selection of engine hardware, hybridization and exhaust aftertreatment.

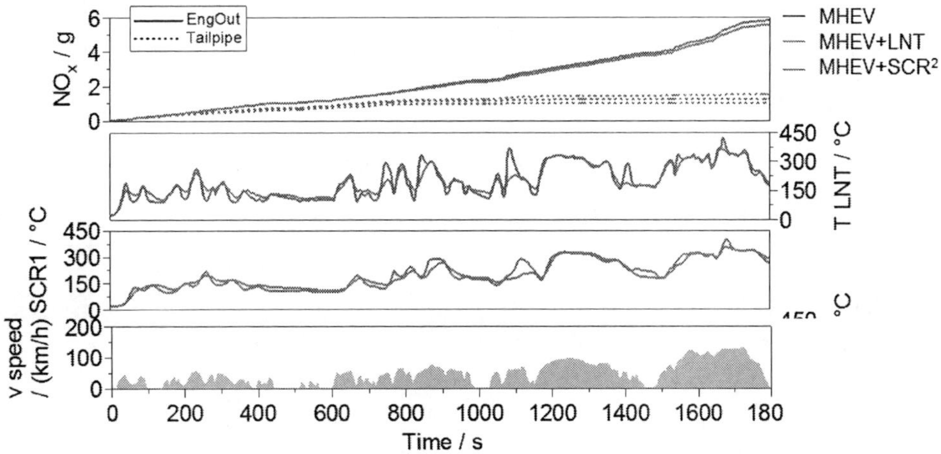

Figure 8 Modal analysis of different exhaust aftertreatment concepts in WLTC

The introduction of an LNT deteriorates the fuel economy slightly by 2.5 % due to intermittent regeneration of the LNT. The presence of an actively controlled LNT nevertheless reduces AdBlue consumption by 50 %, as a significant share of the NOx-conversion is achieved in the LNT and the SCR catalysts only needs to convert a reduced NOx mass.

The introduction of an LNT increases slightly the emissions of unburned hydrocarbons, as the necessary regeneration events of the LNT are not completely HC-neutral. The MHEV+SCR^2 configuration shows a further deterioration of HC-emissions, as the coated DPF in the MHEV+LNT configuration is replaced with an SDPF, which limits the HC oxidation to the LNT. However, all configurations achieve an uncritical level of hydrocarbon emissions.

3.3 Real Driving Emissions (RDE) capability

Contrary to the assessment in the WLTC, where the introduction of an LNT offered only relatively small benefits especially under low speed driving conditions, the benefits of increasing the exhaust gas temperature window, in which a high NOx-conversion is possible, enabled by the LNT becomes obvious.

The combination of an LNT with an underfloor SCR catalyst allows a NOx reduction of 40 % in both investigated RDE drive cycles compared to a system with an SDPF as the only NOx reduction device. However, the achieved emission levels are still not compliant to CN6b regulation and just fall within the EU6d final emission legislation for urban driving.

The transition to a twin SCR-system with twin dosing helps to close this gap. In both RDE cycles NOx-Emissions below 30 mg/km can be achieved for the whole cycle and a conformity factor below 1 can also be realized in urban driving conditions.

Both systems relying on an LNT behave in a very robust manner to different driving styles. Whereas the baseline system without an LNT, shows increased NOx emission levels with higher driving dynamics both investigated combined systems show a reduction of tailpipe NOx emissions, as at least one NOx reduction element operates at a temperature allowing high NOx conversion.

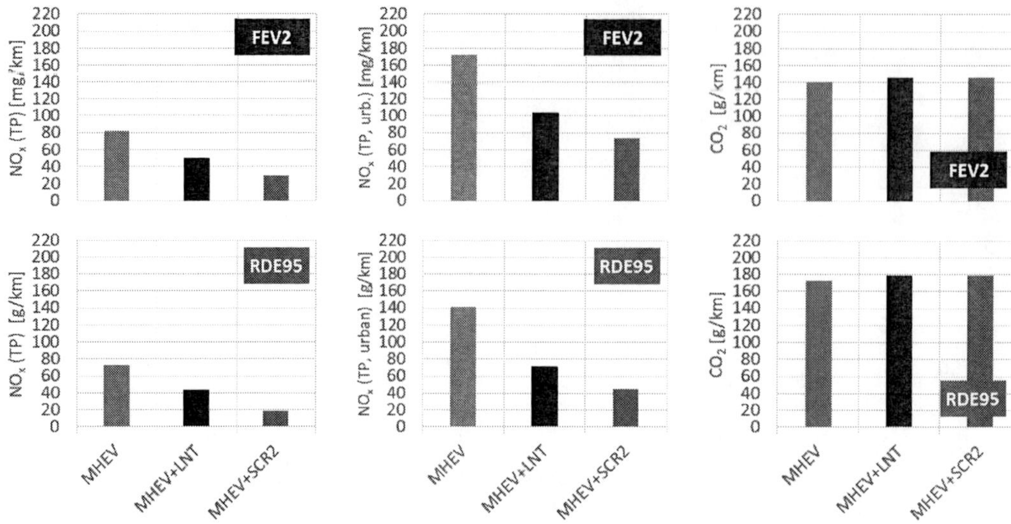

Figure 9 Evaluation of different exhaust aftertreatment concepts under RDE conditions

3.4 Potential with increased degree of electrification

The previously shown results rely only on a mild level of hybridization with a 48 V electric machine. Diesel powertrains relying on a high-voltage systems have been investigated in the past, but only in the context of a simplified engine and exhaust aftertreatment configuration [4]. Within this study an engine architecture with a full feature set and a highly capable exhaust aftertreatment as part of a P2 hybrid powertrain (HEV) is compared to a P0 hybrid configuration with the most advanced aftertreatment (MHEV+SCR²).

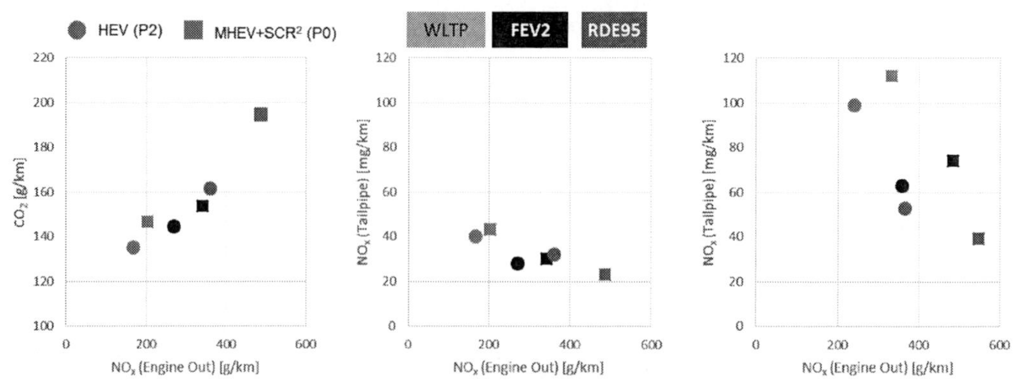

Figure 10 Comparison of emission performance of a P0 and a P2 configuration

The increased capability of the P2 hybrid configuration to recuperate brake energy and to assist the internal combustion engine during highly transient operation enable a significant reduction of engine-out emissions when compared to the previously discussed P0 configuration. For normal driving conditions, this reduction of engine out NOx-Emissions is converted into a reduction of tailpipe emissions, especially in urban driving conditions. It is especially noteworthy to mention, that this is achieved with a simpler exhaust aftertreatment system with only one active SCR catalyst.

Under more dynamic driving conditions, the P2 configuration emits slightly higher NOx emissions, as the simpler exhaust aftertreatment compared to the P0 configuration operates at a slightly lower conversion efficiency.

Aside from the beneficial impact of the P2 hybrid configuration, the P2 hybrid configuration also offers a significant fuel economy benefit dependent on the driving cycle between 5 and 9 %.

4 SUMMARY AND CONCLUSION

Electrification of Diesel powertrains offer new potentials to simultaneously reduce fuel consumption and emissions enabling compliance with future pollutant and emission legislation. Within this study a stepwise approach was taken to identify potentials of different degrees of hybridization and different exhaust gas aftertreatment configurations.

The introduction of a simple and cost-effective MHEV system allows a significant fuel consumption and emission reduction. However, without further improvements of the exhaust gas aftertreatment system future emission legislation cannot be fulfilled. The introduction of more sophisticated, combined NOx reduction systems enable to minimize NOx emissions from Diesel engines with the potential to meet future emission legislation. It is especially noteworthy, that the configuration featuring an LNT, an SDPF and an underfloor SCR catalyst achieved almost 100 % conversion efficiency after the system achieved light-off. The vast majority of NOx-Emissions were emitted in the first part of the cycles. This observation drives the development direction for future activities towards achieving an earlier light-off of the exhaust gas aftertreatment system, e.g. by the use of heating modes or the introduction of electrically heated catalysts.

Further strengthening the degree of hybridization by implementing a P2 configuration with increased battery capacity and a stronger electric machine, allows to further improve the trade-off between emissions and fuel economy.

5 REFERENCES

[1] 2020–2030 CO2 standards for new cars and light-commercial vehicles in the European Union, http://www.theicct.org/

[2] M. Wittler, P. Glusk, A. Nase, M. Daniel - Future Engine Strategies - Survival of the ICE Beyond 2025, 24th Aachen Colloquium Automobile and Engine Technology

[3] http://www.theicct.org/blogs/staff/improving-conversions-between-passenger-vehicle-efficiency-standards

[4] T. Koerfer, B. Holderbaum, D. Blanco-Rodriguez, G. Vagnoni, M. Eisenbarth - Tailored Diesel engines for electrical propulsion systems or tailored electrification for future Diesel powertrain, 11th International MTZ Conference on Future Powertrains

The effects of wet ethanol injection and Miller cycle on a heavy-duty diesel engine operating at full load

V B Pedrozo, T D M Lanzanova, H Zhao
Centre for Advanced Powertrain and Fuels Research (CAPF),
Brunel University London, UK

ABSTRACT

The combination of the dual-fuel combustion strategy with low carbon fuels such as ethanol can help reduce the transport sector's oil dependence and greenhouse gas (GHG) emissions. The use of wet ethanol can further improve these benefits due to energy savings during the production process of ethanol. In this framework, an experimental analysis compared the port fuel injection of anhydrous ethanol and wet ethanol containing 20%, 35%, and 50% of water on a volume basis as potential premixed fuels for dual-fuel engine operation. The investigation was carried out on a single cylinder heavy-duty (HD) engine equipped with a high pressure common rail diesel injection system. A variable valve actuation (VVA) system installed on the intake camshaft allowed for the exploration of Miller cycle via late intake valve closing events. The study was performed at a high load condition of 2.4 MPa net indicated mean effective pressure (IMEP), which corresponds to maximum load of the engine. The results showed that wet ethanol injection increased the net indicated efficiency while minimising the levels of nitrogen oxides (NOx). Miller cycle effectively reduced NOx emissions for the diesel-only and dual-fuel combustion modes at the expense of a decrease in net indicated efficiency. Overall, the dual-fuel operation attained up to 2.6% higher net indicated efficiency and 46% lower NOx emissions than the conventional diesel combustion mode. Thus, the implementation of advanced wet ethanol-diesel dual-fuel combustion technology can be effective in realising low carbon and low emission operations in a commercial heavy-duty diesel engine.

1 INTRODUCTION

The dual-fuel combustion process can provide higher net indicated efficiency and lower NOx emissions than conventional diesel combustion (CDC) [1,2].These improvements can minimise the engine running costs of vehicles and machines equipped with selective catalytic reduction (SCR) aftertreatment systems due to a decrease in the consumption of aqueous urea solution. Furthermore, the use of low carbon fuels like ethanol can effectively decrease the carbon footprint for on and off-road internal combustion engines dependent on oil. Thus, this alternative combustion mode would likely reduce the GHG emissions as well as the operational costs of diesel engines.

The dual-fuel operation is usually achieved by the use of two fuels to control the in-cylinder charge reactivity and local equivalence ratio [3]. A low reactivity fuel, such as gasoline, natural gas, or ethanol is port fuel injected to create a well-mixed charge with the inducted air and exhaust gas recirculation (EGR), if applied. The high reactivity fuel (i.e. diesel) is directly injected into the combustion chamber and serves as thc ignition trigger.

Although the dual-fuel strategy allows for clean engine operation and high net indicated efficiency, challenges still remain at high load conditions. This limitation is mainly a result of excessive in-cylinder pressure rise rates (PRR) caused by uncontrolled autoignition of the low reactivity fuel [4]. Dual-fuel studies have investigated strategies to mitigate the onset of knock and expand the maximum operating load, such as the use of high levels of EGR [5] and/or a piston with a lower compression ratio [6]. However, these modifications can decrease the net indicated efficiency due to increased pumping losses when a higher boost pressure (e.g. more fresh air) is needed for lean high load operation. Consequently, the dual-fuel combustion benefits might be partially or completely lost in comparison with optimised conventional diesel combustion.

Advanced dual-fuel combustion control using variable valve actuation systems and fuels with high resistance to autoignition can possibly help minimise knocking at elevated loads. The modification of the intake valve closing (IVC) events can lower the effective compression ratio and is usually referred to Miller cycle [7]. The approach has been often employed to reduce the mean in-cylinder gas temperature during compression stroke and allow for longer ignition delay [8]. However, the Miller cycle strategy can lead to relatively lower net indicated efficiency than the engine operation with external cooled EGR for a given reduction in NOx emissions [9].

Alternatively, the port fuel injection of fuels such as natural gas and ethanol has been used to extend the dual-fuel operating range with lower EGR rates and high compression ratios. Hanson et al. [10] achieved engine loads up to 2.2 MPa brake mean effective pressure (BMEP) without EGR when using natural gas as a premixed fuel. In other study, Splitter et al. [11] showed that the introduction of 85% ethanol content in gasoline (E85) can decrease the EGR requirements and successfully increase the maximum engine load attained by gasoline-diesel dual-fuel operation. These results demonstrate the importance of the in-cylinder charge temperature and fuel reactivity gradient for expanding the limits of the dual-fuel operation.

The use of wet ethanol in dual-fuel engines can enable thermodynamic benefits as well as represent a more cost-effective alternative than anhydrous ethanol, gasoline, or natural gas. Firstly, the presence of water introduces thermal and dilution effects on the ethanol autoignition process [12]. The high heat of vaporisation and heat capacity of water helps reduce the in-cylinder charge temperature and thus retard the compression ignition timing [13,14]. In addition, the fuel-air mixture is somewhat diluted with increased water injection, which can slow down the ethanol burn rates [15,16]. Secondly, energy savings can be attained during the production process for wet ethanol with higher water fractions [17,18]. This is due to a reduction in the energy spent for the distillation and dehydration phases of anhydrous ethanol, allowing for higher net energy efficiency and lower carbon footprint [19,20].

Wet ethanol injection has been mostly explored to minimise engine running costs in the previous studies. Researchers have demonstrated the use of wet ethanol in conventional spark ignition engines [16,18,21] as well as in homogeneous charge compression ignition (HCCI) engines with high intake air temperatures [22,23]. Recent wet ethanol research has also been conducted on dual-fuel combustion strategies, primarily focused on reactivity controlled combustion ignition (RCCI) engine modelling [14] and mid-load experimental analysis [24]. Moreover, wet ethanol fumigation was carried out on a medium-duty diesel engine operating at high loads, revealing a reduction in NOx emissions and an increase in thermal efficiency [4]. Nevertheless, comparisons between CDC and dual-fuel operation with anhydrous and wet ethanol injection have not been performed yet.

Therefore, to the best of our knowledge, this study is the first attempt to experimentally investigate the potential of wet ethanol injection and Miller cycle to enable high efficiency and clean dual-fuel operation at full engine load. Exhaust emissions and net indicated efficiency were compared to those of a CDC baseline. The dual-fuel combustion characteristics were presented and discussed for anhydrous ethanol as well as wet ethanol containing up to 50% of distilled water on a volume basis. Three different ECRs achieved via late intake valve closing (LIVC) events were explored during the analysis.

2 EXPERIMENTAL SETUP

Base hardware specifications are outlined in Table 1. The engine features a VVA system on the intake camshaft, incorporating a hydraulic tappet on the valve side of the rocker arm [25]. This system allowed for the use of a Miller cycle strategy via LIVC events.

Table 1 – Single cylinder heavy-duty diesel engine specifications.

Parameter	Value
Displaced volume	2.026 dm^3
Stroke	155 mm
Bore	129 mm
Piston type	Re-entrant bowl
Geometric compression ratio	16.8:1
Maximum in-cylinder pressure	18 MPa
Diesel injection system	Common rail, injection pressure of 50–220 MPa
Ethanol injection system	PFI, injection pressure of 0.3 MPa relative

A schematic diagram of the single cylinder heavy-duty engine experimental setup is shown in Figure 1. Fresh intake air was supplied to the engine via an external supercharger with closed loop control of the boost pressure. A throttle valve located upstream of a large-volume surge tank provided fine control over the intake manifold air pressure. The air flow rate was measured with a thermal mass flow meter. An electronically controlled exhaust back pressure valve located downstream of the exhaust surge tank was used to set the required exhaust manifold pressure. Boosted air temperature was controlled using water cooled heat exchanger.

During dual-fuel operation, ethanol was injected through a port fuel injector (PFI). An injector driver controlled the PFI pulse width, which was adjusted according to the desired ethanol energy fraction. The $\dot{m}_{ethanol}$ represents the actual anhydrous ethanol mass flow rate and was calculated by subtracting the water mass flow rate from the total wet ethanol flow rate measured with a Coriolis flow meter. The water mass and volume fractions in the wet ethanol were determined using the fuel density measured by the flow meter and an alcoholometry table [26]. An alcoholmeter was used to confirm the water content. The diesel fuel was supplied to the engine using a high pressure common rail injection system. Two Coriolis flow meters were used to determine the diesel flow rate $\dot{m}_{diesel}$ by measuring the total fuel supplied to and from the high pressure pump and diesel injector.

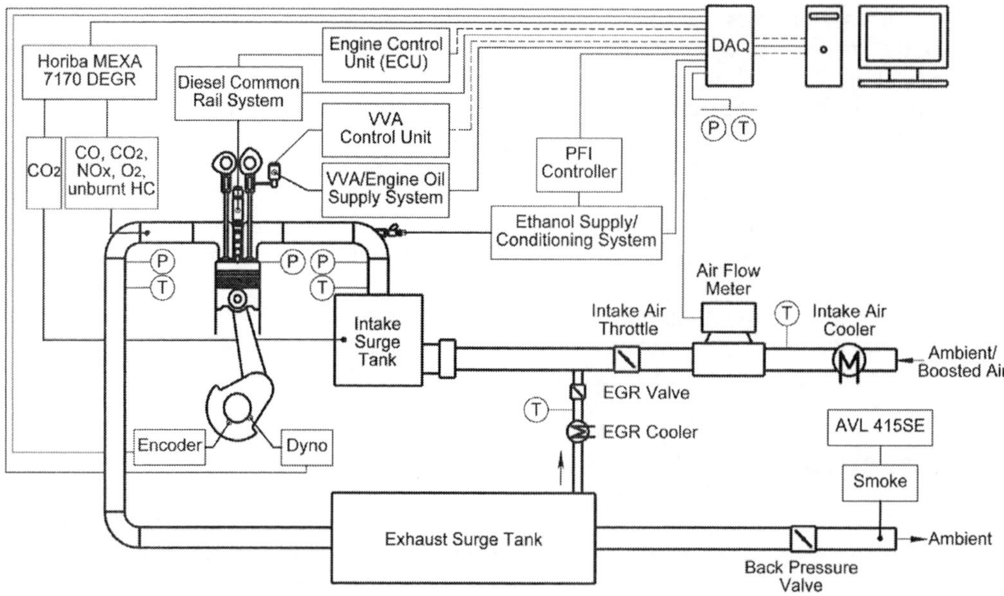

Figure 1 – Schematic diagram of the engine experimental setup.

Gaseous emissions such as NOx, carbon monoxide (CO), and unburnt hydrocarbons (HC) were taken using a Horiba MEXA-7170 DEGR gas analyser system. The in-cylinder pressure was measured using a piezoelectric pressure sensor. Intake and exhaust pressures were measured with two piezoresistive absolute pressure sensors. The intake valve lift profile was continuously monitored by a linear differential variable reluctance transducer located on the top of the valve spring retainer. The diesel injection signal was acquired by measuring the electric current sent by the engine control unit to the injector solenoid using a current probe.

Two National Instruments data acquisition (DAQ) cards were used to acquire the signals from the measurement device. A high speed DAQ card received the crank angle resolved data synchronised with an optical encoder of 0.25 crank angle degrees (CAD) resolution. A lower speed DAQ card acquired the low frequency data, such as engine speed, torque, as well as temperatures and pressures at relevant locations.

3 DATA ANALYSIS

Crank angle based in-cylinder pressure traces were averaged for 200 consecutive cycles for each operating point and used to calculate the IMEP and the apparent net heat release rate (HRR). A constant ratio of specific heats (γ) of 1.33 was assumed throughout the engine cycle. The mass fraction burnt (MFB) was calculated by integrating the HRR. Combustion phasing (CA50) was determined by the crank angle of 50% MFB. The mean in-cylinder gas temperature was computed using the ideal gas law [27].

The in-cylinder pressure rise rate (PRR) was represented by the average of the maximum pressure variations of 200 cycles of cylinder pressure versus crank angle. Cycle-to-cycle variability was measured by the coefficient of variation of IMEP (COV_IMEP) over the sampled cycles. The stoichiometric air/fuel ratio was determined by the conservation of mass of each chemical element in the reactants [27]. The global fuel/air equivalence ratio (Φ) was calculated using the intake air and fuel flow rates.

The ethanol energy fraction (EF) was defined as the ratio of the energy content of the anhydrous ethanol (E100) to the total fuel energy by

$$EF = \frac{\dot{m}_{ethanol}\, LHV_{ethanol}}{(\dot{m}_{ethanol}\, LHV_{ethanol}) + (\dot{m}_{diesel}\, LHV_{diesel})}$$

where $LHV_{ethanol}$ and LHV_{diesel} are the lower heating value of ethanol and diesel, respectively.

Net indicated efficiency represents the ratio of the net indicated power produced by the engine (P_{ind}) in kW to the rate of fuel energy delivered as

$$Net\ indicated\ efficiency\ [\%] = \frac{3.6\, P_{ind}}{(\dot{m}_{ethanol} LHV_{ethanol} + \dot{m}_{diesel} LHV_{diesel})} \times 100$$

The variation in net indicated efficiency was calculated as

$$Variation\ in\ net\ indicated\ eff.[\%] = \left(\frac{Net\ indicated\ efficiency}{Net\ indicated\ efficiency_{baseline}} - 1\right) \times 100$$

where the $Net\ indicated\ efficiency_{baseline}$ is the net indicated efficiency of the most efficient conventional diesel combustion (CDC) case.

4 TEST CONDITIONS

Table 2 summarises the baseline engine operating conditions and highlights the parameters that were optimised. The ethanol energy fraction was swept during the optimisation process. Diesel fuel was introduced using a late single injection between -10 and 5 CAD after top dead centre (ATDC).

Table 2 – Engine operating conditions.

Parameter	Baseline	Optimisation
Engine speed	1200 rpm	
Load	2.4 MPa IMEP	
Diesel injection pressure	180 MPa	
Diesel injection strategy	Single injection	
Diesel injection timing	Varied	Swept
Ethanol energy fraction	0.00	Swept
Intake pressure	0.30 MPa	
Exhaust pressure	0.29 MPa	
EGR rate	0%	
ECR	16.8:1	Swept
Intake air temperature	318 K	

In some cases, a Miller cycle strategy via LIVC events was used to delay the initiation of the compression process and reduce the effective compression ratio (ECR) from 16.8:1 to 15.7:1 and 14.4:1. The ECR was calculated using the effective in-cylinder volume at IVC obtained from the intersection of an extrapolated polytropic compression curve and the average intake manifold gas pressure [8,28].

EGR was not employed in this study because the intake manifold and recycled exhaust gas temperatures could not be controlled with the current gas-to-water charge coolers setup. Stable engine operation was quantified by a COV_IMEP below 3%. A maximum PRR of 3.0 MPa/CAD was considered as the upper bound for calibration. Moreover, the maximum in-cylinder gas pressure (Pmax) was limited to 18 MPa.

5 RESULTS AND DISCUSSION

5.1 The effect of Miller cycle and start of injection on emissions and net indicated efficiency of full load CDC operation

Diesel-only combustion is characterised by an exhaust emissions/net indicated efficiency trade-off where lower NOx emissions are usually attained at the expense of higher smoke and reduced net indicated efficiency. Figure 2 depicts the net indicated emissions of soot (ISsoot) and NOx (ISNOx) as well as the variation in net indicated efficiency for full load CDC operation with varied SOIs and ECRs. The $Net\ indicated\ efficiency_{baseline}$ was equivalent to 45.1% for the case with the most advanced SOI at an ECR of 16.8:1. Later diesel injection timings and lower ECRs of 15.7:1 and 14.4:1 reduced NOx and increased soot emissions. At these conditions, the resulting ISsoot levels exceeded the Euro VI limit of 0.010 g/kWh [29]. This was possibly attributed to the lower combustion temperatures and oxygen availability [9]. At an ECR of 16.8:1, a delay of 8.25 CAD in the SOI reduced the net indicated efficiency by 9.3% for 20% lower NOx emissions. Alternatively, the most efficient case at an ECR of 14.4:1 decreased the net indicated efficiency in 4.7% for 42% reduction in ISNOx. This demonstrates the potential of Miller cycle to minimise NOx emissions with relatively lower fuel consumption than a retarded combustion process. However, the NOx emissions for the CDC operation remained considerably higher than the Euro VI emission target of 0.4 g/kWh [29]. This gives the opportunity of applying alternative combustion strategies such as the dual-fuel operation.

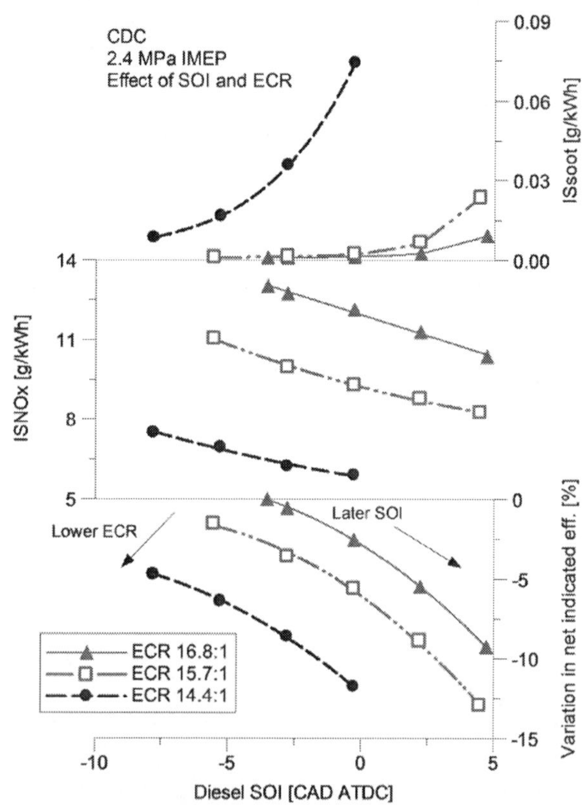

Figure 2 – ISsoot, ISNOx, and variation in net indicated efficiency for full load CDC operation with different SOIs and ECRs.

5.2 The effect of anhydrous ethanol injection on full load dual-fuel operation

Experiments were performed to explore the effect of anhydrous ethanol injection on combustion. The investigation was carried out using the baseline engine operating conditions showed in Table 2 except for the use of different EFs. The diesel injection timing was held at 4.75 CAD ATDC.

Figure 3 shows that the anhydrous ethanol fuel autoignited prior to the diesel SOI. This was a result of the high in-cylinder gas temperatures and pressures at this particular load. Increased EF led to higher first peak heat release and shorter diesel mixing-controlled combustion, which elevated both the Pmax and the levels of PRR. At the baseline ECR of 16.8:1, the dual-fuel operation with a constant diesel SOI was limited to an ethanol energy fraction of 0.19 and a PRR of 1.9 MPa/CAD. The pressure traces for the dual-fuel combustion with higher EFs were not recorded due to excessive PRRs.

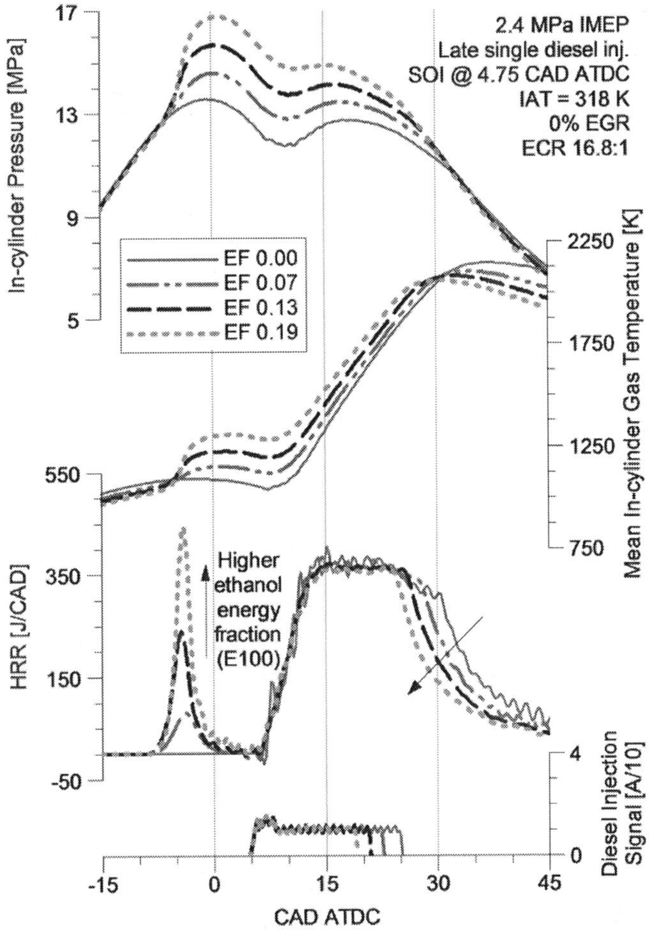

Figure 3 – The effect of ethanol energy fraction on anhydrous ethanol-diesel dual-fuel operation at an ECR of 16.8:1.

5.3 The effect of Miller cycle on full load dual-fuel operation

The subsequent dual-fuel combustion experiments were performed to explore the effect of Miller cycle via LIVC events. The baseline IVC timing was delayed by applying the same LIVC events used for the CDC operation in the subsection 5.1, resulting in ECRs of 15.7:1 and 14.4:1.

Figure 4 reveals that the use of a lower ECR effectively retarded the autoignition of the ethanol fuel and slowed down the reaction rates when operating with an EF of ~0.19. This was primary a result of the lower mean in-cylinder gas temperatures attained during the compression stroke. However, the diesel injection duration became longer for the cases with lower ECRs, indicating a decrease in net indicated efficiency. This was likely caused by the richer dual-fuel combustion and higher heat transfer losses [30] obtained with a Miller cycle strategy at a constant boost pressure. The Φ was increased from 0.67 to 0.87 when reducing the ECR from 16.8:1 to 14.4:1 and holding the diesel SOI at 4.75 CAD ATDC.

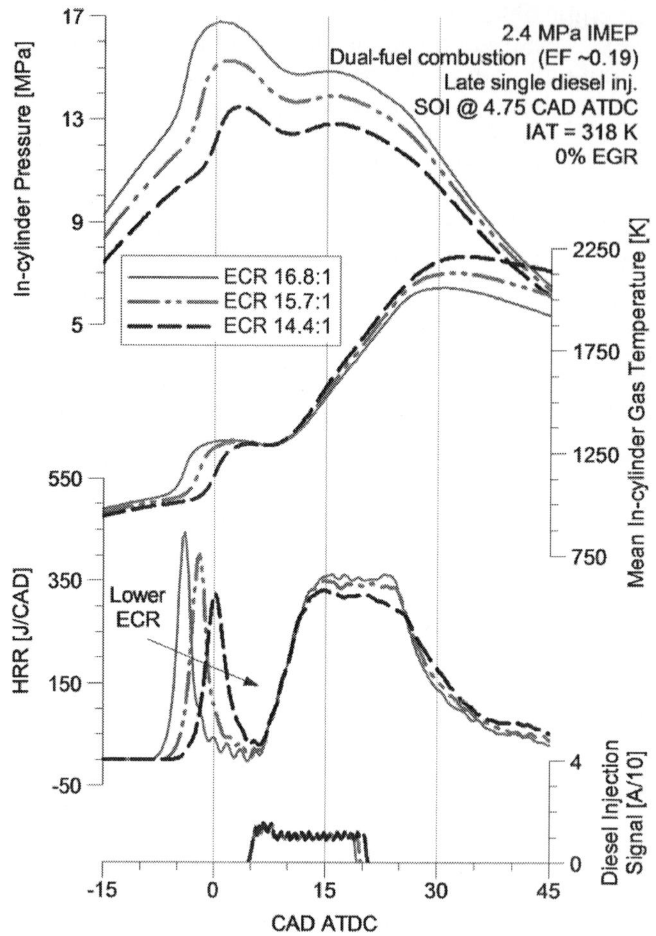

Figure 4 – The effect of ECR on full load dual-fuel operation with an ethanol energy fraction of ~0.19.

5.4 The effect of anhydrous ethanol injection and Miller cycle on emissions and net indicated efficiency of full load dual-fuel operation

The diesel SOI and the EF were swept to determine the actual potential of full load dual-fuel operation with anhydrous ethanol injection and lower ECRs to decrease NOx emissions and increase net indicated efficiency. The optimum SOIs were limited by the Pmax of 18 MPa and PRRs of 3.0 MPa/CAD, as shown in Figure 5. Despite the lower compression pressures and temperatures achieved with a Miller cycle strategy, the maximum EF remained approximately 0.20. Figure 6 depicts the ISsoot, ISNOx, and variation in net indicated efficiency for full load dual-fuel operation with different EFs and ECRs. The plot shows that NOx emissions can be reduced and net indicated efficiency increased at higher ethanol energy fractions.

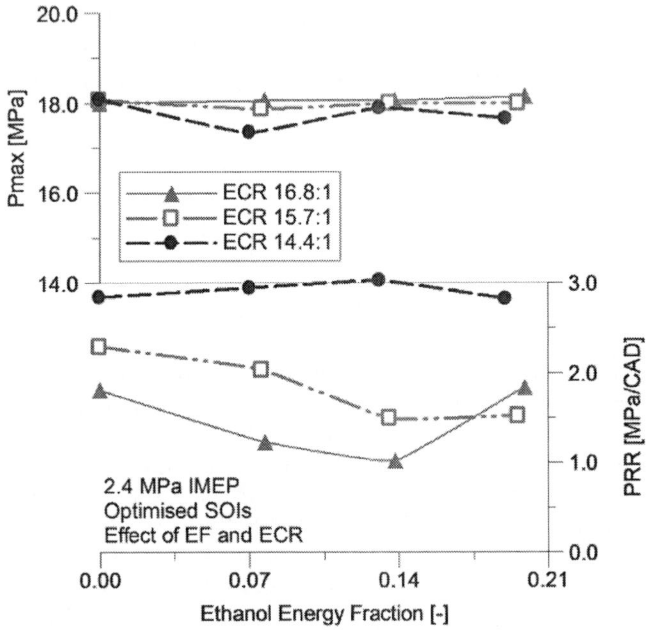

Figure 5 – Maximum in-cylinder pressure and pressure rise rate for anhydrous ethanol-diesel dual-fuel operation with different EFs and ECRs.

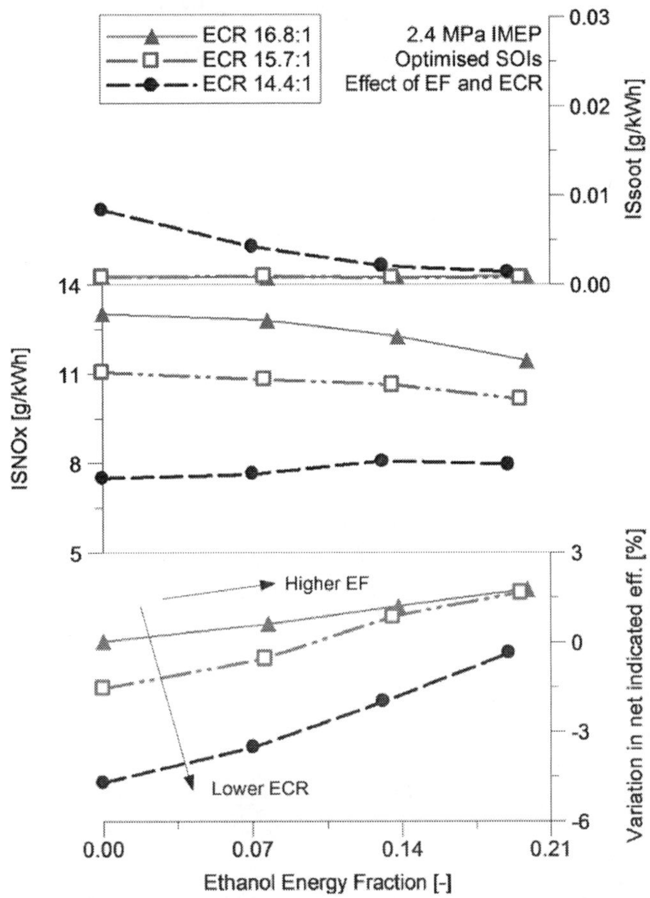

Figure 6 – ISsoot, ISNOx, and variation in net indicated efficiency for anhydrous ethanol-diesel dual-fuel operation with different EFs and ECRs.

The reduction in ISNOx achieved 12% at the ECR of 16.8:1, decreasing from 13.0 g/kWh in the CDC baseline to 11.4 g/kWh in the dual-fuel operating mode. This improvement was attained with an increase in net indicated efficiency of 1.8%. The use of a lower ECR in the dual-fuel operation further decreased the levels of ISNOx. Dual-fuel combustion combined with Miller cycle allowed for up to 39% NOx reduction with a small decrease in net indicated efficiency of 0.4% compared to the CDC baseline. Soot emissions were consistently low at the ECRs of 16.8:1 and 15.7:1. The reduced combustion temperatures and lower oxygen availability increased the smoke numbers at an ECR of 14.4:1. Nevertheless, the use of a higher EF of ~0.20 helped maintain the ISsoot below 0.002 g/kWh, which is significantly lower than the measurements for the CDC operation with Miller cycle.

5.5 The effect of wet ethanol injection on full load dual-fuel operation

The last parameter explored in this study was the water-in-ethanol content. This investigation was carried out at the baseline ECR of 16.8:1. The experiments were performed using anhydrous ethanol (E100) and wet ethanol containing 20%, 35%, and 50% of distilled water on a volume basis. These ethanol-water mixtures were named E80W20, E65W35, and E50W50, respectively.

Figure 7 shows the effect of wet ethanol injection on the dual-fuel combustion process with an ethanol energy fraction of ~0.19. The diesel SOI was set at 4.75 CAD ATDC.

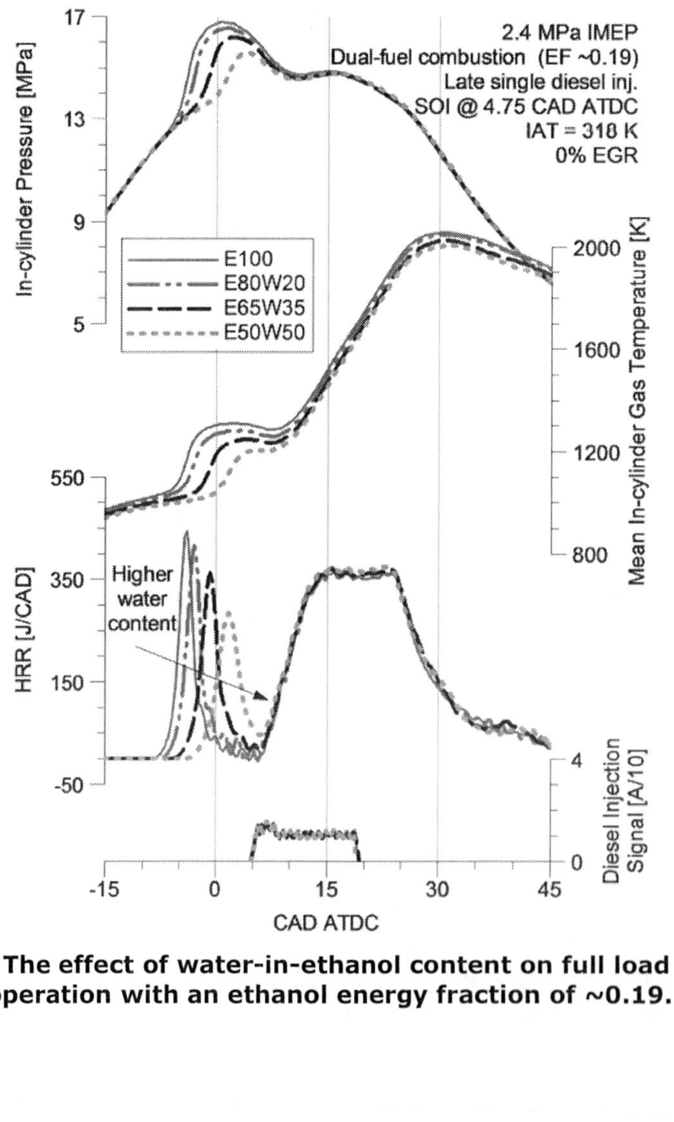

Figure 7 – The effect of water-in-ethanol content on full load dual-fuel operation with an ethanol energy fraction of ~0.19.

The results revealed that wet ethanol with higher water content were effective in decreasing the compression temperatures, retarding the SOC, and slowing down the ethanol compression ignition combustion. The later and slower reaction rates reduced the first peak heat release. The trend was similar to that observed when decreasing the ECR in Figure 4. The diesel mixing-controlled combustion was practically unaffected as the water-in-ethanol content was increased. However, the use of E65W35 and E50W50 somewhat impaired the combustion process between 35 and 45 CAD ATDC. This was probably caused by lower end-of-combustion temperatures.

To demonstrate the thermal and dilution effects of water, a sweep of ethanol energy fraction with E50W50 is depicted in Figure 8. The E50W50-diesel dual-fuel operation led to different combustion behaviour than that attained with anhydrous ethanol injection in Figure 3. The autoignition timing and the position of first peak heat release were delayed as more E50W50 was injected. These findings can be attributed to the charge cooling and heat capacity effects as well as the intake air oxygen displacement introduced by a higher concentration of water.

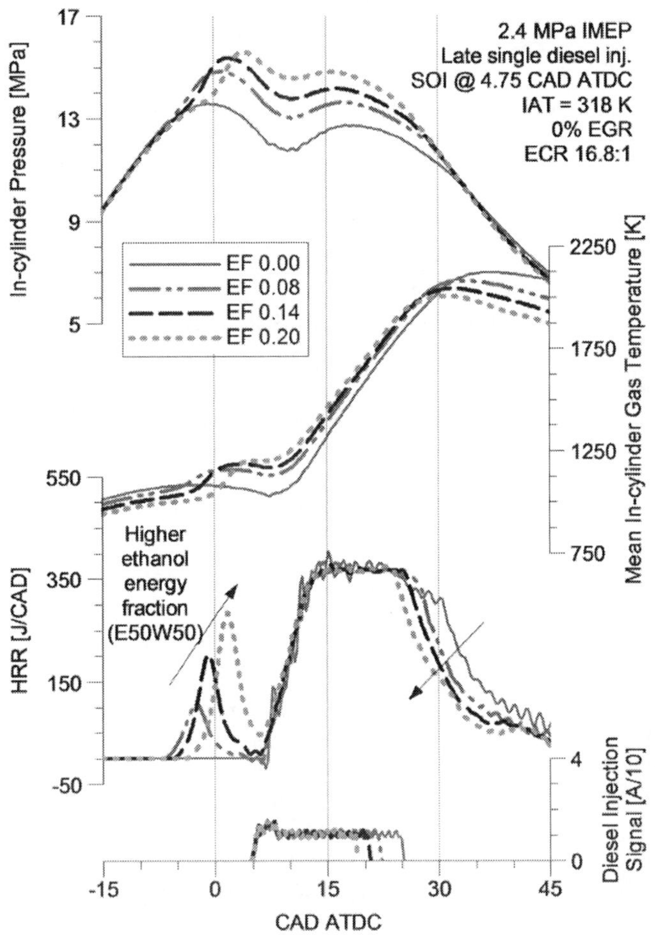

Figure 8 – The effect of ethanol energy fraction on full load E50W50-diesel dual-fuel operation at an ECR of 16.8:1.

5.6 The effect of anhydrous and wet ethanol injection on emissions and net indicated efficiency of full load dual-fuel operation

To explore the potential of wet ethanol injection to curb NOx and increase net indicated efficiency, the diesel SOIs and the EFs were optimised up against the

Pmax and/or PRR limit. Although the water-in-ethanol content introduced thermal and dilution effects, Figure 9 shows that the maximum ethanol energy fraction with E50W50 could not be increased by a large extent and was limited to 0.26. This was attributed to excessive PRRs produced by a simultaneous combustion of ethanol and diesel at higher ethanol energy fractions.

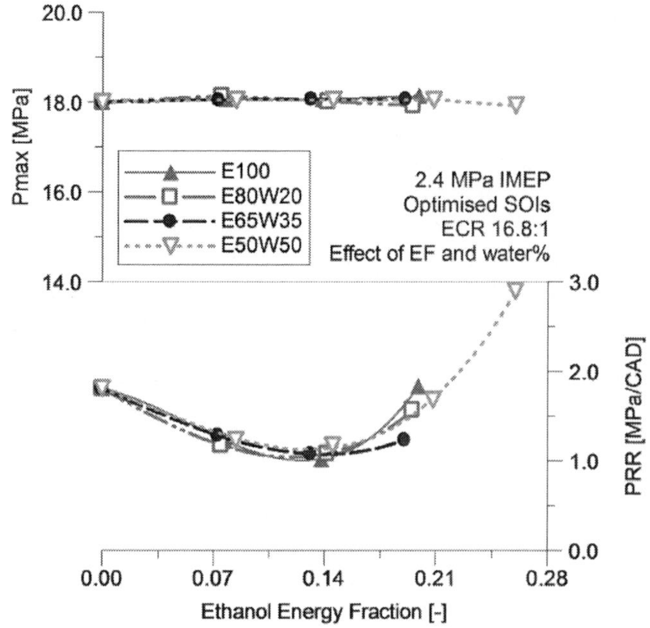

Figure 9 – Maximum in-cylinder pressure and pressure rise rate for full load dual-fuel operation with anhydrous and wet ethanol injection.

Figure 10 reveals how wet ethanol injection affected the exhaust gas temperature (EGT) and exhaust emissions. Increased water percentages helped reduce EGT and NOx emissions. This was probably a result of the higher specific heat capacity, charge cooling, and dilution effects introduced by the presence of water [12-16]. Compared to E100, the use of E50W50 decreased the ISNOx levels by 30% when operating the engine with an EF of 0.20-0.21. Soot emissions were effectively lower than the Euro VI emission limit of 0.010 g/kWh [29] and increased slightly at higher EFs. The later was linked to a decrease in combustion temperature, which is supported by the lower levels of NOx and a reduction of up to 63 K in the EGT compared to the CDC baseline.

The dual-fuel operation led to higher net indicated emissions of CO (ISCO) and unburnt HC (ISHC) as the EF was increased. This was likely caused by higher levels of premixed fuel found in the crevice volumes of the stock diesel combustion system and near the cylinder liner [31]. The water-in-ethanol content showed little influence on the CO and unburnt HC emissions at EFs below 0.14. However relatively higher levels of ISCO and ISCH were attained with wet ethanol as the EF was increased to ~0.20, which supports a dependence on in-cylinder gas temperature.

Figure 11 depicts the exhaust emissions/net indicated efficiency trade-off for the dual-fuel operation with anhydrous and wet ethanol injection. The levels of ISsoot remained below 0.002 g/kWh for all cases. The most efficient engine operation was obtained by the E65W35-diesel dual-fuel combustion with an ethanol energy fraction of 0.19, which increased the net indicated efficiency by 2.6%. The injection of E50W50 reduced ISNOx by up to 46% to 7 g/kWh while increasing the net indicated efficiency in 1.4% over the diesel-only baseline highlighted by red circles.

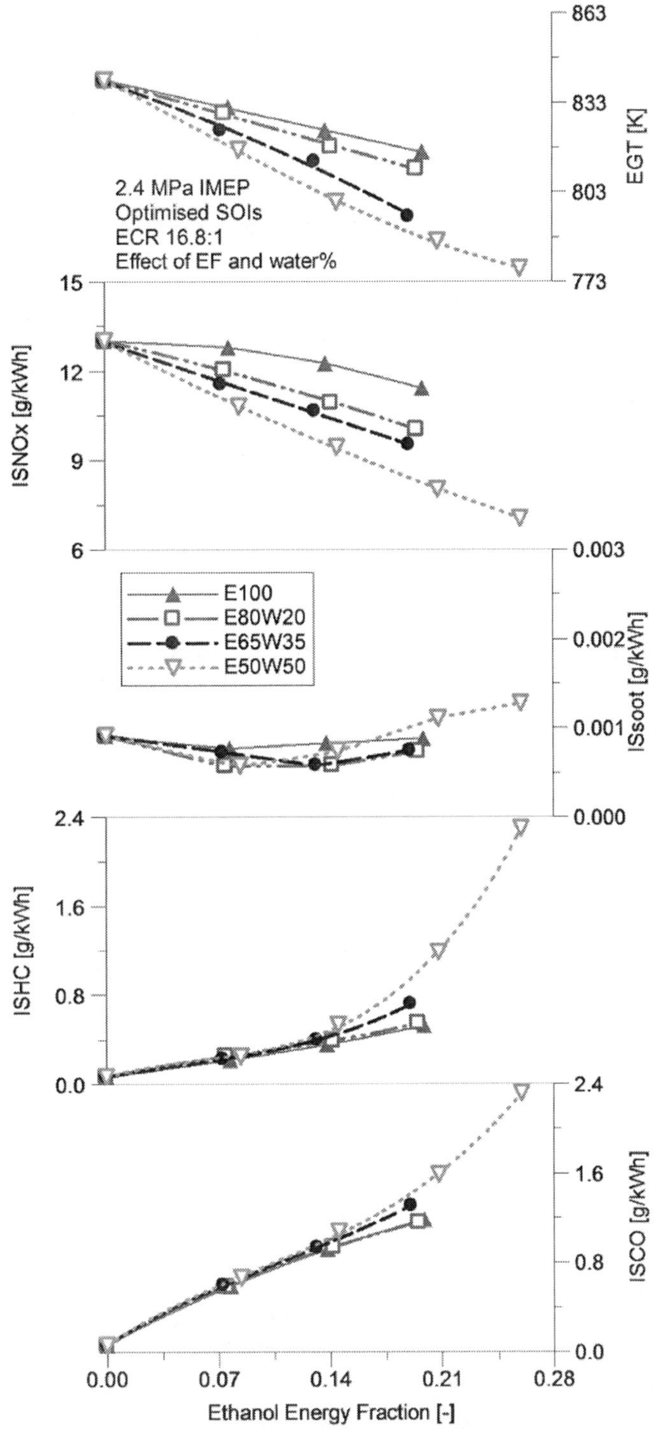

Figure 10 – EGT and exhaust emissions for full load dual-fuel operation with anhydrous and wet ethanol injection.

Figure 12 shows a comparison between the CDC baseline and the optimum dual-fuel case (see green circles in Figure 11). The dual-fuel operation with wet ethanol injection (E50W50) achieved 38% lower NOx emissions and 2.4% higher net indicated efficiency at an ethanol energy fraction of 0.21. These results indicate that the dual-fuel strategy can decrease the engine running costs of SCR equipped vehicles and machines, as supported by our previous study at mid-loads [1].

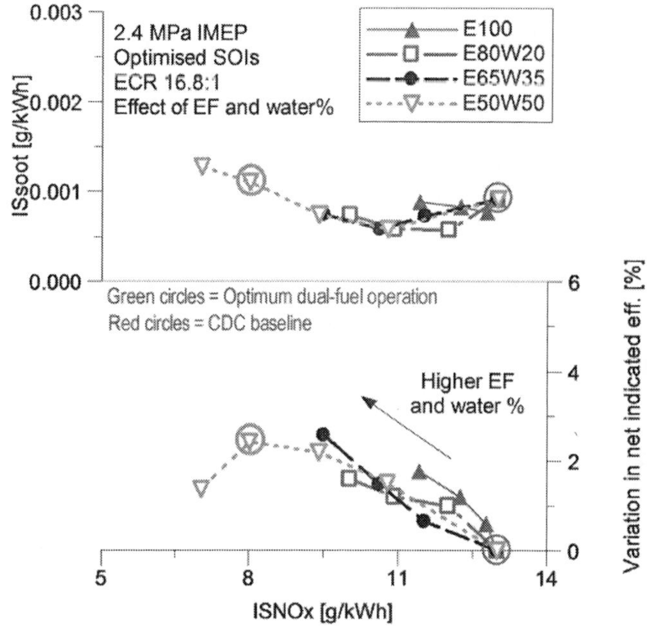

Figure 11 – Trade-off between ISNOx, ISsoot, and net indicated efficiency for full load dual-fuel operation with anhydrous and wet ethanol injection.

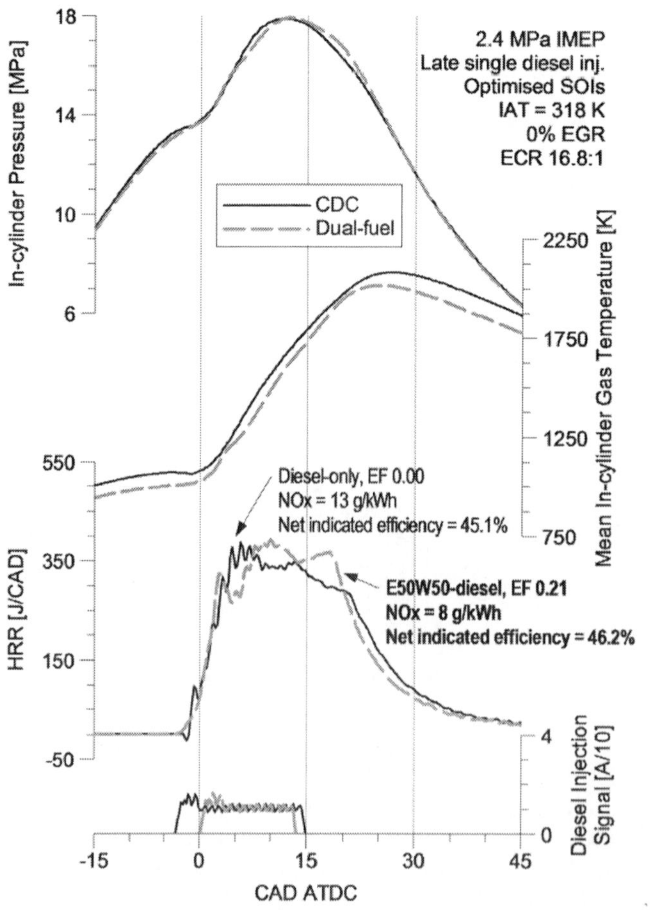

Figure 12 – Comparison between the CDC baseline and the optimum dual-fuel operation with wet ethanol injection (E50W50).

6 CONCLUSIONS

In this study, engine experiments were performed to characterise the potential of wet ethanol injection and Miller cycle to attain high net indicated efficiency and clean dual-fuel operation. The investigation was carried out on a single cylinder heavy-duty diesel engine at a constant speed of 1200 rpm and a high load of 2.4 MPa IMEP. The use of anhydrous ethanol (E100) and wet ethanol containing 20%, 35%, and 50% of water on a volume basis was explored and discussed. Late intake valve closing events were employed to obtain three different effective compression ratios of 16.8:1, 15.7:1, and 14.4:1. Combustion characteristics, exhaust emissions, and net indicated efficiencies were compared to a conventional diesel combustion baseline. The primary findings can be summarised as follows:

- Conventional diesel combustion presented an exhaust emissions/net indicated efficiency trade-off where lower NOx emissions led to higher levels of soot and lower net indicated efficiency. The use of late IVC timings in Miller cycle allowed for better results than retarded diesel injection timings.
- Full load dual-fuel operation yielded autoignition of the premixed anhydrous ethanol fuel prior to the diesel SOI at the baseline ECR of 16.8:1. Excessive PRRs and the maximum in-cylinder pressure of the engine limited the highest ethanol energy fraction to ~0.20.
- The introduction of a Miller cycle strategy via LIVC events decreased the ECR and thus the in-cylinder gas pressure and temperature prior to the start of combustion. This resulted in relatively longer ignition delays, effectively retarding the ethanol autoignition timing.
- E100-diesel dual-fuel combustion allowed for an increase in net indicated efficiency of 1.8% with 12% lower NOx emissions at a constant ECR of 16.8:1. Alternatively, the use of an ethanol energy fraction of 0.19 at an ECR of 14.4:1 reduced the levels of NOx by 39% with a decrease in net indicated efficiency of 0.4% over the CDC baseline.
- The port fuel injection of wet ethanol containing high water content (e.g. E50W50) helped delay the autoignition event and slow down the ethanol burn rate. Higher levels of wet ethanol injection also led to a similar trend, retarding the position of the first peak heat release when operating the engine with constant diesel injection timing. Nevertheless, PRR and Pmax remained a challenge for a dual-fuel operation with wet ethanol injection, limiting the maximum EF to 0.26.
- NOx emissions were reduced when increasing the water-in-ethanol content. Up to 46% lower ISNOx levels than the diesel-only case were achieved with wet ethanol injection. In addition, the E65W35-diesel dual-fuel operation achieved 2.6% higher net indicated efficiency than the CDC baseline. The increase in net indicated efficiency and the NOx reduction were attained at the expense of higher CO and unburnt HC emissions.

Full load dual-fuel operation with wet ethanol injection and Miller cycle can be used to achieve higher net indicated efficiency and lower NOx emissions than the conventional diesel combustion. These improvements can help reduce GHG emissions as well as minimise the total cost of ownership of diesel engines through reduced fuel energy demand and lower EGR requirements/aqueous urea solution consumption in the SCR system.

Lower carbon footprint for wet ethanol combined with competitive fuel costs can further enhance the benefits introduced by the dual-fuel operation. Ultimately, the advantages of the dual-fuel combustion have to be weighed against the additional complexity and upfront cost of a port fuel injection system and extra fuel tank, as well as the vehicle operator requiring additional wet ethanol fuel.

ACKNOWLEDGEMENTS

V. Pedrozo and T. Lanzanova would like to acknowledge CAPES Foundation (Brazil) and CNPq (Brazil) for supporting their PhD studies at Brunel University London under supervision of Prof. Zhao.

REFERENCE LIST

[1] Pedrozo VB, May I, Zhao H. Exploring the mid-load potential of ethanol-diesel dual-fuel combustion with and without EGR. Applied Energy 2017;193:263–75. doi:10.1016/j.apenergy.2017.02.043.

[2] Pedrozo VB, May I, Dalla Nora M, Cairns A, Zhao H. Experimental analysis of ethanol dual-fuel combustion in a heavy-duty diesel engine: An optimisation at low load. Applied Energy 2016;165:166–82. doi:10.1016/j.apenergy.2015.12.052.

[3] Reitz RD, Duraisamy G. Review of high efficiency and clean reactivity controlled compression ignition (RCCI) combustion in internal combustion engines. Progress in Energy and Combustion Science 2015;46:12–71. doi:10.1016/j.pecs.2014.05.003.

[4] Goldsworthy L. Fumigation of a heavy duty common rail marine diesel engine with ethanol-water mixtures. Experimental Thermal and Fluid Science 2013;47:48–59. doi:10.1016/j.expthermflusci.2012.12.018.

[5] Han X, Divekar P, Reader G, Zheng M, Tjong J. Active Injection Control for Enabling Clean Combustion in Ethanol-Diesel Dual-Fuel Mode. SAE International Journal of Engines 2015;8:890–902. doi:10.4271/2015-01-0858.

[6] Benajes J, Pastor J V., García A, Boronat V. A RCCI operational limits assessment in a medium duty compression ignition engine using an adapted compression ratio. Energy Conversion and Management 2016;126:497–508. doi:10.1016/j.enconman.2016.08.023.

[7] Martins MES, Lanzanova TDM. Full-load Miller cycle with ethanol and EGR: Potential benefits and challenges. Applied Thermal Engineering 2015;90:274–85. doi:10.1016/j.applthermaleng.2015.06.086.

[8] He X, Durrett RP, Sun Z. Late Intake Valve Closing as an Emissions Control Strategy at Tier 2 Bin 5 Engine-Out NOx Level. SAE International Journal of Engines 2008;1:2008-01 - 0637. doi:10.4271/2008-01-0637.

[9] Benajes J, Molina S, Martín J, Novella R. Effect of advancing the closing angle of the intake valves on diffusion-controlled combustion in a HD diesel engine. Applied Thermal Engineering 2009;29:1947–54. doi:10.1016/j.applthermaleng.2008.09.014.

[10] Hanson R, Ickes A, Wallner T. Use of Adaptive Injection Strategies to Increase the Full Load Limit of RCCI Operation. Volume 1: Large Bore Engines; Fuels; Advanced Combustion, ASME; 2015, p. V001T03A015. doi:10.1115/ICEF2015-1115.

[11] Splitter D, Hanson R, Kokjohn S, Reitz R. Reactivity Controlled Compression Ignition (RCCI) Heavy-Duty Engine Operation at Mid-and High-Loads with Conventional and Alternative Fuels. SAE Technical Paper 2011:2011-01 - 0363. doi:10.4271/2011-01-0363.

[12] Sjöberg M, Dec JE. Effects of EGR and its constituents on HCCI autoignition of ethanol. Proceedings of the Combustion Institute 2011;33:3031–8. doi:10.1016/j.proci.2010.06.043.

[13] Zhao H. HCCI and CAI engines for automotive industry. Cambridge: Woodhead Publishing Limited; 2007.

[14] Dempsey AB, Das Adhikary B, Viswanathan S, Reitz RD. Reactivity Controlled Compression Ignition Using Premixed Hydrated Ethanol and Direct

Injection Diesel. Journal of Engineering for Gas Turbines and Power 2012;134:082806–082806 - 11. doi:10.1115/1.4006703.

[15] Christensen M, Johansson B. Homogeneous Charge Compression Ignition with Water Injection. SAE Technical Paper 1999. doi:10.4271/1999-01-0182.

[16] Brewster S, Railton D, Maisey M, Frew R. The Effect of E100 Water Content on High Load Performance of a Spray Guide Direct Injection Boosted Engine. SAE Technical Paper 2007. doi:10.4271/2007-01-2648.

[17] Flowers DL, Aceves SM, Frias JM. Improving Ethanol Life Cycle Energy Efficiency by Direct Utilization of Wet Ethanol in HCCI Engines. SAE Technical Paper 2007;01. doi:10.4271/2007-01-1867.

[18] Lanzanova TDM, Dalla Nora M, Zhao H. Performance and economic analysis of a direct injection spark ignition engine fueled with wet ethanol. Applied Energy 2016;169:230–9. doi:10.1016/j.apenergy.2016.02.016.

[19] Saffy H a, Northrop WF, Kittelson DB, Boies AM. Energy, carbon dioxide and water use implications of hydrous ethanol production. Energy Conversion and Management 2015;105:900–7. doi:10.1016/j.enconman.2015.08.039.

[20] López-Plaza EL, Hernández S, Barroso-Muñoz FO, Segovia-Hernández JG, Aceves SM, Martínez-Frías J, et al. Experimental and Theoretical Study of the Energy Savings from Wet Ethanol Production and Utilization. Energy Technology 2014;2:440–5. doi:10.1002/ente.201300180.

[21] Munsin R, Laoonual Y, Jugjai S, Imai Y. An experimental study on performance and emissions of a small SI engine generator set fuelled by hydrous ethanol with high water contents up to 40%. Fuel 2013;106:586–92. doi:10.1016/j.fuel.2012.12.079.

[22] Martins M, Fischer I, Gusberti F, Sari R, Nora MD. HCCI of Wet Ethanol on a Dedicated Cylinder of a Diesel Engine. SAE Technical Paper 2017. doi:10.4271/2017-01-0733.

[23] Mack JH, Aceves SM, Dibble RW. Demonstrating direct use of wet ethanol in a homogeneous charge compression ignition (HCCI) engine. Energy 2009;34:782–7. doi:10.1016/j.energy.2009.02.010.

[24] Fang W, Fang J, Kittelson DB, Northrop WF. An Experimental Investigation of Reactivity-Controlled Compression Ignition Combustion in a Single-Cylinder Diesel Engine Using Hydrous Ethanol. Journal of Energy Resources Technology 2015;137:031101–031101 - 7. doi:10.1115/1.4028771.

[25] Schwoerer J, Kumar K, Ruggiero B, Swanbon B. Lost-Motion VVA Systems for Enabling Next Generation Diesel Engine Efficiency and After-Treatment Optimization. SAE Technical Paper 2010. doi:10.4271/2010-01-1189.

[26] International Organisation of Legal Metrology (IOLM). International Recommendation No22 - Alcoholometry. First Ed. Paris: 1973.

[27] Heywood JB. Internal Combustion Engine Fundamentals. First Ed. McGraw-Hill, Inc.; 1988.

[28] Ickes A, Hanson R, Wallner T. Impact of Effective Compression Ratio on Gasoline-Diesel Dual-Fuel Combustion in a Heavy-Duty Engine Using Variable Valve Actuation. SAE Technical Paper 2015. doi:10.4271/2015-01-1796.

[29] The European Parliament and the Council of the European Union. Commission Regulation (EU) No 582/2011. Official Journal of the European Union 2011;167.

[30] Splitter DA, Reitz RD. Fuel reactivity effects on the efficiency and operational window of dual-fuel compression ignition engines. Fuel 2014;118. doi:10.1016/j.fuel.2013.10.045.

[31] Kokjohn SL, Hanson RM, Splitter D a, Reitz RD. Fuel reactivity controlled compression ignition (RCCI): a pathway to controlled high-efficiency clean combustion. International Journal of Engine Research 2011;12:209–26. doi:10.1177/1468087411401548.

Effects of injection timing and EGR ratio on the combustion process of the mixture of fatty acid methyl esters and diesel in a common rail high pressure diesel engine

Y Qian, Y Zhang, D Han, D Ju, L Shi, X Lu
Key Laboratory for Power Machinery of M.O.E,
Shanghai Jiao Tong University, P. R. China

ABSTRACT

As biodiesel is deemed to be one kind of the main alternates for diesel fuel, researches have been conducted to find feasible ways of efficient and clean combustion for biodiesel in conventional diesel engine. This paper conducted an experimental study on the combustion process for the mixture of fatty acid methyl esters and high-quality diesel in a modified diesel engine. The combustion and emission characteristics of commercial diesel, methyl decanoate/diesel (1:4 in volume) and methyl hexanoate/diesel (also 1:4 in volume) are compared. Results shown that under the same injection timing, fuels blended with methyl hexanoate demonstrated the higher CO emissions. Retarding the injection timing of the fuel blends generally resulted in an increase in the total number of particulate matter (PM) and a decrease in average particle size. The indicated thermal efficiency firstly increased and then decreased with the retarding of injection timing and nonetheless the value was kept over 42%. Besides, as the exhaust gas recirculation (EGR) ratio is increased, CO emissions increased while HC emissions remained almost unchanged. At the same EGR ratio, fuel blended with methyl hexanoate demonstrated a higher level of CO and HC emissions. PM emissions of diesel was influenced significantly by EGR while for fuel blends, the impact was slight. At the higher EGR ratios, there was a remarkable reduction for indicated thermal efficiency, among which methyl hexanoate/diesel showed the lowest indicated thermal efficiency at different EGR ratios whilst the value was higher than 42%. According to the experimental results, low NOx and particle emissions and relatively high efficiency for diesel blending fatty acid methyl esters can be achieved via various strategies including proper injection timing and high EGR ratio.

Keywords: Methyl hexanoate; Methyl decanoate; Injection timing; EGR; Diesel; Combustion and emissions

1. INTRODUCTION

Driven by the ever-increasing energy demand and environmental pollutions, researchers are now focusing on the development of biomass energy, a main kind of renewable energy, and its applications in conventional diesel engines [1-2]. One important kind of biomass energy is biodiesel. It can be derived from a variety of vegetable oil, animal fat and waste cooking oil. Biodiesel can be directly used in some conventional diesel engines due to the relatively high cetane number [3-4]. However, some physicochemical properties of biodiesel, including density, viscosity and oxygen content, differ from those of commercial diesel and this leads to the distinct effects on combustion and emission performances [5]. For this reason, wide and deep investigations into the combustion and emission characteristics of biodiesel in diesel engines are conducted by many researchers.

Studies on NOx emissions of biodiesel have reported that compared with commercial diesel, biodiesel generates higher NOx emissions. Biodiesel consists of a large number of oxygenated fatty acid methyl/ethyl esters. Higher oxygen content provides favorable conditions for NOx formation. For another, soot emissions are the effective radiators. The lower soot level during biodiesel combustion reduces the radiative heat loss to the environment thus maintains a high in-cylinder temperature, favoring the NOx formation [6-9]. Schönborn et al [10, 11] found that esters with shorter chain length tended to generate the higher NOx emissions. Biodiesel with shorter chain length increases the ignition delay and consequently augments the fraction of premixed combustion. If a higher proportion of fuel is burned in premixed combustion, the combustion temperature would be higher correspondingly, which is responsible for the higher amounts of NOx. Meanwhile, Pham et al. [12] claimed an increase in the total number and mass concentration of particle emissions as the fuel fatty acid chain length increases. The effects of unsaturation of a biodiesel on PM emissions are considered complicate. Fuel with unsaturation bond tends to decompose into C2H4 and C2H2 during combustion. Those alkenes and alkynes are primary precursors of soot formation, therefore the higher unsaturation degree a fuel has, the higher PM emission there will be [13-16]. On the other hand, higher unsaturated fatty acid fuel can prolong ignition delay, allowing more time for fuel-air mixing and ultimately reducing PM emissions [17-18]. Besides, several researchers confirm that the higher oxygen concentration in biodiesel fuel can help to reduce PM emissions [19-20].

Apart from distinct physicochemical properties of fuel itself, engine operating parameters such as injection timing, injection pressure and EGR also play an important role in biodiesel combustion and emissions [21]. Labecki et al. [22] investigated rapeseed oil and its blends with diesel fuel on engine performance and found that at 20% EGR ratio, more than 65% of NOx emissions could be reduced as the concentration of rapeseed oil in the blend increases. The high concentration of rapeseed oil in diesel may cause high soot, CO and THC emissions. Can et al [5] blended soybean biodiesel in 20 vol. % with commercial diesel to study the effects of EGR on fuel combustion, engine performances as well as pollutant emissions. Results revealed that both the maximum in-cylinder pressure and peak value of heat release rate increased under the combined effects of biodiesel fuel and EGR applications. Higher premixed combustion fractions were seen together with the improvements in NOx-smoke trade-off performance. However, the CO and HC emissions would be higher.

Biodiesel generally has higher viscosity compared with commercial diesel fuel, which leads to relatively worse volatility, atomization and PM emissions for biodiesel combustion. Soot emissions can even reach a quite high level in some cases [23-26]. Hence, it makes significant sense to investigate the prospective applications of short chain fatty acid methyl esters, which have low viscosity and high volatility, on common rail high pressure diesel engines. Moreover, methyl hexanoate and methyl decanoate have been widely adopted and studied as biodiesel surrogate component. However, previous researches focused on the fundamental ignition characteristics and chemical kinetic mechanism, the investigation on diesel engines is scarce. Therefore, the methyl hexanoate and methyl decanoate were chosen in this paper to study their combustion and emission characteristics on diesel engine. In this paper, fatty acid methyl esters were blended with diesel at the ratio of 20 vol. % for experimental studies. The main attention was focused on analyzing the effects of injection timing and EGR ratio on combustion and emission characteristics of the blended fuel.

2. EXPERIMENTAL SETUP

2.1 Engine and instrumentation

The experiments were performed on an electronic-control common rail in-line four-cylinder diesel engine, the main specific engine parameters are listed in Table 1. For reducing the amount of fuel samples used in the experiment, the intake and exhaust systems of the fourth cylinder were rebuilt as test cylinder, while the rest cylinders worked in conventional diesel mode. The precise control of injection quantity and injection phase in fourth cylinder was enabled by an impressed electronic control unit (ECU). The measure point of exhaust emissions was arranged at approximately 1m downstream from the exhaust pipe of test cylinder and the emissions were sampled before flowing into the total exhaust pipe. The layout of experimental platform is shown in Fig. 1. The in-cylinder gas pressure was measured by a pressure transducer (Kistler 6125B). The charge output from this transducer was converted to amplified voltage using an amplifier (Kistler 5015A) and then input to combustion analyzer via data acquisition system (D2T Orisis). The in-cylinder heat release rate and combustion temperature could be obtained on the basis of the in-cylinder pressure and the first law of thermodynamics. The real-time measurement of PM with its particle size ranging from 5 nm to 1000 nm was measured by DMS 500 fast particle sample analyzer manufactured by Cambustion company.

Table 1 Engine parameters

Engine parameters	values
Bore × Stroke [mm]	113×130
Displacement [L]	5.3
Compression ratio	18
No. of nozzle holes × nozzle diameter [mm]	7×0.176
Intake valve open [°CA ATDC]	337.5
Intake valve close [°CA ATDC]	-144.5
Exhaust valve open [°CA ATDC]	112.5
Exhaust valve close [°CA ATDC]	-334.5°

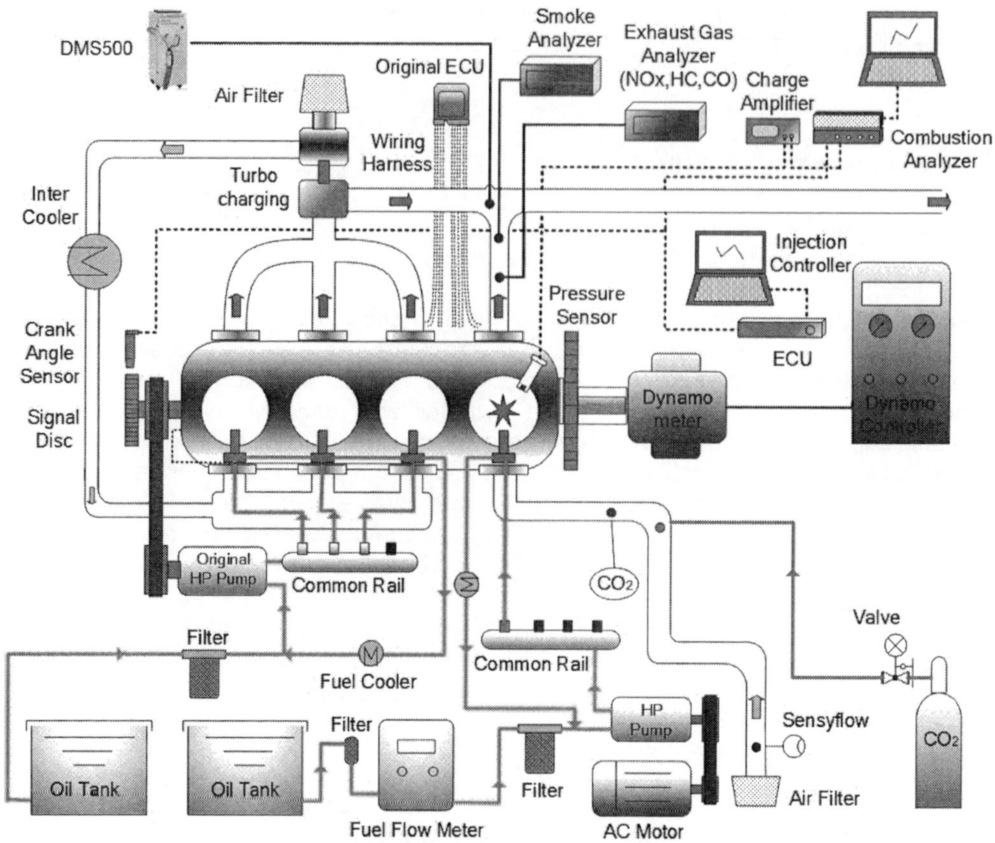

Fig. 1 Schematic of experimental setup

2.2 Test fuels

The fuels used in the experiment are Europe VI standard diesel, methyl hexanoate and methyl decanoate. The basic physical and chemical properties of the three fuels are listed in Table 2. Due to the significant difference between methyl hexanoate and diesel in physical and chemical properties, it is not suitable to use pure methyl hexanoate on diesel engine directly. Therefore, it blends with diesel according to the volume proportion of 1:4, namely the volume fraction of methyl hexanoate is 20% (ignoring the volume loss of the mixed solution), the methyl decanoate does so too. The physical parameters of the blended fuels are shown in Table 3 and Fig. 2.

Table 2 Properties of diesel, methyl hexanoate and methyl decanoate

Name	diesel	methyl hexanoate	methyl decanoate
Molecular formula	-	C7H14O2	C11H22O2
Molecular weight	180~200	130.2	186.3
Density [g/cm^3]	0.836	0.885	0.871
Boiling point [°C]	-	150	224
Purity [%]	-	>99.5	>99.5
Lower heat value[MJ/kg]	42.65	30.41	34.51
Cetane number	52.0	-	47.2
C [wt %]	85.9	64.6	71.0
H [wt %]	13.6	10.8	11.8
O [wt %]	<0.5	24.6	17.2

Table 3 Properties of diesel and the mixed fuels

	Viscosity[mm²/s][1]	Density[g/cm³]	Lower Heating value[MJ/kg][2]
Diesel	2.86	0.836	42.65
Methyl hexanoate/Diesel	1.94	0.842	40.20
Methyl decanoate/Diesel	2.42	0.840	41.02

1. The viscosity is measure according to ASTM D445-17a.

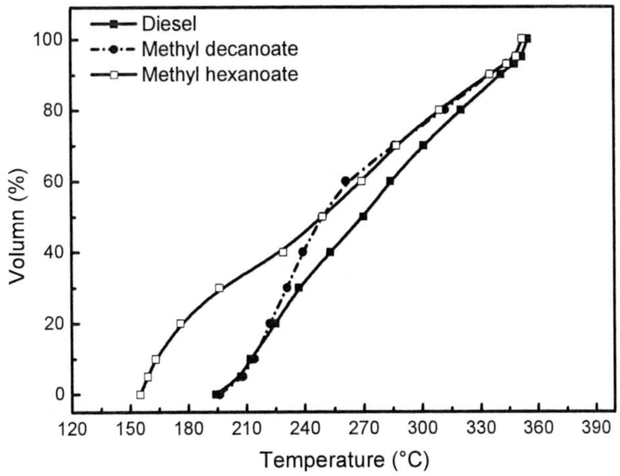

Fig. 2 Distillation curves of diesel and the other two mixture fuels

The distillation range of the test fuels were measured according to ASTM D86-17. As shown in Fig. 2, the distillate speeds of fractions in blended fuels are different due to blending with different fatty acid methyl esters. To be more specific, the blended fuel containing methyl hexanoate is distilled first due to the lowest boiling temperature of methyl hexanoate among testing fuels. Besides, the initial distillate timing of blended fuel containing methyl decanoate is similar to diesel. Generally, methyl hexanoate/diesel has the highest volatility, followed by methyl decanote/diesel, and the diesel has the worst volatility among three fuels.

2.3 Experiment procedure

The operating conditions for the engine are listed in Table 4. During the experiment, the intake temperature was maintained at 25 °C ±3. Besides, coolant and lubrication oil temperatures were kept at 80 °C ±1 and 85 °C ±3, respectively. The data obtained in this experiment were all steady state data. The heat release rate and in-cylinder temperature are calculated by a D2T combustion analyzer. CO emissions were measured by non-dispersive infra-red analyzer (NIDR CAI 602P), total hydrocarbon (THC), and NOx emissions were measured by a heated flame ionization detector (HFID CAI 600) and heat chemiluminescent analyzer (HCLD CAI 600), respectively. The particulate matter distribution was measured by a fast particulate analyzer (DMS500). The intake and exhaust temperature was measured by Omega K type thermocouple. The in-cylinder gas pressure signal was recorded every 0.5°CA and the value finally used were an average of recorded data from 100 continuous cycles. In addition, various amount of pure CO2 was inducted into intake pipe as exhaust gas recirculation.

Table 4 Engine operating conditions

	Engine Speed (rpm[1])	Injection Number	Injection Pressure (MPa)	Injection Quantity (mg/cycle)	Injection Timing (°CA ATDC[2])	EGR Ratio (%)
Part I	1500	Single	120	45	-1,-3,-5,-7, -9,-11,-13	0
Part II	1500	Single	120	45	-7	0,20, 40,60

1.rpm: rotate per minute
2.ATDC: after top dead center

2.4 Definition of EGR ratio

EGR ratio is defined by the CO2 concentration in intake and exhaust gas [27]:

$$\text{EGR\%} = \frac{CO_{2\,int}}{CO_{2\,ext}} \times 100\%$$

where $CO_{2\,int}$ is the volume fraction of CO_2 in the intake pipe and $CO_{2\,ext}$ is the volume fraction of CO_2 in the exhaust pipe, respectively.

3. EXPERIMENTAL RESULTS AND DISCUSSION

3.1 Effect of injection timing on the combustion and emissions of binary blend fuels

As one of the most important combustion parameters, the injection timing has a significant impact on the formation of the in-cylinder mixed gas. Too early or late injection will deteriorate the combustion process, which means increasing emissions and reducing thermal efficiency. Therefore, it is important to determine the appropriate fuel injection phase.

Fig. 3 shows the in-cylinder pressure and heat release rate of the three fuels (diesel, methyl decanoate/diesel, methyl hexanoate/diesel) under three injection timings of -3, -7, -11°CA ATDC. With the advance of injection timing, the combustion phase is advanced. The in-cylinder pressure and heat release rate curves keep moving forward. The peak value of the in-cylinder pressure increases as the injection timing advances, while it is the opposite for the peak value of the heat release rate. This is because with the advance of the injection timing, the fuel is injected earlier into the cylinder, which enables more fuel to burn before top dead center (TDC), so that the in-cylinder pressure and heat release rate increase earlier. On the other hand, the higher in-cylinder temperature at the later

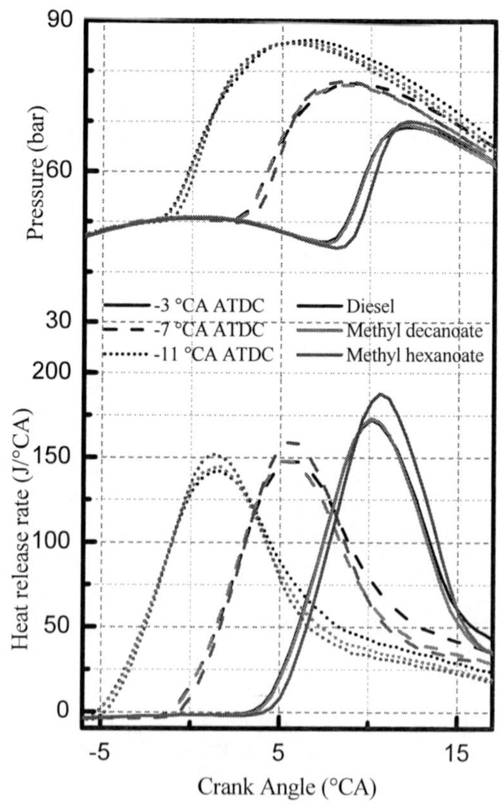

Fig. 3 Effects of the injection timing on the in-cylinder pressure and heat release rate

injection timing is conducive to the evaporation of the fuel, which further makes the heat release process more concentrated and the peak heat release rate higher.

In addition, under the same injection timing, the peak values of the heat release rate of methyl hexanoate/diesel is higher than those of diesel and methyl decanoate/diesel. The lower boiling point and viscosity of methyl hexanoate are beneficial to the atomization and evaporation of the fuel in the spray process, which improves the mixing of fuel and air, thus methyl hexanoate/diesel has a higher peak heat release rate.

Fig. 4 shows the ignition delay and the in-cylinder maximum temperature under different injection timings. As can be seen from Fig. 4 (a), with the delay of the injection timing, the ignition delay remains unchanged firstly and then increases gradually. Because the injection pulse width in this experiment is about 6 °CA, as the injection timing delays, the fuel injection process continues until after the TDC, in which the in-cylinder temperature and pressure are gradually decreased, leading to a shorter residence time of the fuel at high temperature, so the ignition delay is increased. It is worth to note that the ignition delay of methyl hexanoate/diesel is longer than that of diesel and methyl decanoate/diesel when the injection timing is retarded to -7°CA ATDC. This can be interpreted that the shorter carbon chain length of methyl hexanoate/diesel leads to a lower compression activity. For the favorable in-cylinder pressure and temperature conditions, the ignition delay does not show any differences when the injection timing is earlier. With the retard of the injection timing, the in-cylinder temperature and pressure conditions for the fuel atomization become worse, and the influence of the cetane number on the combustion process is strengthened, resulting in a longer ignition delay and a retarded combustion phase.

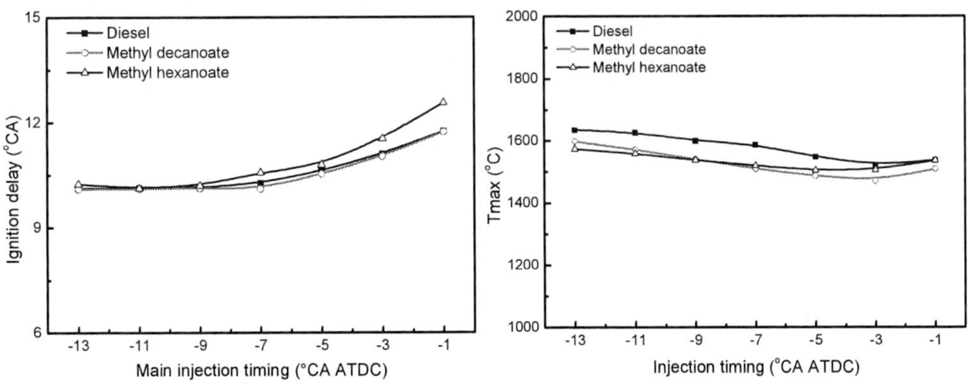

(a) (b)

Fig. 4 Effects of the injection timing on the ignition delay and in-cylinder maximum temperature

In Fig. 4 (b), the maximum mass average temperature gradually increases with the advance of the injection timing. Early injection timing means that more fuel can be burned in the compression stroke, so that the combustion temperature is higher. In addition, diesel has the highest in-cylinder maximum temperature among the three fuels at the same injection timing. More heat can be released before TDC with the same injection quantities due to its high lower heating value (LHVs), so the maximum temperature of diesel is higher. It is noteworthy that the maximum mass average temperature of methyl hexanoate/diesel is higher than that of methyl decanoate/diesel at the late injection timing. The better volatility and the longer ignition delay of methyl hexanoate/diesel are favorable for the atomization and mixing processes, which contributes to the increase of the peak value of heat release rate, so the maximum mass average temperature is higher.

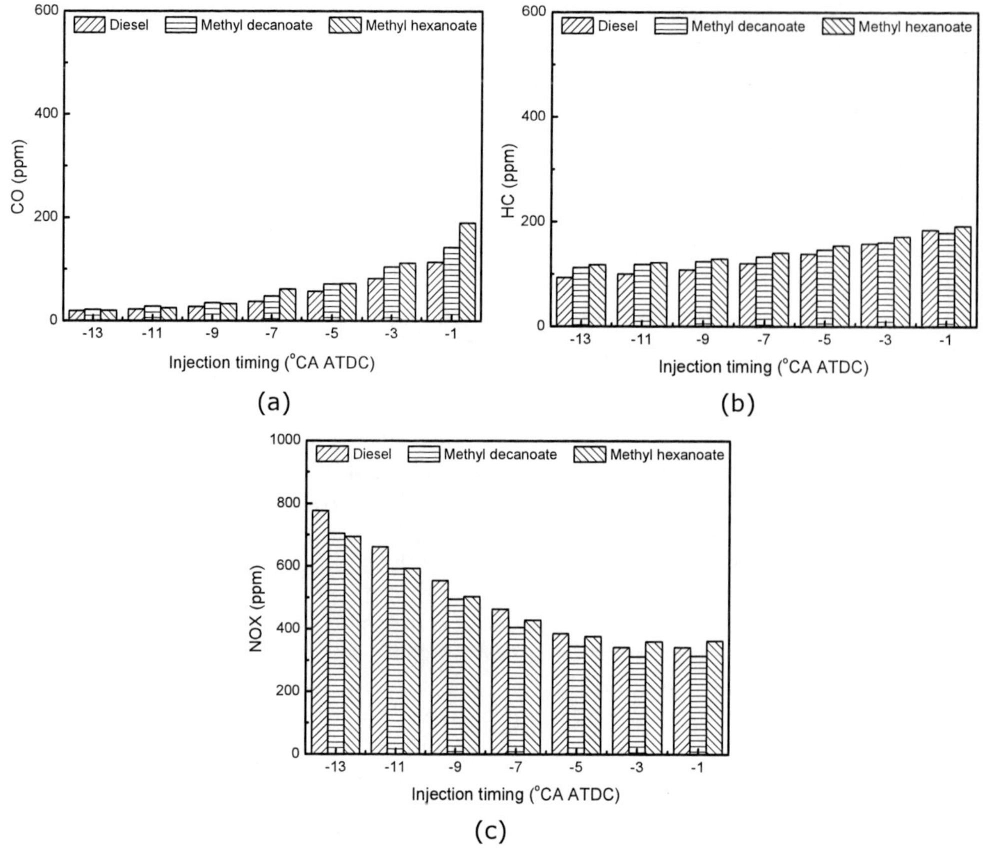

Fig. 5 Effects of injection timing on the main gas emissions

Fig. 5 illustrates the effects of injection timing on the main gas emissions. As shown in Fig. 5 (a), the CO emissions increase with the delay of injection timing, nevertheless it is below 200ppm all through. Obviously, with the delay of injection timing, the in-cylinder combustion temperature decreases, leading to the incomplete oxidation of CO emissions. As a result, CO emissions increase gradually as the injection timing delays. It is noteworthy that because of the higher LHVs of diesel, its combustion temperature is higher and its CO emissions are lower than those of other two blended fuels. In Fig. 5 (b), the HC emissions increase slightly with the delay of injection timing, and the specific values are less than 200ppm. Similar to CO emissions, with the delay of injection timing, the decrease of the in-cylinder combustion temperature results in a difficulty in the oxidation of HC emissions. Therefore, the HC emissions increase gradually. In addition, the HC emissions of diesel are lower than those of other two fuels. Diesel has a smaller penetration distance owing to its lower density, so the propensity of fuel impingement decreases, resulting in a low HC in the boundary layer. Also, the higher in-cylinder temperature of diesel is benefit to the oxidation of HC emissions. Consequently, the longer ignition delay and lower in-cylinder maximum temperature of the two blended fuels are responsible for the higher CO and HC emissions. It can be observed from Fig.5 (c) that with the delay of injection timing, the NOx emissions decrease continually. Due to the delay of combustion phase resulting from the delay of injection timing, more combustion takes place in the expansion stroke, when the in-cylinder temperature decreases rapidly, thus the duration of high in-cylinder temperature shortens and NOx emissions decrease. It is notable that when the injection timing is advanced, the NOx emissions of diesel are higher than those of methyl decanoate/diesel and methyl hexanoate/diesel, and with the delay of injection timing, the NOx emissions of methyl hexanoate/diesel are higher than those of commercial diesel. The formation of NOx is strongly related to the combustion temperature, the oxygen concentration and the in-cylinder high temperature duration. Because of the lower LHVs, the in-cylinder maximum combustion temperature of two blends is lower (as shown in Fig.4), which prohibit the NOx formation. Diesel's higher in-cylinder combustion temperature result in the higher NOx emissions. With the delay of injection timing, the ignition delay of methyl hexanoate/diesel increases and the peak heat release rate rises, thus the in-cylinder maximum combustion temperature increases, leading to the higher NOx emissions than diesel and methyl decanoate/diesel.

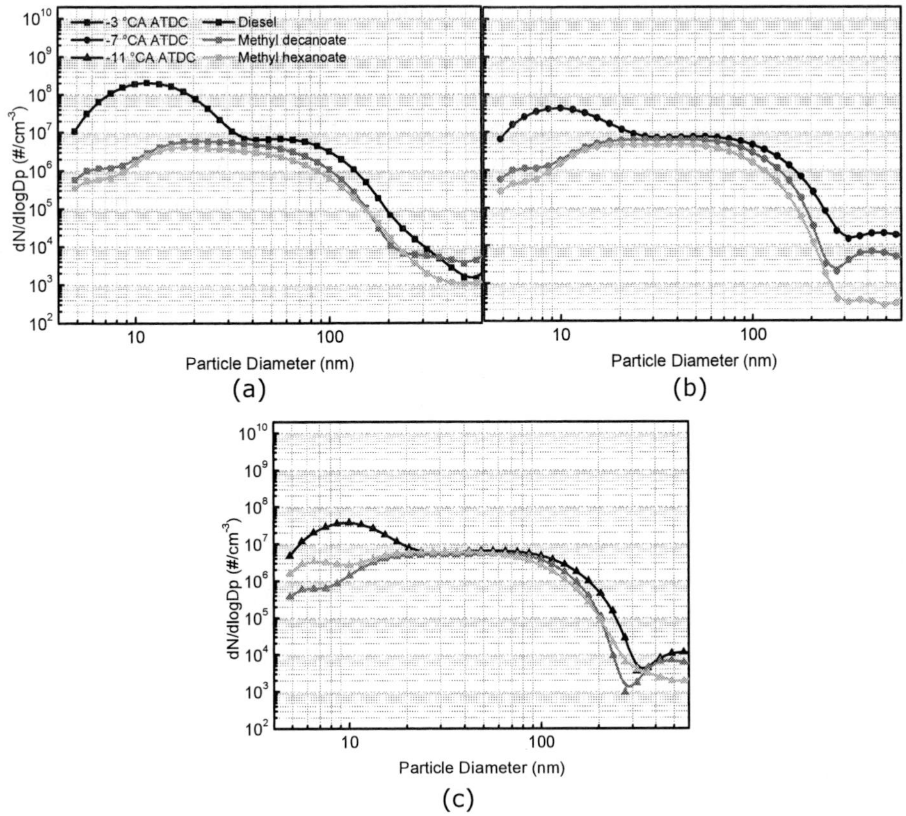

Fig. 6 Effects of injection timing on particle number size distribution

Generally, particles can be divided into nucleation mode and accumulation mode according to its size. Nucleation particle is formed by volatile components and sulfate. Accumulation particle takes soot as kernel, to which organics (e.g. unburned hydrocarbon) and inorganics (e.g. sulfate) are attached on the surface. Fig. 6 illustrates the particle number size distributions of the three fuels at the different injection timings. The injection timing in Fig. 6(a), (b), (c) is -3, -7 and -11°CA ATDC, respectively. It can be seen that at different injection timings, the peak number concentration of the nucleation mode particles of diesel is higher than those of other two blended fuels. Besides, with the delay of injection timing, the peak number concentration of nucleation mode particles of pure diesel increases gradually. Because of the common rail pressure of 120MPa, the atomization of fuel is preferable, which reduces the local oxygen deficit zone in combustion process, thus the number concentration of accumulation mode particles decreases. On the other hand, owing to that diesel molecule hardly contains oxygen atom, the oxidation to fine soot particles is less effective than the other two fuels. With the delay of injection timing, the longer ignition delay results in the improvement of the mixture formation of fuel and air, and the volatile components in boundary layer increase, which leads to the increase of the number concentration of nucleation mode particles.

Fig. 7 shows the effects of injection timing on the indicated mean effective pressure (IMEP) and indicated thermal efficiency. As shown in Fig. 7(a), at the same injection timing, the IMEP of diesel is about 10% higher than that of the other two blended fuels, and as the injection timing delays, the IMEP of diesel firstly increases slightly and then decreases gradually. In addition, when the injection timing is earlier than -7°CA ATDC, the IMEP of methyl decanoate/diesel is higher than that of methyl hexanoate/diesel. The fuel with higher LHVs will release more heat at the same injection quantity. For LHVs, diesel> methyl decanoate > methyl hexanoate, so the order of IMEP is diesel> methyl decanoate/diesel > methyl hexanoate/diesel. With the retard of injection timing, the ignition delay of methyl hexanoate/diesel becomes longer, which leads to a more concentrated heat release process. Therefore, the IMEP of methyl hexanoate/diesel is higher than that of methyl decanoate/diesel.

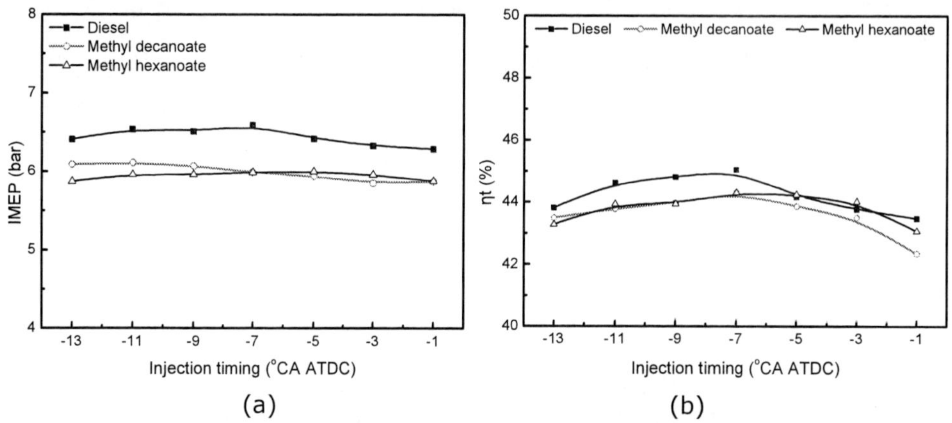

Fig. 7 Effects of injection timing on IMEP and indicated thermal efficiency

The indicated thermal efficiency of the three fuels increases initially and then decreases as the injection timing delays, as seen in Fig. 7(b). The working capacity of in-cylinder mixture increases firstly and then decreases. On the other hand, with the further delay of injection timing, the CO and HC emissions increase owing to the deterioration of combustion process, which reduces the combustion efficiency. When the injection timing is earlier, the combustion becomes more complete owing to the better mixing process, and the indicated thermal efficiency of diesel is obviously higher than those of the other two fuels. The difference in the indicated thermal efficiency of three fuels becomes smaller as the injection timing delays. But when the injection timing is closer to TDC, the indicated thermal efficiency of methyl hexanoate/diesel, which is close to that of diesel, is gradually higher than that of methyl decanoate/diesel. As shown in Fig. 3 and Fig. 4(b), the higher heat release rate and the higher in-cylinder maximum temperature of methyl hexanoate/diesel which result from the longer ignition delay are responsible for the higher thermal efficiency. It is noteworthy that the IMEPs of the diesel fuel are always much higher than other two fuels regardless of injection timing, however the thermal efficiencies of diesel fuel are not higher at late injection timings. The high LHVs of diesel is the dominant factor for this phenomenon.

3.2 Effects of EGR on conventional emissions of binary blend fuels

EGR are widely used as an important strategy to achieve low temperature combustion technology. However, EGR exerts a significant impact on the conventional gas emissions and the PM emissions. Especially it can significantly reduce NOx emissions.

Fig. 8 shows the effects of EGR ratio on the main gas emissions. It can be seen that the CO emissions are relatively low under the low and medium EGR ratio conditions (less than 40%) and apparently increase at high EGR ratio. With the decrease of the carbon chain length, the CO emissions gradually increase. Under the higher EGR ratio, the in-cylinder temperature apparently decreases, which leads to the difficulty in the oxidization of CO. Thus, the CO emissions gradually increase. The effect of EGR ratio on HC emissions is not obvious. The difference of HC emissions among three fuels is mainly caused by the fuel reactivity. The mixture of low reactivity fuel and air has a longer ignition delay, so there are more unburned hydrocarbons in the boundary layer, which easily forms HC emissions. As Fig. 8(c) showed, with the increase of EGR ratio, the in-cylinder temperature and oxygen concentration gradually decrease, which apparently restraints the NOx formation. The NOx emissions decrease to about 100pm under the 60% EGR ratio.

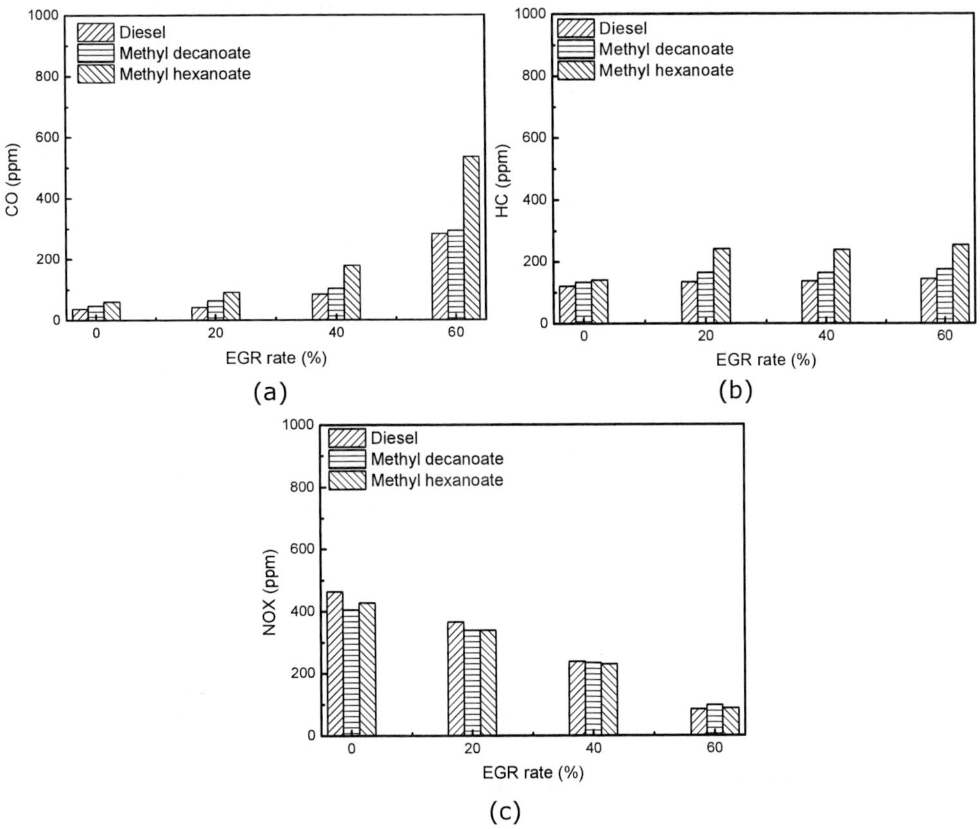

Fig. 8 Effects of EGR on the main gas emissions

Fig. 9 (a), (b) and (c) show the particle number size distributions at the 0%, 40% and 60% EGR ratios, respectively. In general, diesel has a higher particle number concentration at each EGR ratio and the peak number concentration of nucleation mode particles is much higher than that of accumulation mode particles. The effects of EGR ratio on the particle number size distributions of the other two fuels are not significant. At the same EGR ratio, the particle number concentration of diesel is higher than that of the other two fuels. This phenomenon is attributed to that the molecules of the two fatty acid methyl esters contain different proportions of oxygen atoms, which is conducive to reducing the particle emissions. Especially under the high EGR ratio, the severe situation of in-cylinder local oxygen deficit enlarges the advantage of the oxygen content in the fatty acid methyl ester molecules, thus the PM emissions of the other two blended fuels are obviously lower than that of diesel.

Thanks to the highest oxygen content in methyl hexanoate molecule, the particle number concentration of methyl hexanoate/diesel decreases significantly. Furthermore, with the increase of EGR ratio, the peak number concentration of nucleation mode particles decreases initially and then increases. There are two main reasons. The EGR prolongs the ignition delay to promote the mixing of fuel and air, which is effective in reducing the PM formation; however, on the other hand, excessive CO2 tremendously reduces the in-cylinder oxygen concentration, leading to a relatively high PM emission.

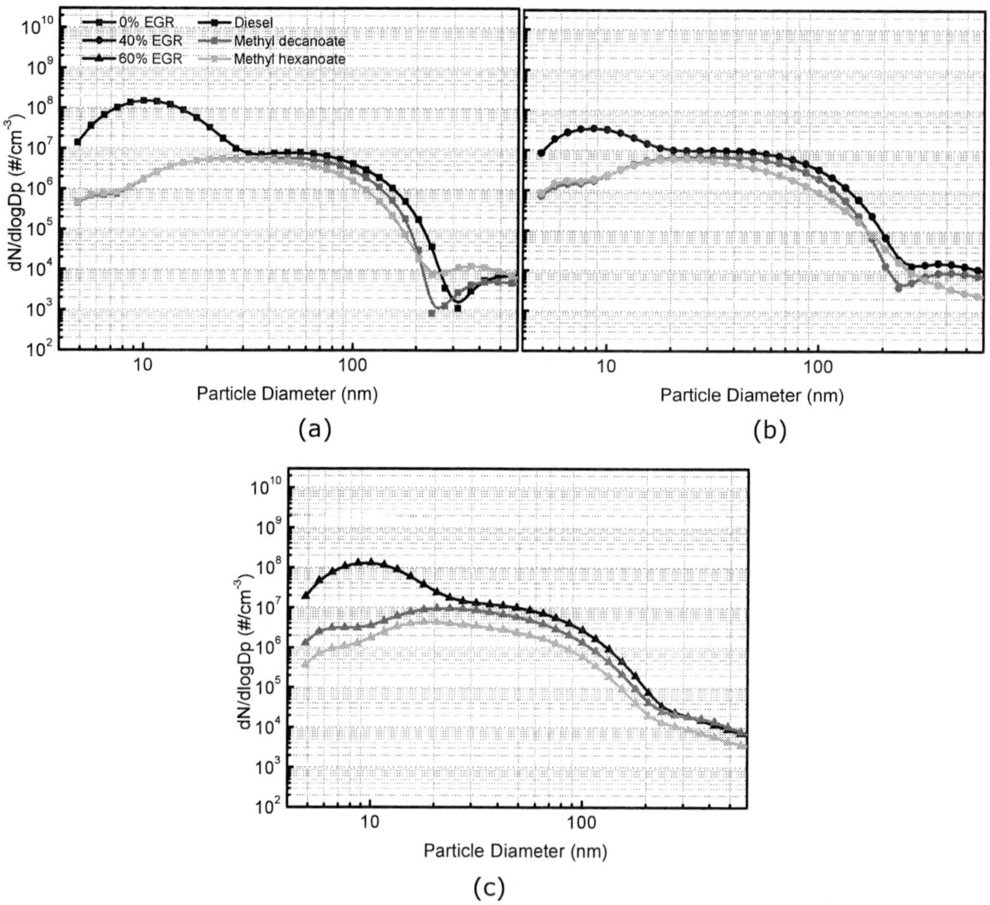

Fig. 9 Effects of EGR on particle number size distribution

Fig. 10 illustrates the IMEP and the indicated thermal efficiency under different EGR ratios. On the whole, with the increase in EGR ratios, the IMEP and the indicated thermal efficiency of three fuels gradually decrease, which is attributed to the reduction of the working capacity of mixtures with the delay of combustion phase. Because methyl hexanoate/diesel has the lowest reactivity and LHVs, which results in a more delayed combustion phase, the IMEP and the indicated thermal efficiency of methyl hexanoate/diesel are the lowest among the three fuels at the same EGR ratio. It is worth noting that the IMEP and the indicated thermal efficiency of diesel increase when the EGR ratio increases to 20%. When the EGR ratio further increases, although the homogeneity of the mixture constantly increases, most of the combustion takes place in the expansion stroke, which leads to the reduction of working capacity of mixture, so that the IMEP and the indicated thermal efficiency decrease gradually.

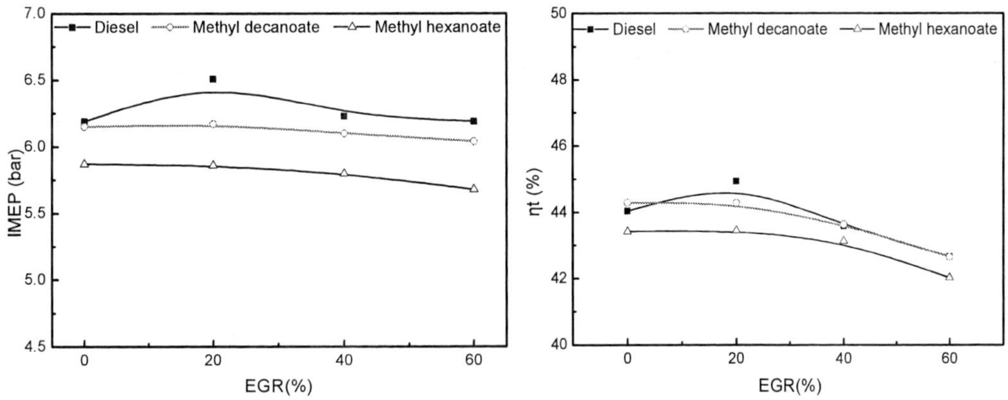

(a) (b)

Fig. 10 Effects of EGR on IMEP and indicated thermal efficiency

4. CONCLUSIONS

In order to enlarge the species of biodiesel surrogate component and research the combustion and emission characteristics of short carbon chain length fatty acid methyl ester on diesel engine, the article focused on the combustion and emissions of methyl hexanoate/diesel, methyl decanoate/diesel and diesel under different injection timing and EGR ratio conditions. Based on the above researches, the conclusions can be reached as follows:

1. The methyl hexanoate/diesel features a higher peak heat release rate and a longer ignition delay as a result of the lower reactivity and better volatility. The maximum mass average temperatures of the two binary blend fuels are lower than those of diesel, and the maximum temperature of methyl hexanoate/diesel is higher than that of methyl decanoate/diesel at the late injection timings.
2. The IMEP and the indicated thermal efficiency of diesel are higher than those of the other two blended fuels. When the injection timing is later than -7°CA ATDC, the IMEP and the indicated thermal efficiency of methyl hexanoate/diesel is gradually higher than that of methyl decanoate/diesel.
3. Compared with diesel, methyl hexanoate/diesel and methyl decanoate/diesel have the higher CO and HC emissions and lower NOx emissions. Overall, with the decrease of the carbon chain length, the CO, HC and NOx emissions gradually increase.
4. Because of the oxygen presence, the particle number concentrations of two binary blend fuels are lower than that of diesel. The methyl hexanoate/diesel has the lowest particle number concentration owing to the highest oxygen content in methyl hexanoate molecule. The difference of the particle number size distributions among three fuels becomes larger when the EGR ratio is higher.

ACKNOWLEDGMENTS

This work was supported by the Foundation for Distinguished Young Scholars of China (Grant No.51425602) and the National Basic Research Program of China (Grant No. 2013CB228405).

REFERENCES

[1] Rakopoulos DC, Rakopoulos CD, Giakoumis EG. Impact of properties of vegetable oil, bio-diesel, ethanol and n-butanol on the combustion and emissions of turbocharged HDDI diesel engine operating under steady and transient conditions. Fuel. 2015;156:1-19.

[2] Mofijur M, Rasul MG, Hyde J. Recent Developments on Internal Combustion Engine Performance and Emissions Fuelled With Biodiesel-Diesel-Ethanol Blends. Procedia Engineering. 2015;105:658-64.

[3] Ashraful AM, Masjuki HH, Kalam MA, Rizwanul Fattah IM, Imtenan S, Shahir SA, et al. Production and comparison of fuel properties, engine performance, and emission characteristics of biodiesel from various non-edible vegetable oils: a review. Energy Conversion & Management. 2014;80:202-28.

[4] Demirbas A. Biodiesel: A Realistic Fuel Alternative for Diesel Engines. El Campo Boletín De Información Agraria. 2008;107:págs. 99-103.

[5] Can Ö, Öztürk E, Solmaz H, Aksoy F, Çinar C, Yücesu HS. Combined effects of soybean biodiesel fuel addition and EGR application on the combustion and exhaust emissions in a diesel engine. Applied Thermal Engineering. 2016;95:115-24.

[6] Sahoo PK, Das LM, Babu MKG, Arora P, Singh VP, Kumar NR, et al. Comparative evaluation of performance and emission characteristics of jatropha, karanja and polanga based biodiesel as fuel in a tractor engine. Fuel. 2009;88:1698-707.

[7] ÇElİKten I, Koca A, Arslan MA. Comparison of performance and emissions of diesel fuel, rapeseed and soybean oil methyl esters injected at different pressures. Renewable Energy. 2010;35:814-20.

[8] Liu Y, Liu H, Huo M, Lee CFF, Li J. Spray and Combustion Characteristics of Butanol-Biodiesel-Diesel in a Constant Volume Combustion Chamber. Biochemistry. 2012;7:1490-505.

[9] Aydİn H, İlkİlİç C. Exhaust emissions of a CI engine operated with biodiesel from rapeseed oil. Energy Sources Part A Recovery Utilization & Environmental Effects. 2011;33:1523-31.

[10] Schönborn A, Ladommatos N, Williams J, Allan R, Rogerson J. The influence of molecular structure of fatty acid monoalkyl esters on diesel combustion. Combustion & Flame. 2009;156:1396-412.

[11] Schönborn A, Ladommatos N, Allan R, Williams J, Rogerson J. Effect of the Molecular Structure of Individual Fatty Acid Alcohol Esters (Biodiesel) on the Formation of Nox and Particulate Matter in the Diesel Combustion Process. Sae International Journal of Fuels & Lubricants. 2008;1:849-72.

[12] Pham PX, Bodisco TA, Stevanovic S, Rahman MD, Wang H, Ristovski ZD, et al. Engine Performance Characteristics for Biodiesels of Different Degrees of Saturation and Carbon Chain Lengths. Sae International Journal of Fuels & Lubricants. 2013;6:188-98.

[13] Ladommatos N, Parsi M, Knowles A. The effect of fuel cetane improver on diesel pollutant emissions. Fuel. 1996;75:8-14.

[14] Schulz H, Melo GBD, Ousmanov F. Volatile organic compounds and particulates as components of diesel engine exhaust gas. Combustion & Flame. 1999;118:179-90.

[15] Graboski MS, Mccormick RL, Alleman TL, Herring AM. Effect of Biodiesel Composition on Engine Emissions from a DDC Series 60 Diesel Engine: Final Report; Report 2 in a Series of 6. Biomass Fuels. 2003.

[16] Fontaras G, Kousoulidou M, Karavalakis G, Tzamkiozis T, Pistikopoulos P, Ntziachristos L, et al. Effects of low concentration biodiesel blend application on modern passenger cars. Part 1: Feedstock impact on regulated pollutants, fuel consumption and particle emissions. Environmental Pollution. 2010;158:1451-60.

[17] Lapuerta M, Armas O, Rodriguez-Fernandez J. Effect of the Degree of Unsaturation of Biodiesel Fuels on NOx and Particulate Emissions. Sae International Journal of Fuels & Lubricants. 2008;1:1150-8.
[18] Haas MJ, Scott KM, Alleman TL, McCormick RL. Engine Performance of Biodiesel Fuel Prepared from Soybean Soapstock: A High Quality Renewable Fuel Produced from a Waste Feedstock. Energy & Fuels. 2001;15:1207-12.
[19] Wu F, Chen WW, Shuai S. A Study On Emission Performance of a Diesel Engine Fueled With Five Typical Methyl Ester Biodiesels. Atmospheric Environment. 2009;43:1481-5.
[20] Nguyen VH, Pham PX. Biodiesels: Oxidizing enhancers to improve CI engine performance and emission quality. Fuel. 2015;154:293-300.
[21] Sayin C, Gumus M. Impact of compression ratio and injection parameters on the performance and emissions of a DI diesel engine fueled with biodiesel-blended diesel fuel. Applied Thermal Engineering. 2011;31:3182-8.
[22] Labecki L, Ganippa LC. Effects of injection parameters and EGR on combustion and emission characteristics of rapeseed oil and its blends in diesel engines. Fuel. 2012;98:15-28.
[23] Zhu L, Cheung CS, Huang Z. A comparison of particulate emission for rapeseed oil methyl ester, palm oil methyl ester and soybean oil methyl ester in perspective of their fatty ester composition. Applied Thermal Engineering. 2016;94:249-55.
[24] Agarwal AK, Gupta T, Kothari A. Particulate emissions from biodiesel vs diesel fuelled compression ignition engine. Renewable & Sustainable Energy Reviews. 2011;15:3278-300.
[25] Man XJ, Cheung CS, Ning Z. Effect of Diesel Engine Operating Conditions on the Particulate Size, Nanostructure and Oxidation Properties when Using Wasting Cooking Oil Biodiesel. Energy Procedia. 2015;66:37-40.
[26] Datta A, Mandal BK. A comprehensive review of biodiesel as an alternative fuel for compression ignition engine. Renewable & Sustainable Energy Reviews. 2016;57:799-821.
[27] Shan X, Qian Y, Zhu L, Lu X. Effects of EGR rate and hydrogen/carbon monoxide ratio on combustion and emission characteristics of biogas/diesel dual fuel combustion engine. Fuel. 2016;181:1050-7.

On board fast measurement of vehicle NOx emissions

M Peckham, M Duckhouse, H Bradley, M Irwin, M Hammond
Cambustion Ltd, UK

ABSTRACT

An on-board version of a fast response chemiluminescence analyzer with a 3 millisecond response time enables the recording of high temporal resolution transient NOx emissions during cold start and pull-away, gear changes, speed bumps and other real world transient conditions. Tests were performed on a 1.6 litre turbo GDI vehicle. The data has been correlated with on-board engine data available from the vehicle's electronic control unit to study the underlying causes of these short-duration events. Exhaust gas was sampled from after the catalyst but before the silencer (muffler) to preserve all high frequency detail and to allow accurate integration of pollutant concentration and engine exhaust mass flow. Various aspects of the vehicle's calibration are seen to contribute to the short-duration bursts of tailpipe emissions which would not be visible with slower response analyzers.

DEFINITIONS/ABBREVIATIONS

CAN	Controller area network
ECU	Engine control unit
g/km	Grams per kilometer
GPS	Global positioning system
HC	Hydrocarbon
HEGO	Heated exhaust gas oxygen
Km/h	Kilometers per hour
OBD	On board diagnostics
PN	Particle number
ppm	Parts per million
RDE	Real driving emissions
TWC	Three way catalyst
WLTP	Worldwide harmonized light vehicles test procedure

1 INTRODUCTION

Despite steadily reducing vehicle emissions limits, the European Commission, European Environment Agency and national governments of its member states are concerned over the lack of improvement in urban air quality. Much of this concern centres on the discrepancy between legislated laboratory-based test cycles such as the WLTP [1] and real world driving conditions [2]. To address this all vehicle manufacturers selling to the EU will, from September 2017, be required to carry out real world driving emissions (RDE) testing using portable emissions measurement systems (PEMS) to quantify NOx, CO, CO_2 and PN [3]. This is a fast-moving area of legislation and conference speakers often predict rapid tightening of RDE regulations in short timescales. Note that the measurement of RDE hydrocarbons has been dropped as a requirement largely due to the difficulties associated with fueling the FID analyzer.

At the time of introduction, real driving emissions legislation will exclude data from the cold start portion of the cycle, where significant emissions occur [4] but this is expected to be introduced soon after [5]. Initial legislation will also define a conformity factor, to allow for the technical difficulties involved in improving real world driving emissions in the short term. This conformity factor will reduce from 2.1 to 1.5 by January 2021 [6].

1.1 Real Driving Emissions Route

The real driving emissions test must contain elements of urban (up to 60km/h), rural (between 60 and 90km/h) and motorway (above 90km/h) driving with tight control over test conditions achieved through limits on stopped time and percentage of total trip distance spent at each condition. All conditions are classified by instantaneous vehicle speed [3].

Table 1. Summary of RDE route conditions [3]

Mode	Vehicle speed	Portion of Cycle	Other Conditions
Stopped	Less than 1km/h	10% (of urban by time)	Several stops of 10s or longer
Urban	Up to 60km/h	Between 29 and 34% (by distance)	Average speed between 15 and 30km/h
Rural	60 to 90km/h	Between 23 and 44% (by distance)	
Motorway	90 to 145km/h	Between 23 and 43% (by distance)	Above 100km/h for at least 5 minutes

The minimum distance for urban, rural and motorway operation respectively should be 16km. The total trip duration should be between 90 and 120 minutes [3] and must be conducted on a working day as defined by the Union in Council Regulation [7].

1.2 Portable Emissions Measurement

The European Commission also defines strict regulations for allowable PEMS equipment. Data should be continuously recorded throughout the test with the system not taking electrical power from a source that draws energy, either directly or indirectly, from the test vehicle. The weight of the equipment should be kept to a minimum so as to reduce the impact on the vehicle's emissions by its implementation. Additionally some method of determining exhaust mass flow, either by direct measurement or calculation, must be included as well as an indication of vehicle speed, from the ECU, via GPS or directly measured [3]. Measurement of exhaust mass flow allows conversion of ppm emissions to grams. Ambient conditions should also be logged.

1.3 Calculating Exhaust Mass Flow

Typical PEMS systems calculate the exhaust mass flow from a pitot tube measuring the exhaust mass flow at the tailpipe. However, notwithstanding the time response differences between mass flow and fast emissions data, for unobtrusive and readily available exhaust mass flow, it is possible to calculate mass from other parameters; indeed this is specifically mentioned in the RDE legislation [3]. Calculation of exhaust mass flow requires two of the following three parameters to be known:

1. Air mass flow rate [kg/s]
2. Fuel mass flow rate [kg/s]
3. Air-to-fuel ratio

For the purposes of the tests described within this paper, the air mass flow and engine-out Air-to-Fuel ratio recorded from the vehicle's own OBD port, were used to calculate the exhaust mass flow. Sometimes, the inlet air mass flow rate is not reported via OBD and then a non-obtrusive 3rd-party air mass flow meter can be fitted immediately downstream of the intake air filter. This can also be used to correlate air mass flow with intake manifold pressure.

1.4 Calculating Instantaneous Mass Emissions

Instantaneous mass emissions [g/s] must be determined by multiplying instantaneous concentration [ppm] by instantaneous exhaust mass flow rate [kg/s]. Values must be corrected with respect to the ratio of densities of exhaust components [kg/m^3] to the density of exhaust gas [kg/m^3] as described in the following table [3].

Table 2. Raw exhaust gas u values depicting the ratio between the densities of exhaust component or pollutant [kg/m^3] and the density of the exhaust gas [kg/m^3] (6)

Fuel	ρ_e [kg/m^3]	Component or pollutant i					
		NO_x	CO	HC	CO_2	O_2	CH_4
		ρ_{gas} [kg/m^3]					
		2,053	1,250	(1)	1,9636	1,4277	0,716
		u_{gas} (2) (6)					
Diesel (B7)	1,2943	0,001586	0,000966	0,000482	0,001517	0,001103	0,000553
Ethanol (ED95)	1,2768	0,001609	0,000980	0,000780	0,001539	0,001119	0,000561
CNG (3)	1,2661	0,001621	0,000987	0,000528 (4)	0,001551	0,001128	0,000565
Propane	1,2805	0,001603	0,000976	0,000512	0,001533	0,001115	0,000559
Butane	1,2832	0,001600	0,000974	0,000505	0,001530	0,001113	0,000558
LPG (5)	1,2811	0,001602	0,000976	0,000510	0,001533	0,001115	0,000559
Petrol (E10)	1,2931	0,001587	0,000966	0,000499	0,001518	0,001104	0,000553
Ethanol (E85)	1,2797	0,001604	0,000977	0,000730	0,001534	0,001116	0,000559

(1) Depending on fuel.
(2) At l = 2, dry air, 273 K, 101,3 kPa.
(3) u values accurate within 0,2 % for mass composition of: C = 66 – 76 %; H = 22 – 25 %; N = 0 – 12 %.
(4) NMHC on the basis of $CH_{2,93}$ (for THC the u_{gas} coefficient of CH_4 shall be used).
(5) u accurate within 0,2 % for mass composition of: C_3 = 70 – 90 %; C_4 = 10 – 30 %.
(6) u_{gas} is a unitless parameter; the u_{gas} values include unit conversions to ensure that the instantaneous emissions are obtained in the specified physical unit, i.e. g/s.

Using this information the following equation can be applied [3]:

$$m_{gas,i} = u_{gas} \cdot c_{gas,i} \cdot q_{mew,i} \qquad (2)$$

where:

$$m_{gas,i} = mass\ of\ the\ exhaust\ component\ \text{"gas"}\ [g/s]$$

$$u_{gas} = ratio\ of\ the\ density\ of\ the\ exhaust\ component\ \text{"gas"}\ and\ the\ overall\ density\ of\ the\ exhaust\ as\ listed\ in\ \text{Table 2}$$

$$c_{gas,i} = measured\ concentration\ of\ the\ exhaust\ component\ \text{"gas"}\ in\ the\ exhaust\ [ppm]$$

$$q_{mew,i} = measured\ exhaust\ mass\ flow\ rate\ [kg/s]$$
$$gas = respective\ component$$
$$i = number\ of\ the\ measurement$$

1.5 Vehicle On Board Diagnostics

On-board diagnostics (OBD) were first implemented in 1994 to monitor vehicle parameters, in particular to detect malfunctions from deterioration, tampering or improper maintenance which might lead to a significant increase in emissions. These diagnostics are increasingly important as emissions control solutions become more complex [8].

Table 3. Sources of excessive vehicle emissions monitored by OBD [8]

Monitoring Categories		
Catalyst Efficiency	Misfire Detection	Evaporative System
Secondary Air	Fuel System	Exhaust Gas Sensors
Exhaust Gas Recirculation	Crankcase Ventilation	Engine Cooling System
Cold Start Strategies	Variable Valve Timing	Direct Ozone Reduction
Air Metering System	Transmission Controls	Forced Air Systems
Hybrid Systems	Engine Speed/Angle	

Use of a vehicle's OBD system allows simple correlation of emissions measurements, taken during RDE testing, with engine parameters such as speed, load and lambda. An engine calibrator can make an informed choice as to what parameters would be of most interest to log. However, it is also often the case that a calibrator would have an open ECU at their disposal with the ability to log (and adjust) numerous parameters at somewhat faster speeds. Mass flow from the ECU is particularly useful for integrating with the fast emissions concentration to provide (fast) cumulative grams of emissions.

1.6 Fast Emissions Measurement

Analyzers for measuring emissions in real time have been available for some time and they are already used extensively for development of engine and after treatment calibrations particularly during transient operation [9]. The ability to measure HCs, NOx, CO, CO_2 and particulates fast enough to correlate with each exhaust event gives calibrators the additional information required to reduce short duration emissions spikes [2]. Up until now, these instruments have primarily been restricted to measurements on engine and chassis test cells but there is some argument that the same real time data could provide the necessary insight to meet RDE legislation.

For RDE testing it is desirable for the analyzer to have the following features:

1. As low power consumption as possible (to reduce required battery size)
2. Lightweight and Compact
3. Low zero and span drift
4. Quiet
5. Integration with other vehicle/engine parameters (lambda, speed, load) so that transient emissions can be correlated.
6. On-line calculation of cumulative gram emissions
7. Cope with climatic conditions
8. Durable and vibration insensitive
9. Safe
10. Minimal changes to vehicle

2 EXPERIMENTAL SETUP

A benchtop fast NOx analyzer with a 2 millisecond response time, Cambustion CLD500 [10], was adapted to run from a 110Ah conventional 12V car battery in the trunk of a Euro 4 Turbocharged GDI vehicle. As of 2015, Euro 4 vehicles comprise 31.3% of all UK licensed vehicles in use, therefore this is a representative test subject [11]. The vehicle's TWC had been hydrothermally aged to 50,000 miles from a previous project which led to emissions characteristics similar to that of a typical production vehicle. Fast emissions data was logged alongside OBD parameters taken from the vehicle CAN port.

Table 4. Vehicle summary [12]

Vehicle make	Peugeot
Date of first registration	March 2008
Year of manufacture	2008
Cylinder capacity (cc)	1598
CO_2 Emissions (g/km)	167
Fuel type	Petrol

2.1 Modifications

In order to use an emission analyzer designed for test cell applications on board several changes were required primarily to reduce weight and power consumption [13].

1. Single channel system to reduce power consumption and allow battery operation (110Ah 12V)
2. 1,000W Inverter added to allow operation of 230V components from 12V battery
3. Custom 300mm sample probe used to allow access to pre-silencer point through vehicle floor tray
4. Restrictor fitted to sample probe to reduce sample flow and therefore size of vacuum pump required
5. Small diaphragm pump used to reduce power consumption
6. Switchover system installed to allow warm up from mains power before switching to battery, maximizes possible test time and avoids use of battery during highest power consumption
7. Compact cabinet to house components
8. Exhaust gas expelled out under vehicle after sampling

Figure 1. Modified CLD500 with battery in boot of vehicle

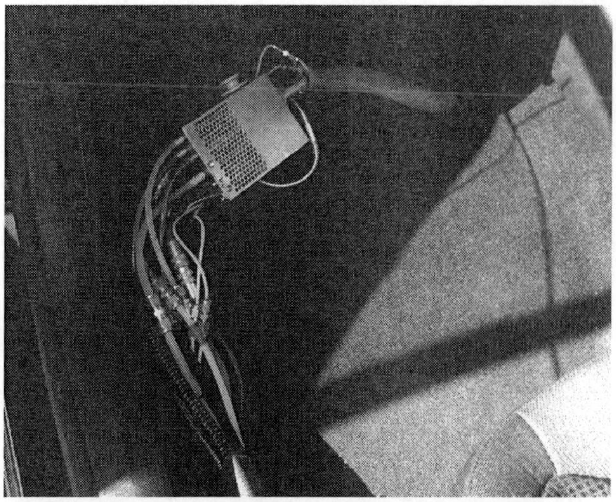

Figure 2. CLD500 sample head in passenger seat foot well

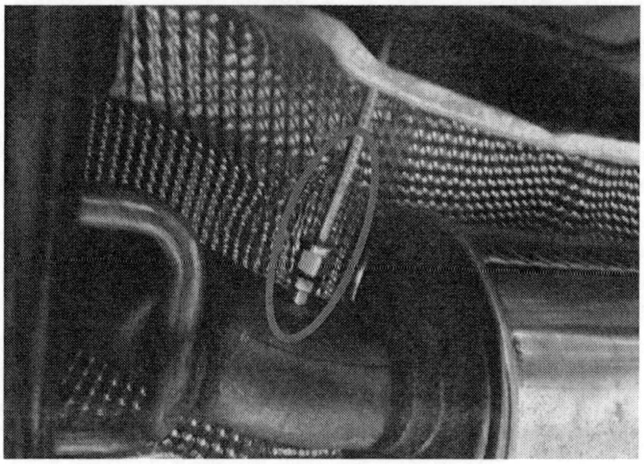

Figure 3. Sample probe position post catalyst

This set up allowed approximately 4 hours of operation from the battery and the ability for customers to switch between the mobile and lab-based analyzer configurations easily. Prior to testing, the analyzer was calibrated using zero and span gases located at the start of the RDE route. A span check was also carried out when the vehicle returned to this point to assess instrument drift. This meant it was not necessary to carry on-board gas bottles, which simplified the safety requirements. It is worth noting that some of these changes (restricted sample flow and longer sample probe) slow the time response of the instrument but due to inherent mixing associated with the post-catalyst sample point this does not present a problem. The data is still at a frequency several orders of magnitude faster than a typical PEMS device. After modification, the time response of the instrument was defined using a flick test [14], giving a result of around 3ms; a very small compromise from the 2ms response time achievable from the lab-based configuration.

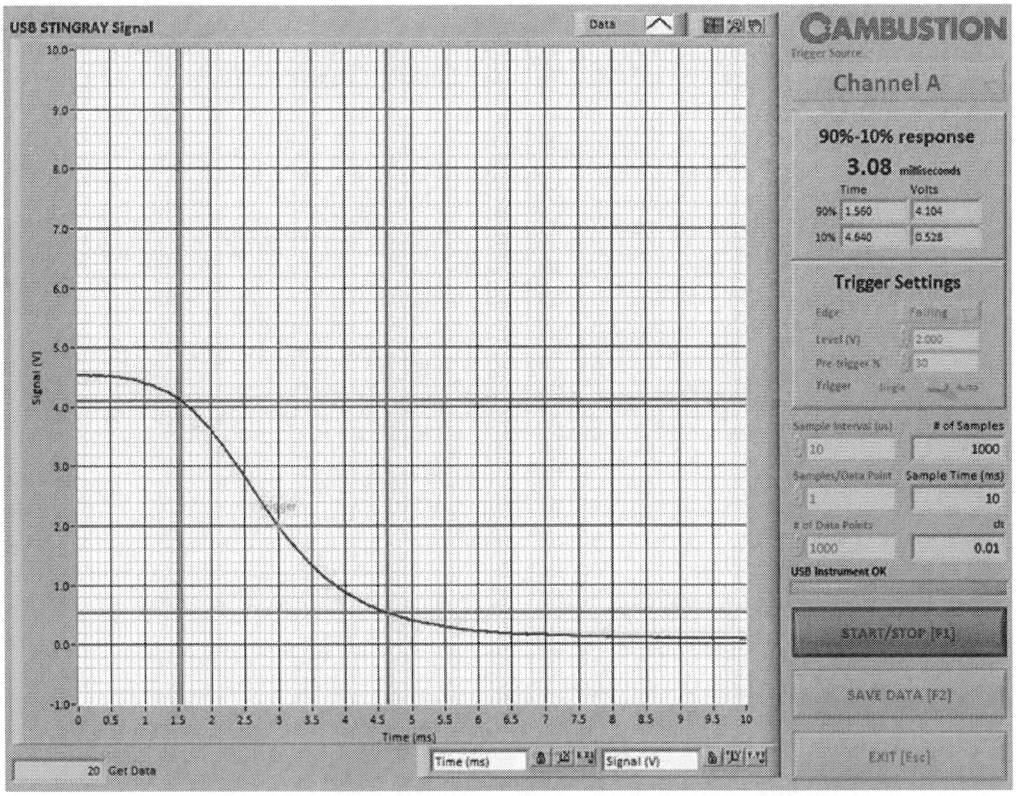

Figure 4. Time response test on modified CLD500

2.2 Time-alignment of data

It is important to note that for the study of transient emissions, the logging of emissions values (ppm) and the other engine parameters with which the data is correlated should be as fast as possible. This can be achieved (up to the limit of the transducers' response times) most readily with data loggers provided with an "open" ECU as used by engine calibrators and researchers.

This (production) vehicle, however, had no such open ECU attached and therefore engine data was recorded at the best rate available from the OBD port - about 10Hz.

The fast emissions data has such a short delay time that, at the catalyst locations, there is no appreciable advantage in correcting for the few milliseconds of delay. However, it is worth noting that the transport delay of gas from the engine exhaust port to the sampling position, varying with exhaust flow rate, can be as long as 300ms (at idle). No time alignment correction was made to the data before calculating "instantaneous" mass emissions because the transients tended to happen at times of high exhaust mass flow (and therefore short transit delays), the transients were significantly longer than a few milliseconds' duration, the exhaust mass flow rate (although compressible) can be considered to be instantly affected by intake air mass flow changes and also the delay time would have been changing *during* the transient and is difficult to predict.

The emissions events of interest during the tests reported within this paper are those which produce a significant contribution to the total cumulative emissions result.

2.3 The RDE Route

Based on the requirements an example RDE route was designed around Cambridge (UK) to allow representative testing of the modified analyzer.

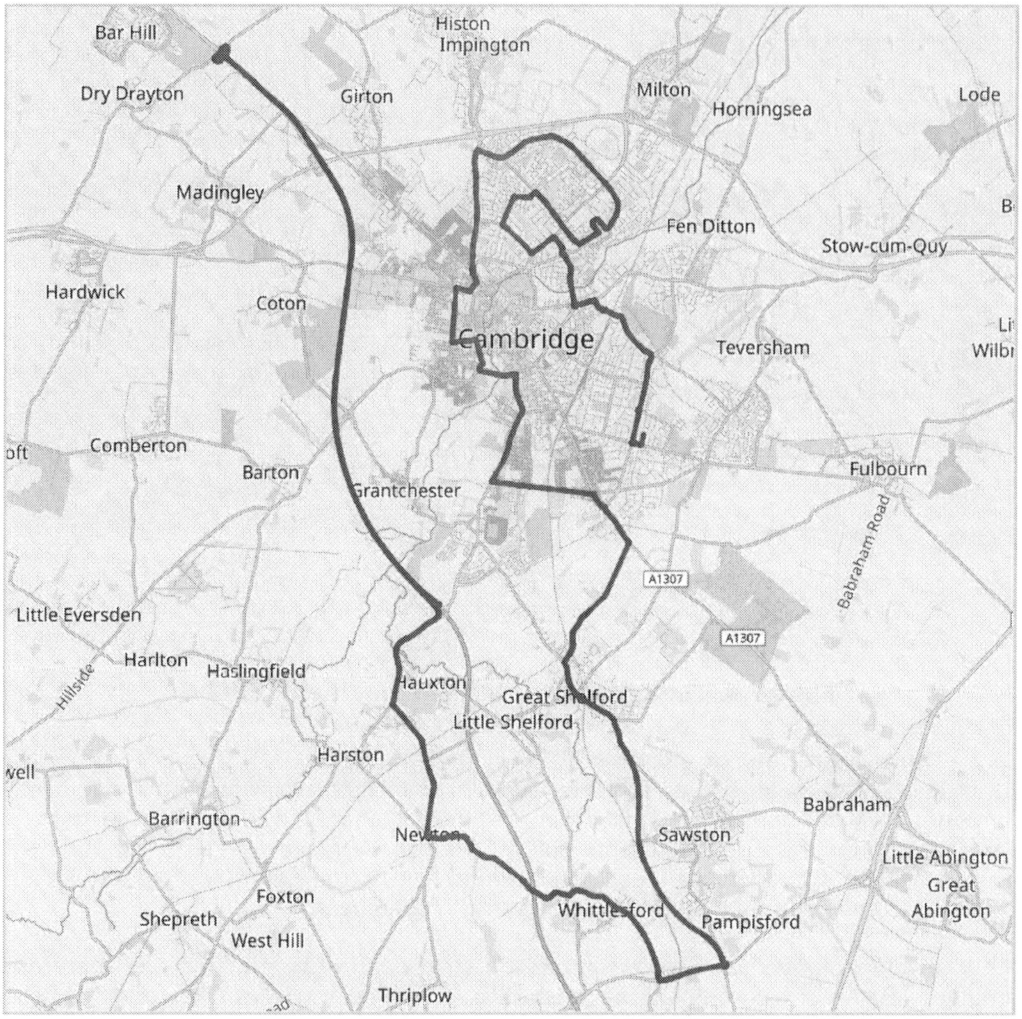

Figure 5. Cambridge RDE route

2.4 LabVIEW Software

A LabVIEW interface was developed to log parameters from the intake air mass flow meter, the Cambustion fast NOx analyzer and the OBD port. Combining this data during logging also simplified the post-processing. The interface was set to automatically calculate exhaust mass flow and plot instantaneous emissions, cumulative emissions and lambda. It also displayed information on the current driving mode (urban, rural or motorway), percentage of stopped time and distance travelled – this made live adaptation of the route to meet certain criteria simple.

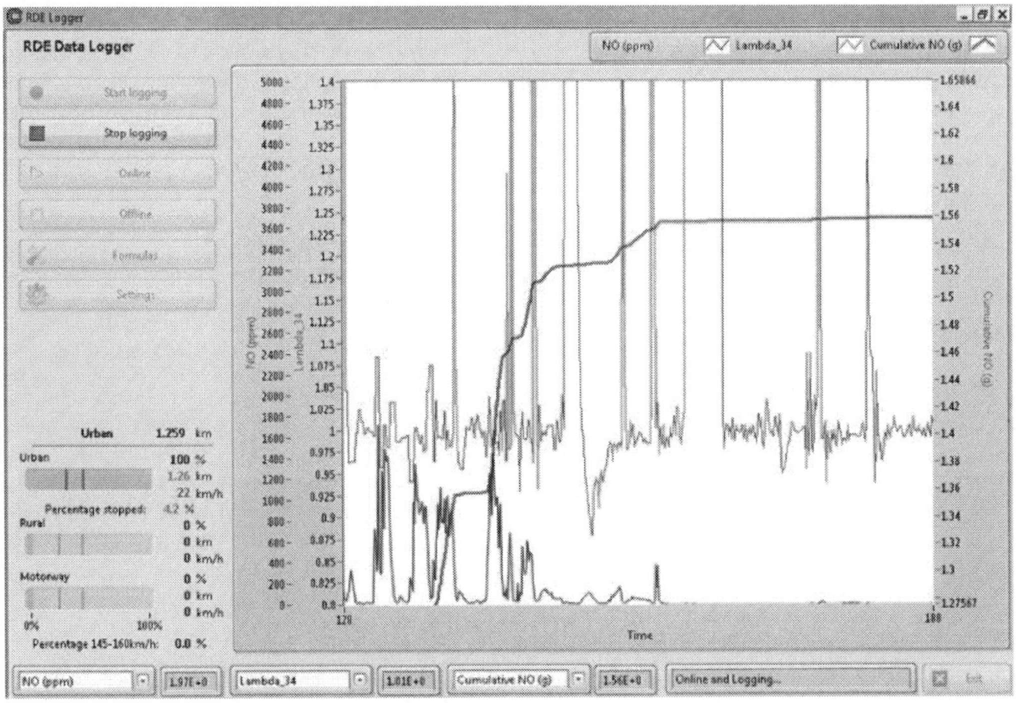

Figure 6. Screenshot of LabVIEW interface for live data interpretation

2.5 Excel Macro for Data Processing

Due to the large quantities of parameters logged at high frequency it was also necessary to develop a series of excel macros to process the data and plot key events within the drive cycle. Users could select parameters such as vehicle speed/load, air fuel ratio and HEGO voltage to plot fast NO emissions against as well as choosing specific portions of the cycle to look at, for example cold start. The macro also gave a summary of the total emissions, emissions per kilometer and percentage of stopped time to assess compliance with official PEMS regulations.

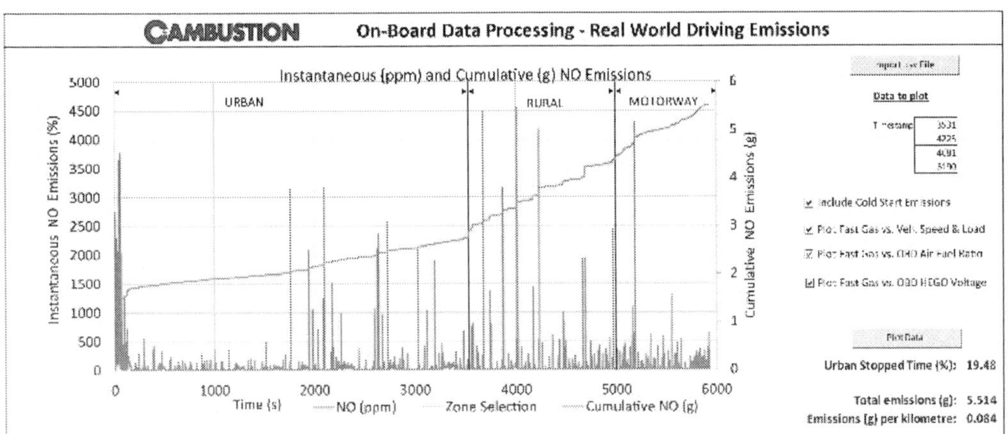

Figure 7. Example of test data processed with Excel macro

3 RESULTS

3.1 Cold-start

As expected, the portion of the cycle which contributes to the largest quantity of emissions is the cold start. As previously mentioned the cold-start portion of RDE testing will not count towards the cumulative limit, although it must be measured, but it is expected to become a legislated requirement in the near future. The data shows that after about 100 seconds, when the catalyst is sufficiently warm, the tailpipe emissions are significantly less and the cumulative rise flattens off, however there are still some events throughout the cycle which cause short duration spikes in the post catalyst emissions. These events will be looked at in the coming sections.

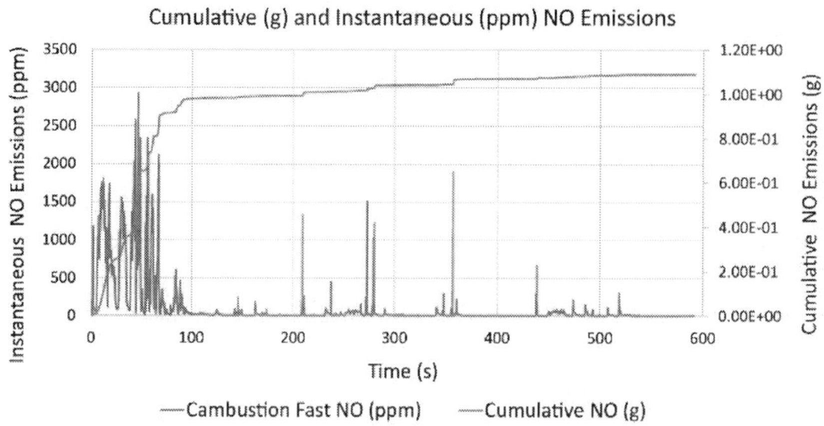

Figure 8. Plot of cumulative and instantaneous NO during cold start

The fact that the tailpipe emissions can be measured very fast and correlated without any significant delay with respect to ECU engine parameters allows calibrators to identify the causes of such breakthrough without having to correct for a smoothed and delayed signal. In general, the emissions of NO downstream of the catalyst during the first portion of the cycle are proportional to engine load, as the formation of NOx is dependent on combustion temperature and residence time.

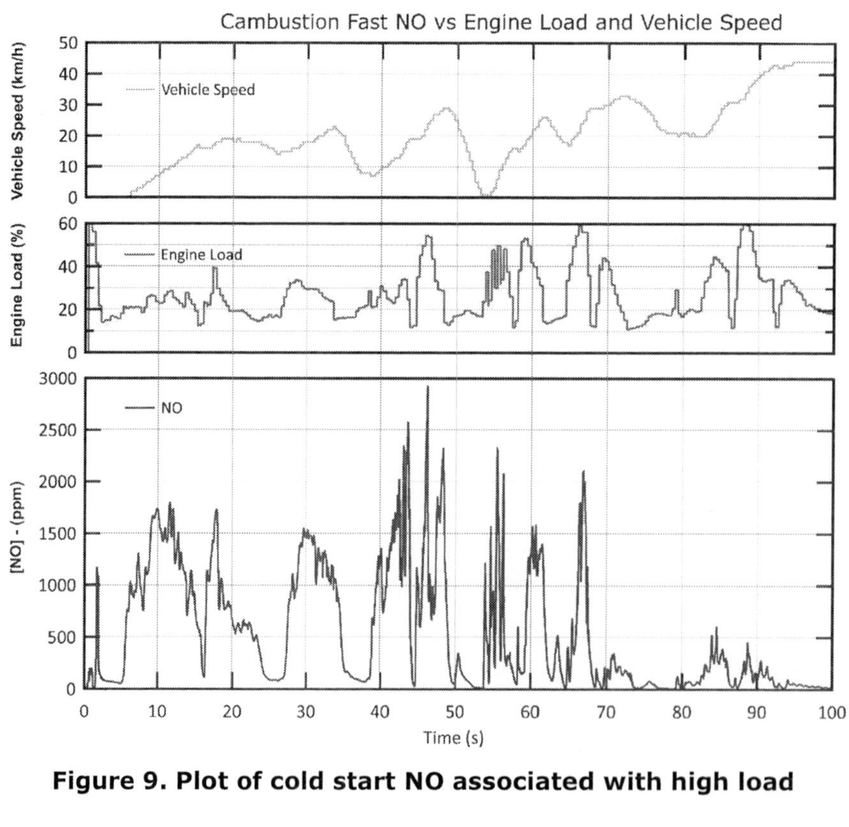

Figure 9. Plot of cold start NO associated with high load

3.2 Speed Bump

According to recent studies, the use of speed bumps in urban environments may lead to high levels of NO emissions associated with decelerating and then accelerating [15]. During the RDE route several such traffic calming measures were encountered and the effect on the NO emissions could be easily seen alongside the causal engine lambda control [16].

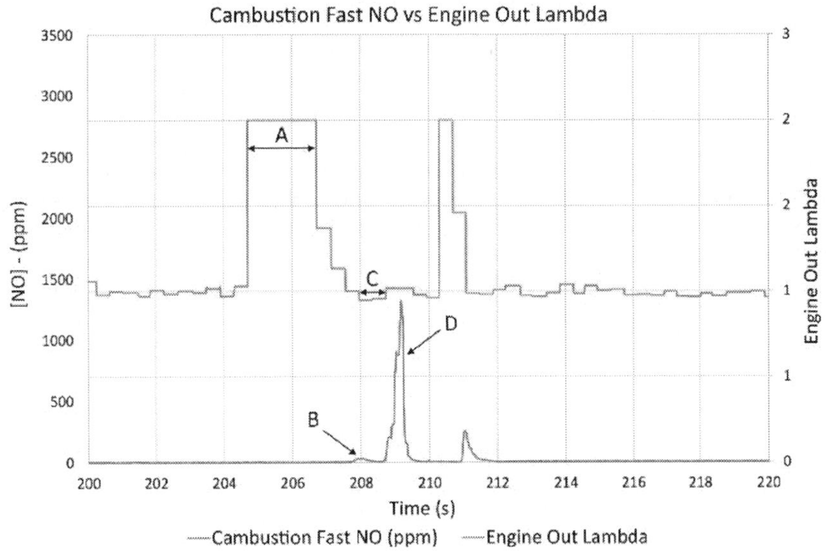

Figure 10. Plot of fast NO transient during speed bump

The following features are shown:

A. Deceleration fuel shut-off
B. Small spike in NO emissions associated with start of combustion
C. Rich purge to reload the catalyst with reductants following decel fuel shut-off
D. But rich purge is insufficient leading to high NO breakthrough under load

3.3 Traffic Lights

Similarly, the changes in speed associated with stopping at a traffic light lead to transient NO emissions when load is applied to the engine.

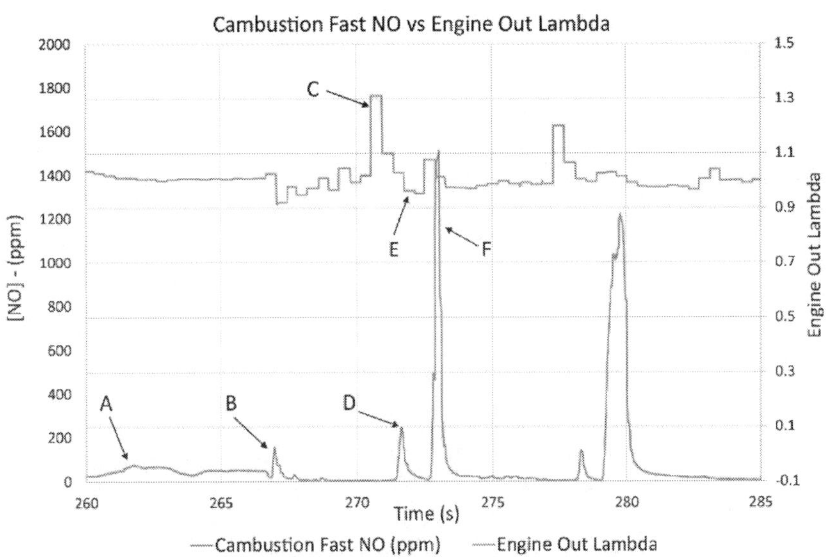

Figure 11. Plot of fast NO transients at traffic lights

The following features are shown:

A. Gradual increase in NO during idle when waiting at red traffic light
B. Small spike in NO emissions associated with start of combustion as traffic light goes green
C. Fuel cut during gear change leads to very lean operation
D. Small spike in NO emissions after gear change
E. Rich purge to try and reload the catalyst with reductants following lean operation
F. But still significant breakthrough in NO as load increases

These features are repeated for the second gear change.

Details of the engine speed and load conditions during this pull away are shown on the graph below. Once again it can been seen that high levels of NO tend to correlate with increases in engine load but the effectiveness of the catalyst rich purge is also a factor in attenuating the tailpipe [NO].

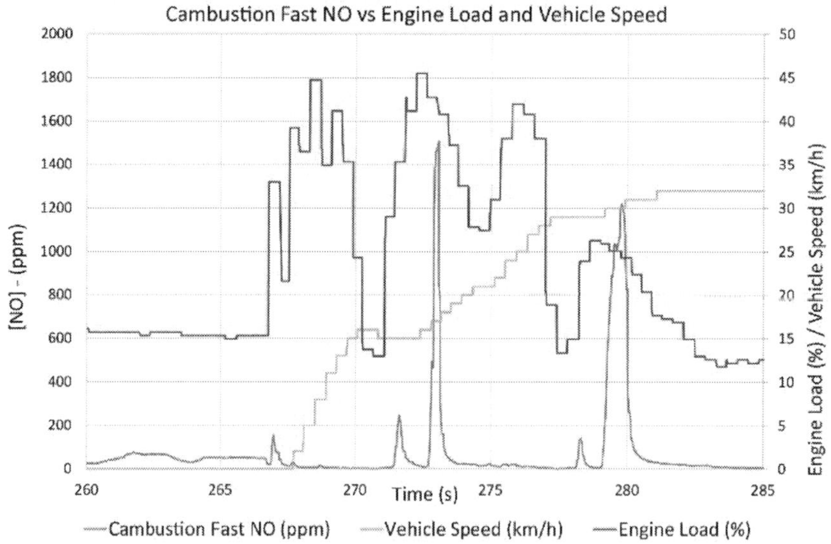

Figure 12. Plot of NO with vehicle speed and engine load

3.4 Two Channel

After improvement of the experimental set-up by reducing the analyzer power consumption further and gaining access to a sampling point upstream of the catalyst, it was possible to sample fast NO emissions at two points simultaneously. Data shown below correlated with first engine out lambda and then vehicle speed and engine load shows similarly transient behavior during gear changes.

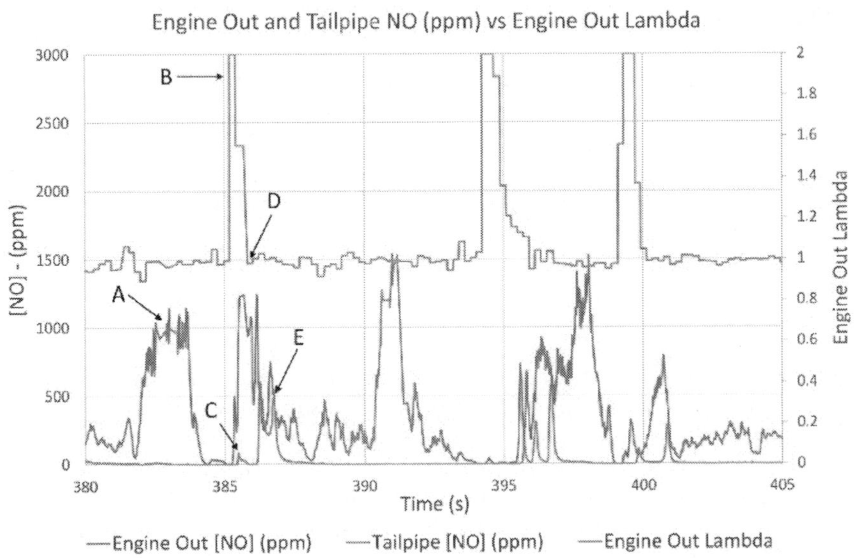

Figure 13. Plot of pre and post catalyst NO with lambda

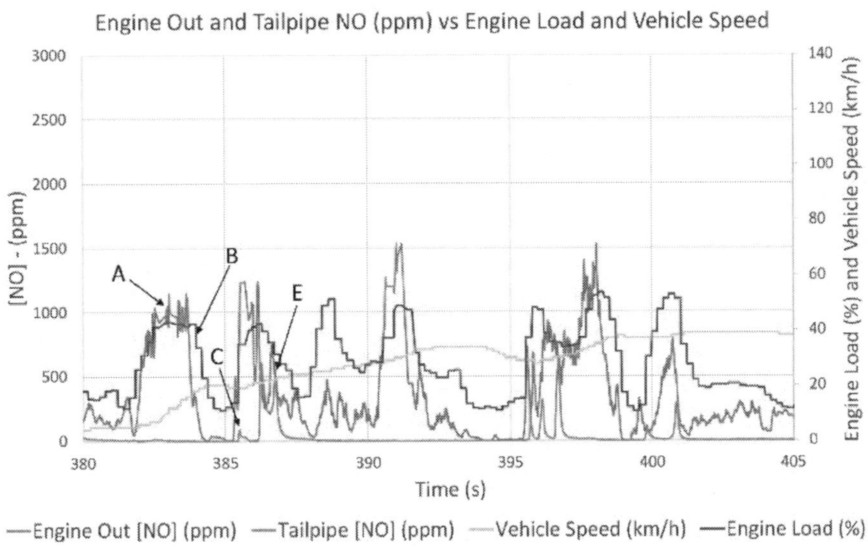

Figure 14. Plot of pre and post catalyst NO with speed and load

The following features are shown:

A. High engine load produces high levels of engine out NO but catalyst is sufficiently warm and conditioned with reductants to give good conversion efficiency. Tailpipe emissions remain low
B. Fuel cut during gear change leads to very lean conditions
C. Lean operation purges the catalyst of reductants leading to NO breakthrough in tailpipe when combustion resumes
D. Short rich purge after gear change to try and reload the catalyst with reductants
E. Rich purge is insufficient and further NO breakthrough can be seen in the tailpipe during acceleration

The same features are repeated for several gear changes.

4 CONCLUSIONS

Fast response gas analyzers have been adapted for on-board use and have been used to log transient tailpipe and engine out emissions. The natural alignment of the emissions data with ECU operating parameters reveals correlations of tailpipe NOx spikes with catalyst light-off, engine load, breakthrough at idle and catalyst rich purge phenomena following decel fuel shut-offs associated with gear changes and vehicle decelerations.

5 FUTURE WORK

Work has already begun adapting a fast response CO/CO2 analyzer for on-board work. Initial single channel engine out results are shown below with some features highlighted.

A. Lambda oscillation around stoichiometry leads to CO production
B. Long deceleration fuel shut-off leads to a drop in CO and CO2 to zero
C. Rich purge leads to significant levels of CO

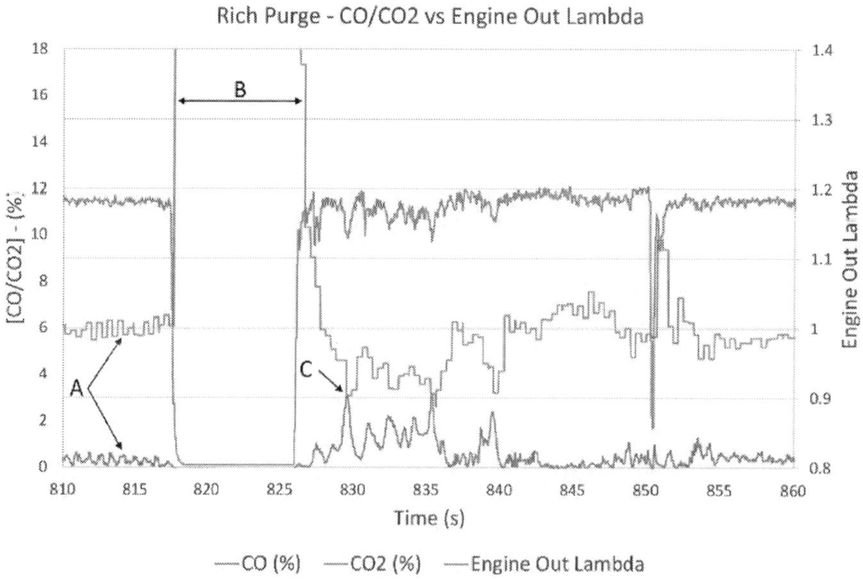

Figure 15. Plot of tailpipe CO/CO2 with lambda

Further work must now be done on two-channel sampling of CO and CO2 to assess catalyst efficiency. In order to ascertain the value of fast response on-board data to engine calibrators it will be necessary to make a direct comparison between legislative PEMs equipment and the Cambustion fast analyzers. Finally, this data will need to be combined with the mass flow measurements, or calculated values to give cycle g/km results.

REFERENCES

1. Mock, P., "World-Harmonized Light-Duty Vehicles Test Procedure (WLTP)," ICCT Policy Update, 2013-11
2. Duckhouse, M., Peckham, M., Campbell, B., Dunnett, J. et al., "A Study of "off-cycle" Emissions From a Euro 6 GDI Vehicle," Aachen Colloquium China, Beijing, 2016-11

3. European Commission, "amending Regulation (EC) No 692/2008 as regards emissions from light passenger and commercial vehicles (Euro 6)," Commission Regulation (EU) 2016/427, 2016-03-31
4. Rodríquez, J. F., "Investigations on the Pollutant Emissions of Gasoline Direct Injection Engines During Cold-Start," Doctor of Philosophy in Mechanical Engineering, Massachusetts Institute of Technology, 2016-06
5. Weissbaeck, M., Dreisbach, R., Enzi, B. et al, "Diesel – The Road Ahead," 38th International Motor Symposium, Vienna, 2017-04
6. European Commission, "Commission welcomes Member States' agreement on robust testing of air pollution emissions by cars," Press release IP/15/5945, 2015-10-28
7. European Council, "determining the rules applicable to periods, dates and time limits," Regulation (EEC, Euratom) No 1182/71, (OJ L 124, 8.6.1971, p.1), 1971-06-03
8. Lyons, A. "On-Board Diagnostics (OBD) Program Overview," California Air Resources Board, 2015-04
9. Peckham, M., Finch, A., Campbell, B., "Analysis of Transient HC, CO, NOx and CO2 Emissions from a GDI Engine Using Fast Response Gas Analyzers," SAE Technical Paper 2011-01-1227, doi:10.4271/2011-01-1227.
10. http://www.cambustion.com/products/cld500
11. https://www.gov.uk/government/collections/vehicles-statistics
12. https://vehicleenquiry.service.gov.uk/ViewVehicle
13. Peckham, M., "CLD500 Fast NOx Analyzer Configuration for Mobile [NO] Sampling, Ford Galaxy Diesel (Euro V) – Summary," Cambustion, 2015-06-15
14. Cheng, W., Summers, T., Collings, N., "The Fast Response Flame Ionizations Detector," Progress in Energy and Combustion Science Volume 24, Issue 2, Pages 89-124, 1998
15. Crombie, H., O'Rourke, D., Robinson, S., "Traffic Management and Enforcement, and Financial Incentives and Disincentives," National Institute for Health and Care Excellence, Evidence Review, 2016-08
16. https://www.youtube.com/watch?v=oodknBoy2ww

Reduced chemical kinetics mechanism for syngas combustion and NOx formation

N Stylianidis, U Azimov
Department of Mechanical and Construction Engineering,
University of Northumbria, UK

ABSTRACT

Chemical kinetics and computational fluid-dynamics (CFD) analysis were performed to evaluate the combustion of syngas derived from biomass solid feedstock in a micro-pilot ignited supercharged dual-fuel engine under lean conditions. For this analysis, a new reduced syngas chemical kinetics mechanism was developed and validated by comparing the ignition delay and laminar flame speed data with those obtained from experiments and other detail chemical kinetics mechanisms available in the literature. The reaction sensitivity analysis was conducted for ignition delay at elevated pressures in order to identify important chemical reactions that govern the combustion process. We found that HO2+OH=H2O+O2 and H2O2+H=H2+HO2 reactions showed very high sensitivity during high-pressure ignition delay times and had considerable uncertainty. The chemical kinetics of NOx formation was analysed for H2/CO/CO2/CH4 syngas mixtures by using premixed laminar flame speed reactors. The new mechanism showed a very good agreement with experimental measurements and accurately reproduced the effect of the equivalence ratio on NOx formation. Finally, the new mechanism was used in a multidimensional CFD simulation to predict the combustion of syngas in a micro-pilot-ignited supercharged dual-fuel engine and results were compared with experiments. The mechanism showed the closest prediction of the in-cylinder pressure and the rate of heat release (ROHR).

1. INTRODUCTION

The environmental problems due to the exhaust gas emissions have forced governments to the adoption of stringiest emission regulations. That has prompted engine manufacturers to the investigation of possible solutions that will reduce the exhaust gas emissions but at the same time will increase or at least retain the efficiency of the engine A promising solution is by using synthesized gases (syngas) as fuels.[1]. Syngas fuels consist of various combustible fuels (CH4,CO and H2) and non-combustible fuels (CO2,N2). The proportion of each constituent varies relatively to the gasification method and the type of feedstock or coal used [2]. For example, syngas fuels derived by coal gasification are mostly consisted of H2 and CO with varying levels of CO2 and water [3], while industrial coke oven gases and biofuels are composed by H2/CO/CO2/N2 and a considerable amount of CH4 [4].

However, the variety of components that included in syngas fuels additional to the effect of external factors such as the turbulence increase the complexities to analyse the chemical reactions that take part during the different stages of the combustion process. A tool that was used for the detail analysis of the combustion chemistry and its interactions with the turbulence is the chemical kinetic mechanism. Chemical Kinetic mechanisms are necessary features for the design and the development of new optimised IC engines. They are imported into computational fluid dynamic (CFD) simulations and coupled with turbulence sub models for an accurate simulation of the combustion chemistry and its interactions with the turbulence.

During the previous years lot of chemical kinetic mechanisms have been developed for the simulations of the combustion process by using different fuels. In a previous research [5], we performed a multidimensional CFD simulation of syngas (CH_4, CO, CO_2, N_2, H_2) combustion in a micro pilot ignited dual fuel by using a constructed reduce chemical kinetic mechanism. The proposed mechanism was validated against experimental results in terms of the pressure tracers and heat release showing very good matching. Keromnes et al [6] studied the combustion and oxidation of hydrogen and syngas mixtures (H_2/CO/CO_2/N_2) at elevated pressures. They developed a new chemical kinetic mechanism, which consists of 51 reactions. The constructed mechanism, was validated in terms of ignition delay time, flame speed and species concentration profiles for pressures up to 70 bar, temperature range between 900-2500K and different equivalent ratios (0.1-4). Frassoldati et al [7-8] investigated numerically and experimentally the combustion characteristics, the flame structure and the NOx and soot emissions of different CO/H_2 syngas mixtures. The authors proposed a detail updated chemical kinetic mechanism in which they incorporated full chemical NOx reactions pathway. The detail mechanism consists of 152 reactions and was validated against experimental data showing a very good agreement. However, Frassoldati mechanism as well as Keromnes mechanism do not include CH_4 chemistry, which is necessary for simulations of syngas mixtures including methane. One of the most famous and well-validated chemical kinetic mechanisms applicable for modelling of a variety of syngas mixtures is GRI mech. 3.0 [9]. GRI mech. 3.0 is a detail chemical kinetic mechanism, which includes full hydrogen, carbon and methane chemistry as well as detailed thermal and prompt NOx and Soot pathways. However, it consists of 325 reactions and 121 species and therefore requires relatively high computational time.

Although the variety of mechanisms that have been developed during the previous years, there is a need of a new robust and computational efficient reduce chemical kinetic mechanism that will be possible to simulate accurately syngas combustion by using a variety of H_2/CO/CO_2/CH_4 fuels and in addition to calculate NOx. Therefore during that research, the reduced chemical kinetic mechanism for H_2/CO/CO_2/CH_4 syngas combustion we have developed on a previous research[5] was optimized by incorporating a 12 NOx chemistry pathway and by changing the rate constants of important hydrogen reactions. Detail analysis of the chemical kinetics and the identification of the important reactions that take part during syngas oxidation was achieved by using sensitivity analysis. The new reduced chemical kinetic mechanism was validated against experimental measurements of ignition delay times, laminar flame speeds and NO concentration profiles. Finally, the reduced mechanism was used in a multidimensional CFD analysis of CH_4/CO/CO_2/H_2 syngas combustion in a micro pilot ignited dual fuel engine. The ignition of the premixed syngas fuel is initiated by the autoignition of a small amount of diesel fuel, in that case n-heptane, which introduce in the engine in a liquid form from the micro pilot injector [10]. For the validation of the reduced mechanism a set of four different syngas mixtures at different pressures, temperatures and eq.ratios where used in order to test the applicability of the mechanism to simulate different syngas compositions. All of the syngas mixtures used during this study, including modelling and initial combustion parameters, are presented in Table 1.

Table 1 Syngas fuels used during this study

No	Fuel Mixture	Composition (%)		Eq. Ratio	Initial Pressure	Initial Temperature	Model	Ref.
Laminar Flame Speed								
F1	H2/CO/CO2/ CH4	TYPE 1	54/11/25/ 10	0.1-1	1 atm	298 K	Laminar Flame speed	11
		TYPE 2	60/10/0/ 30					
		TYPE 3	32/58/0/ 10					
Ignition Delay Time								
F2	H2/CO/CH4 /O2/AR	0.406/0.406/0.075/ 1.113/98		0.5	1.6-32.0 atm	1050-2030 K	Shock Tube/ Constant Volume	12
NOx Analysis								
F3	H2/CO/CO2/ CH4	37.5/37.5/20/5.0		0.71-1.34	1 atm	300 K	Premixed Laminar flame Nox	13
CFD								
F4	H2/CO/CO2/ N2/CH4	13.7/22.3/16.8/ 45.3/1.9		0.52-0.63	225 kPa	298 K	CFD	10

2. CHEMICAL KINETIC MECHANISM

The constructed reduced chemical kinetic mechanism for H2/CO/CO2/CH4 syngas is based on a previous mechanism we proposed on a previous research [5] and is updated by incorporating a 12 reaction NOx pathway proposed by [14,15] and by using new rate constants for important hydrogen reactions. The important reactions affecting the reactivity of the mixture were identified via reaction sensitivity analysis. Detail description of sensitivity analysis can be found in [16]. For the purposes of that study, reaction sensitivity analysis has been performed for the fuel mixture 4 Type1 tested by Azimov et al [10], see Table 1, at temperature 1000 K, eq. ratio 0.63 and pressures 10,30 and 50 bars. The thirteen most sensitive reactions based on the sensitivity analysis were chosen and presented on Figure 1.

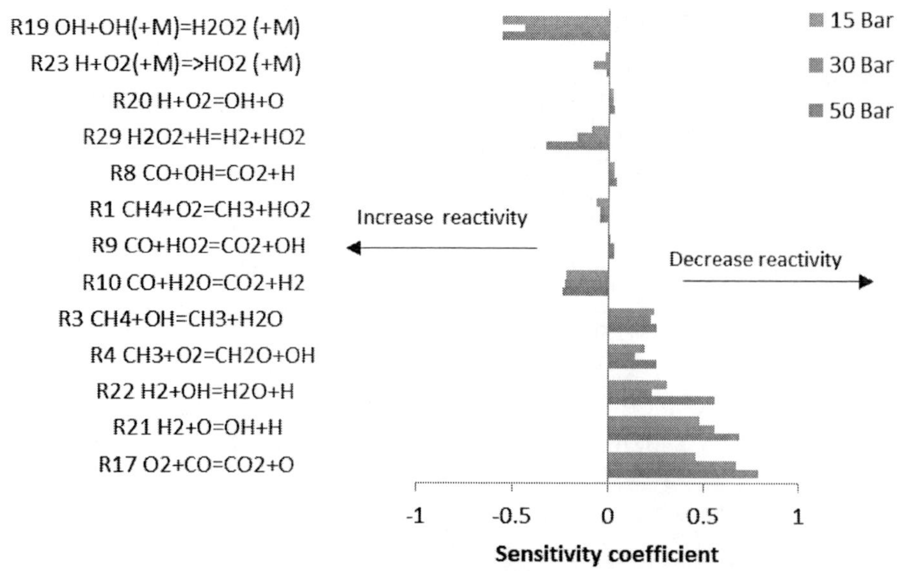

Figure 1. Sensitivity analysis highlighting the 13 ,most sensitive reactions of the mixture H2/CO/CO2/CH4=13.7/22.3/16.8/1.9 at temperature 1000 K, pressures 15,30 and 50 bars and eq.ratio 0.63.

According to Figure 1, reactions such as H2O2(+M) =OH+OH(+M), H2O2+H=H2+HO2 and CO+H2O=CO2+H2 where found to contribute to increase syngas reactivity while reactions such as H2+O=OH+H, O2+CO=CO2+O and CH4+OH=CH3+H2 contribute to the decreasing of the mixtures reactivity. It is obvious that the syngas combustion is mostly driven by hydrogen based reactions such as R19 and R29. R19 H2O2(+M) =OH+OH(+M) is characterized as the central kinetic feature in the operation of the engine because is responsible for the decomposition of H2O2 and the formation of two high reactive OH radicals especially during high pressure conditions. The formation of OH radical species will increase the reactivity of the mixture and in turn will affect important combustion parameters such as the flame speed and ignition delay time. R29 H2O2+H=H2+HO2 on the other hand, was found to be very sensitive during low temperature and high-pressure conditions. The interaction between H2 and HO2 radicals, leads to the formation of H2O2, which in turn will be decomposed via R19 to produce highly reactive OH. Because of the importance of the two reactions on syngas combustion, it was decided that the rate constants of both reactions would be updated in order to reduce the uncertainty factor and to increase the accuracy of the mechanism. For reaction 29, the rate constant proposed by Konnov et al [17], has low uncertainty factor and therefore has been adopted on the proposed mechanism. On the other hand, for reaction 19 the rate constant proposed by Keromnes et al [6] was chosen for that mechanism.

Moreover, R31 was highlighted by other research groups as very important for an accurate simulation of the combustion process because of its complicated temperature dependence and its importance on the consumption of high reactive OH radicals at different temperature ranges [5, 18, 19, 20, and 21]. Therefore, it was decided that a duplicate R31 used in this mechanism by adopting the reaction rate proposed initially by Keyser et al [22] and modified by Keromnes et al [6] for lower temperature range, while for higher temperatures the rate constant proposed by Konnov et al [17] was used.

The reduced chemical kinetic mechanism for H2/CO/CO2/CH4 syngas combustion including the 12reaction NOx pathway is presented on Table 2.

Table 2 Constructed Chemical Kinetic Mechanism

No.	Reactions	A	n	EA	Ref.
1	CH4+O2=CH3+HO2	3.98E13	0.0	56855.5	23
2	CH4+HO2=CH3+H2O2	9.04E12	0.0	24629.4	23
3	CH4+OH=CH3+H2O	1.60E07	1.83	2771.1	24
4	CH3+O2=CH2O+OH	3.30E11	0.0	8934.4	25
5	CH2O+OH=HCO+H2O	3.90E10	0.89	406.1	23
6	CO+O(+M)=CO2(+M)	0.964E11	0.0	3800.0	26
	Low Pressure Limit	0.2070E27	-3.340	7610.0	
	Enhanced Third Body Efficiencies: H2O=12.00, H2=2.00, CO=1.50, CO2=2.00				
7	CO+OH=CO2+H	0.9600E12	0.14	7352.0	26
8	CO+OH=CO2+H	0.7320E11	0.03	-16.0	26
9	CO+HO2=CO2+OH	0.3000E14	0.00	23000.0	26
10	CO+H2O=CO2+H2	0.2000E9	0.00	38000.0	26
11	HCO+M=CO+H+M	0.1200E18	-1.00	17000.0	26
	Enhanced Third Body Efficiencies: H2O=5.00, CO2=3.00, H2=1.90, CO=1.90				
12	HCO+O=CO2+H	0.3000E14	0.0	0.0	26
13	HCO+H=H2+CO	0.1000E15	0.0	0.0	26
14	HCO+OH=H2O+CO	0.5000E14	0.0	0.0	26
15	HCO+HO2=H2O2+CO	0.4000E12	0.0	0.0	26
16	HCO+HO2=>H+OH+CO2	0.3000E14	0.0	0.0	26
17	O2+CO=CO2+O	0.2530E10	0.0	47700.0	26
18	O2+HCO=HO2+CO	0.1000E13	0.0	0.0	26
19	OH+OH(+M)=H2O2(+M)	2.951E14	0.00	4.843E04	6
	Low Pressure Limit	1.202E+17	0.0	4.55E04	
	Troe parameters: 0.5 1E-30 1E30				
	Enhanced Third Body Efficiencies: H2=2.5, H2O=12.00, CO=1.90, CO2=3.80, AR=0.64				
20	H+O2=OH+O	3.52E16	-0.7	17061.4	24
21	H2+O=OH+H	5.06E4	2.67	6287.6	27
22	H2+OH=H2O+H	1.17E9	1.3	3633.5	27
23	H+O2+M=HO2+M	4.6E12	0.4	0.0	6
	Low Pressure Limit	1.737E19	-1.23	0.0	
	Enhanced Third Body Efficiencies: H2=1.3, H2O=10.0, CO=1.9, CO2=3.8				
24	H+H+M=>H2+M	1.30E18	-1	0.0	28
	Enhanced Third Body Efficiencies : H2=2.5, H2O=12.0, CO=1.9, CO2=3.8, AR=0.5				
25	H+OH(+M)=H2O(+M)	3.5E22	-2.0	0.0	6
	Enhanced Third Body Efficiencies H2 =.73, H2O=3.65, AR=.38				
26	HO2+H=>OH+OH	7.08E13	0.0	293.8	28
27	HO2+H=>H2+O2	1.66E13	0.0	821.8	6
28	O+H2O=OH+OH	2.97E06	2.02	1.340E04	6
29	H2O2+H=H2+HO2	1.7E12	0.0	3755	17
30	HO2+HO2=H2O2+O2	1.300E11	0.0	-1.630E03	6
31a	HO2+OH=>H2O+O2	2.89E13	0.0	-496.9	17
31b	HO2+OH=>H2O+O2	2.456E13	0.00	-4.970E02	6

NOx mechanism					
32	N+NO=N2+O	3.5E13	0.0	330.0	29
33	N+O2=NO+O	2.65E12	0.0	6400	29
34	N+OH=NO+H	7.3E13	0.0	1120	29
35	N2O+O=N2+O2	1.4E12	0.0	10810	9
36	N2O+O=2NO	2.9E13	0.0	23150	9
37	N2O+H=N2+OH	4.4E14	0.0	18880	29
38	N2O+OH=N2+HO2	2.0E12	0.0	21060	9
39	N2O(+M)=N2+O(+M)	1.3E11	0.0	59620	29
	Low Pressure Limit	6.2E14	0.0	56100	
	Enhanced Third Body Efficiencies: H2=2.0, H2O=6.0, CH4=2.0, CO=1.5, CO2=2.0				
40	HO2+NO=NO2+OH	2.11E12	0.0	-480.0	9
41	NO+O(+M)=NO2(+M)	1.06E20	-1.41	0.0	9
	Enhanced Third Body Efficiencies: H2=2.0, H2O=6.0, CH4=2.0, CO=1.5, CO2=2.0				
42	NO2+O=NO+O2	3.9E12	0.0	-240.0	9
43	NO2+H=NO+OH	1.32E14	0.0	360.0	9

3. MECHANISM VALIDATION

Experimental measurements obtained from the literature and calculated results by using the GRI mech. 3.0 for ignition delay time, laminar flame speed and NO concentration profiles were used as quantitative measurements for the validation of the proposed mechanism. 0D and 1D simulations were performed by using Digital Analysis of Reactive systems (DARS) [30]. Finally a multidimensional CFD analysis of CH4/CO/CO2/H2 syngas combustion in a micro pilot ignited dual fuel engine was conducted.

3.1 Ignition Delay Time

Ignition delay time is based on the time required from the fuel injection until the ignition occurrence [31]. Some studies determined ignition occurrence based on the slope of the pressure tracers while other studies based on the maximum slope of OH concentration [31-32]. During this study, ignition delay times were calculated based on the slope of the pressure tracers. In order to validate the reduce mechanism the experimental data obtained from Mathieu et al [12] for H2/CO/CH4/O2/AR syngas mixture were used. For the experimental studies, shock tube reactor was used. Therefore, the experimental measurements were simulated by using constant volume reactor, available in DARS, and by assuming adiabatic conditions. The experimental results were measured at a range of T=1010-1920 K, P= 1.6-31 atm and eq.ratio 0.5. The comparison between the calculated and the experimental results is presented on Figure 2. At all of the conditions the reduced mechanism is in a close agreement with the experimental results and reproduces accurately the effects of the temperature and pressure on ignition delay time. In contrast, GRI mech. 3.0 over-predicts the experimental ignition delay times.

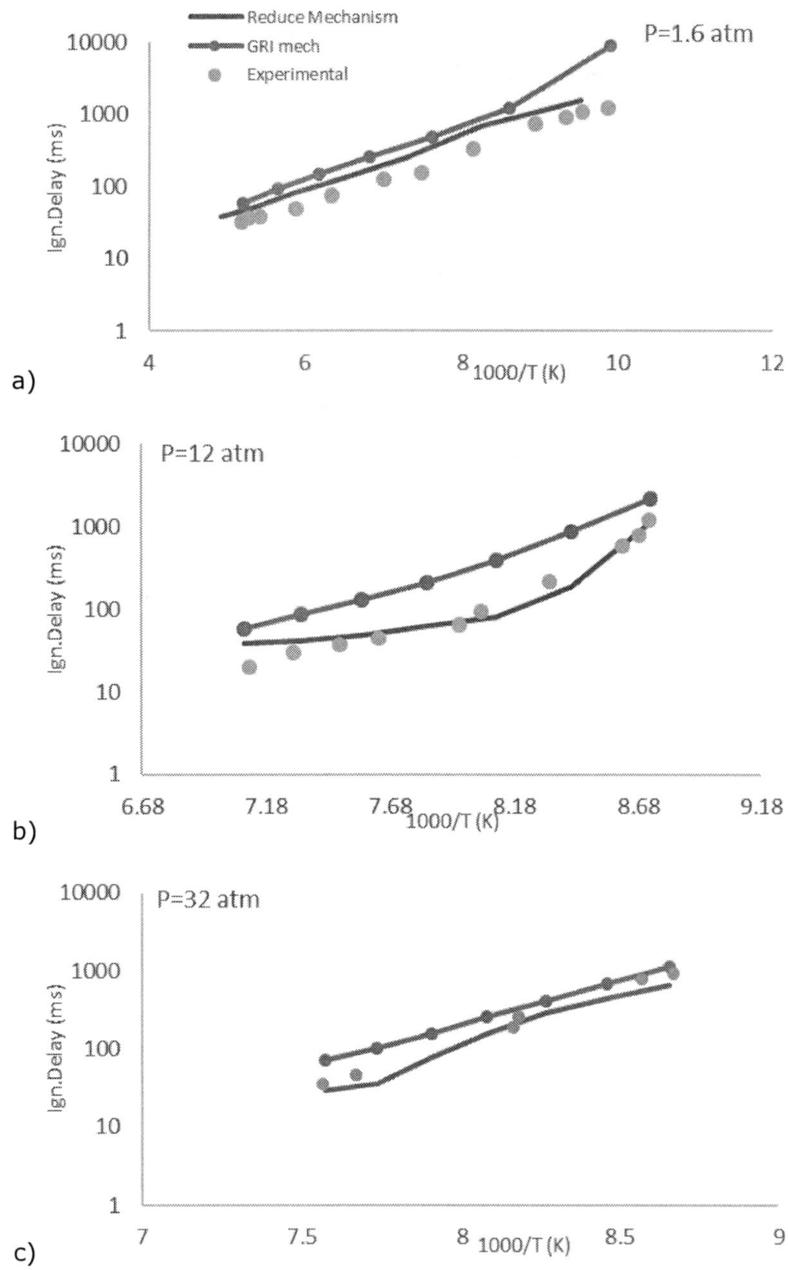

Figure 2. Comparison among the proposed chemical kinetic mechanism, GRI mech. 3.0 and ignition delay time determined for H_2/CO/CH_4 fuel mixtures.

3.2 Laminar flame speed (L.F.S)

Laminar Flame speeds have been calculated by using premixed laminar freely propagating module available in DARS [31]. For Laminar flame calculations the flame front assumed to propagate perpendicular to a hypothetical z axis with the burned mixture moving to z+ direction and unburned mixture to z- direction.

The constructed mechanism was validated against the experimental results obtained by Park et al [11]. The authors measured experimentally the laminar flame speed of H_2/CO/CO_2/CH_4 syngas mixture, see Table 1, on a flat flame burner at pressures 1 atm, temperature 298 K and equivalence ratios between 0.1 and 0.9.

The initial and boundary conditions used during the simulations were taken directly from the experimental investigation by Park et al, hence allowing a direct comparison with the experimental measurements. Figure 3, shows that the calculated results by using the constructed mechanism are in a very good agreement with the experimental results, while GRI mech. 3.0 under-predicts the experimental results especially for Fuel Type 2, Figure2(b), and Fuel Type 3, Figure 2 (c).

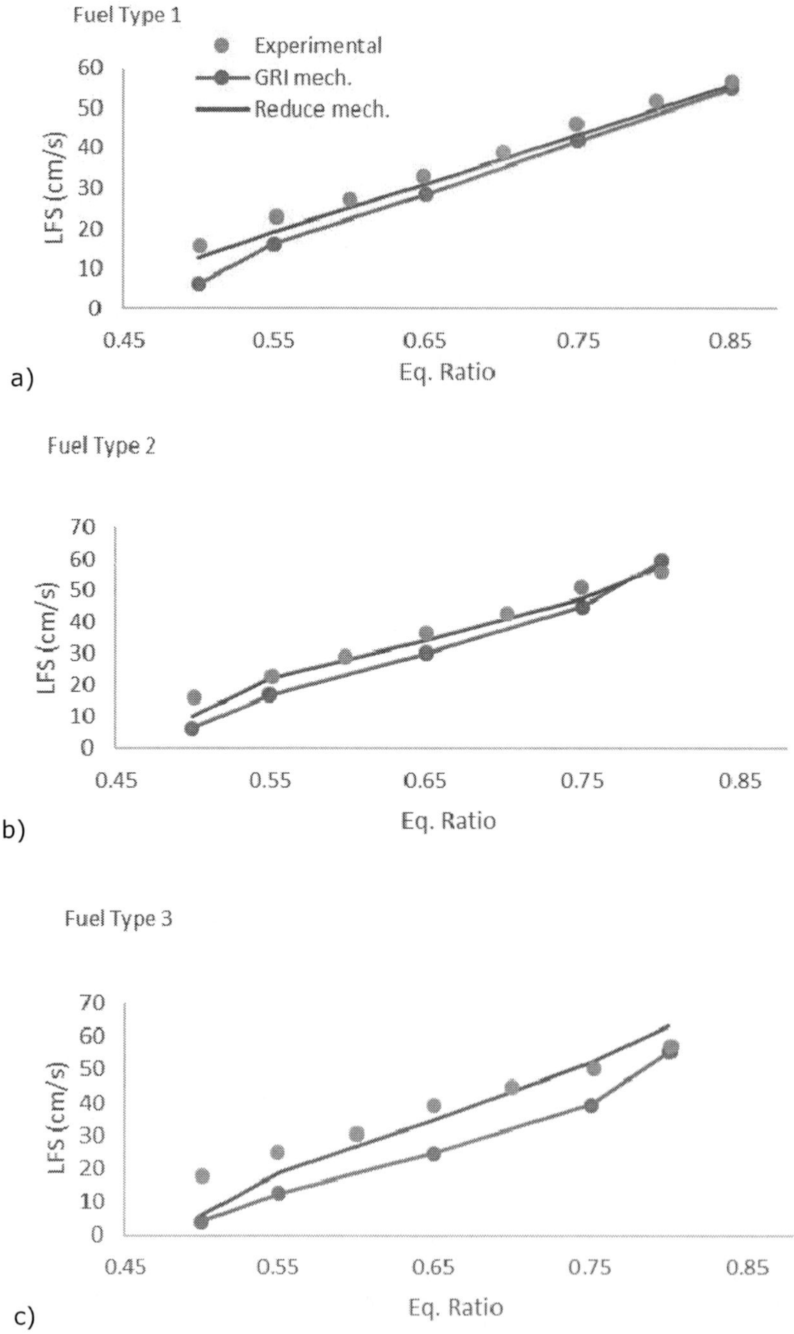

Figure 3. Comparison among the proposed chemical kinetic mechanisms, GRI mech. 3.0 and experimental laminar flame speed determined for H2/CO/CO2/CH4 fuel mixtures.

3.3 NOx Investigation

One of the main advantages of syngas combustion is the reduced NOx and soot emissions. Therefore it is very important the reduced chemical kinetic mechanism to predict accurately NOx formation. Unfortunately, only few experimental studies can be found that investigated the effect of different syngas compositions on NOx formation and therefore can be used for the validation of the proposed mechanism. This is because a lot of experimental studies related to NOx formation focused on the effects of individual gas components and not in the fuel blend. Therefore, there is a need for more experimental data related to NOx formation during syngas combustion. During this study the reduced mechanism was compared against experimental results obtained by Watson et al [13] and numerical results by using the detail GRI mech. 3.0 mechanism.

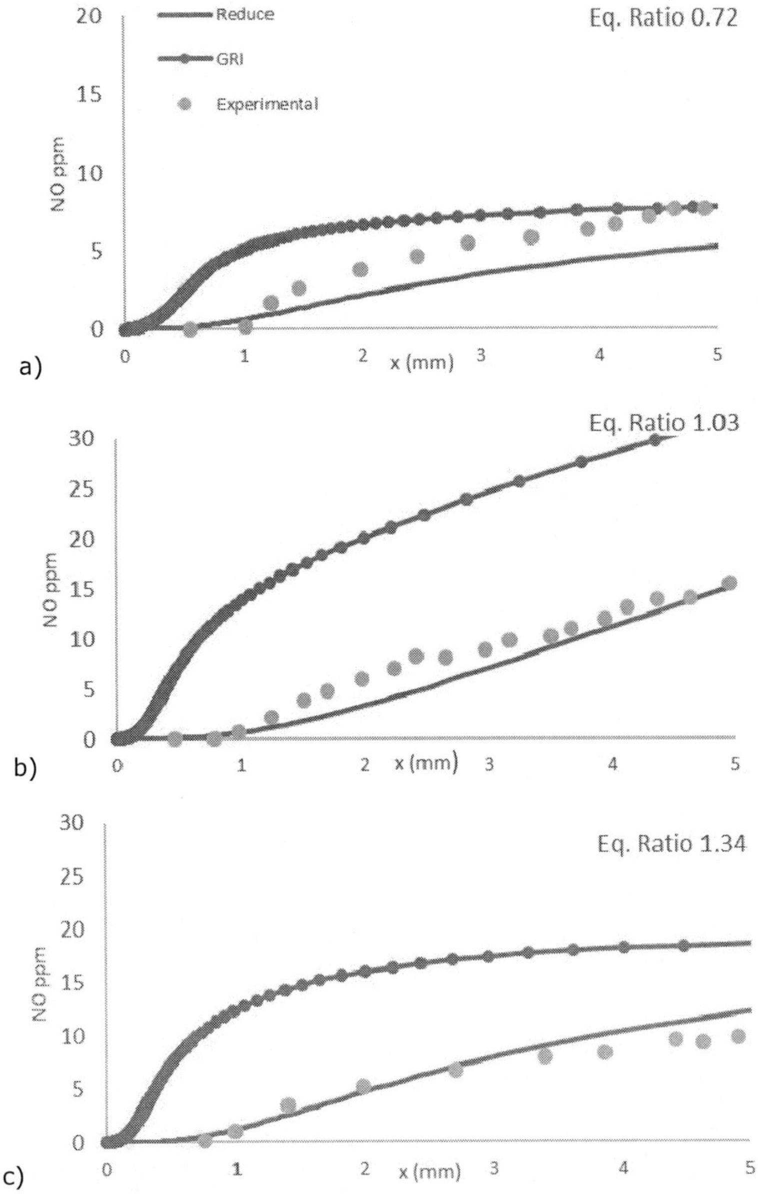

Figure 4. Comparisons between the calculated NO profiles by using the proposed mechanism and GRI mech. 3.0 and the experimental measurements obtained from [13].

Watson et al [13] investigated experimentally NOx formation during the combustion of $CH_4/H_2/CO/CO_2$ syngas mixture. The experiments were conducted at atmospheric pressures in jet wall stagnation flames over a range of eq.ratios 0.7-1.4 and T=300 K. Figure 4 above shows the illustrative comparisons of the simulated and experimental data of NO profiles over the axial distance of the burner for $H_2/CH_4/CO/CO_2$ mixtures. The reduced mechanism is in a good agreement with the experimental results and captures very well the effect of the equivalence ratio on NO formation. The concentration of NO decreases as the equivalence ratio decreases and this is because of the leaner condition of the mixtures. GRI Mech. 3.0 results deviate from the experimental data especially at higher equivalence ratios.

3.4 Multicomponent CFD analysis

Finally, a multidimensional CFD simulation analysis was performed for the investigation of syngas combustion in a micro pilot ignited dual fuel engine. CFD analysis occurred by using Star CD V 4.2 package [33]. During the combustion process a small amount of diesel based fuel (in that case n-heptane) introduced in liquid form inside the combustion chamber. To simulate the pilot-injected diesel spray and the ignition of n-heptane, a single step global reaction, $C_7H_{16} + 11O_2 = 7CO_2 + 8H_2O$, was implemented in the reaction mechanism. The global reactions is based on an eddy breakup (EBU) mixing representation and has specific EBU reaction parameters [34]. According to previous studies, for conventional diesel combustion, both diesel and C_7H_{16} fuels show a similar ROHR [35]. This implies that C_7H_{16} is only used for the initiation of the pilot ignition of the premixed syngas fuel. After the syngas ignition, the combustion proceeds without any C_7H_{16} chemistry. The reason for this simplification is that the injected n-heptane amount was 1.2 mg/cycle, which provides only 2% of the total energy value in the cylinder, and has negligible contribution to the total ROHR [13,15].

In order to formulate and resolve complex gas-phase chemistry problems, STAR CD linked with CHEMKIN [36]. The thermodynamic properties of the computational cells, incorporated in CHEMKIN model, which in turn resolves complex chemistry equations and returns the new thermochemical properties of the species. The coupling between the flow solver technology of STARCD and the chemistry solver technology of CHEMKIN offers a very capable tool for the analysis of complex chemistry problems [30].

By obtaining the new calculated thermodynamic information for all the computational cells, different sub-models activated for the calculation of the flow rate between the cells, the heat transfer and mass transfer. Reynolds averaged equations used to incorporate the turbulence, the chemistry and the n-heptane liquid spray injection into the CFD model. More specifically, high Reynolds $k-\varepsilon$ model used for the turbulence description by using standard wall functions. For an accurate and detail analysis of the combustion process the model is based on the experimental specifications provided by [10,37]. For the cylinder walls and cylinder head constant temperature 450 K used, while for the piston surface the temperature set to 500 K. PISO algorithm implemented into the model for the transient flow calculation, while the properties of the fuel atomization were calculated by using the well-known model Reitz and Diwakar [38,39]. Full model specification such as the cylinder geometry, engine specifications, number of cells and etc. can be found in [31].

The simulated results by using the reduced mechanisms and GRI Mech. 3.0 were compared against experimental data results from Azimov et al [10] for fuel mixture 4, at initial temperature 250 K, initial pressure 225 kPa and equivalence 0.52 and 0.63. The pressure and ROHR plots shown in Figure 5, show a good match between the simulation results by using the reduced mechanism and the experimental data for different equivalence ratios and injection timings. In contrast the simulation results using GRI mech. 3.0, deviate from the experimental measurements at all of the conditions.

Moreover, due to the low number of reactions and species incorporated into the reduced mechanism it requires only 10,800 seconds (3 hours) for a complete CFD simulation, while GRI mech.3.0 requires 86,400 seconds (1 day) for a full simulation.

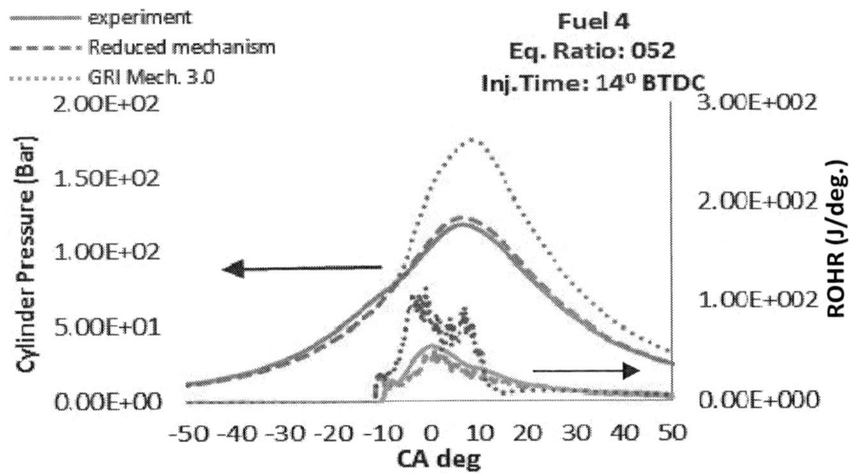

a)

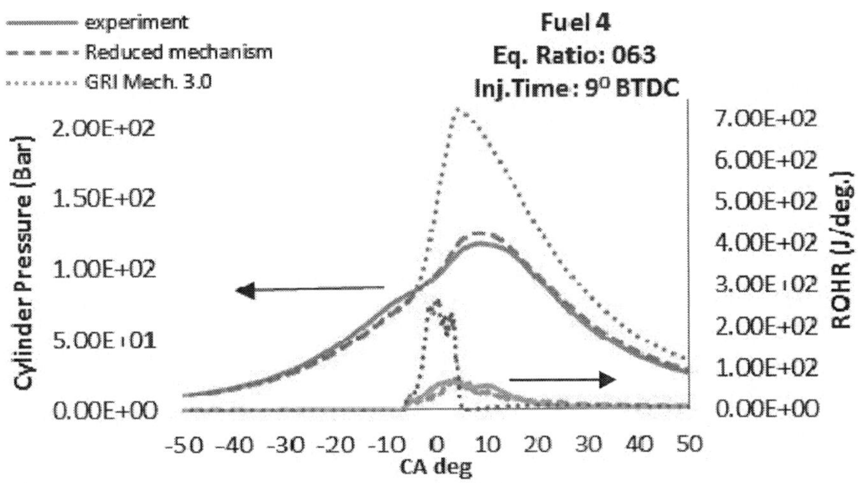

b)

Figure 5. Comparison of experimental and simulated in-cylinder pressures and heat release rates of dual-fuel micro-pilot ignited syngas combustion for fuel 4 Type 1[10]. Computed using the proposed chemical kinetic mechanism.

4. CONCLUSION

The previously published mechanism for syngas combustion proposed by [5], has been updated by incorporating a 12 step NOx chemistry pathway and by using new rate constants for important hydrogen reactions. Reaction sensitivity analysis was conducted to investigate the reactions and the species that play important role on syngas combustion. Hydrogen chemistry was found to be very important for the combustion process. Therefore, reactions R19 and R29 the which are responsible for the formation and the consumption of high reactive OH radicals, were updated by using the rate constants proposed by Konnov et al for R29 and the rate constants proposed by Keromnes et al for R19. Additionally, a duplicate R31 for low and high temperature regions was used in order to address the issue of the complicated temperature dependence, by adopting the rate constants proposed by Keromnes et al [6] for lower temperature range, and by Konnov et al [17] for higher temperatures.

The proposed mechanism was then validated in terms of laminar flame speeds and ignition delay times by comparing the calculated results with experimental measurements and numerical results from the detail GRI mech. 3.0. The reduced mechanism reproduces accurately the ign. delay times and laminar flame speeds at all of the conditions and predicts the NO concentration profiles very well for all of the conditions. The good matching level between the numerical results by using the proposed mechanism and the experimental measurements at all of the combustion parameters and all of the fuel mixtures that were used proves the applicability of the reduced mechanism to simulate accurately the combustion by using different syngas compositions.

Finally, a CFD analysis of syngas combustion occurred by using the constructed mechanism and GRI mech.3.0. The calculated pressure and ROHR by using the reduced mechanism are in a good agreement with the experimental data at all of the equivalence ratios and injection timings. Additionally, the computational time required for a complete CFD simulation by using the reduced mechanism is lower than GRI mech.3.0 due to the lower number of reactions and species included in the mechanism.

By summarizing all of the results, it can be concluded that the reduced mechanism offers an accurate, robust and computational efficient solution for the simulation of a variety of multicomponent syngas mixtures.

5. REFERENCES

1. L. Wang, C.L. Weller, D.D. Jones, M.A. Hanna, Contemporary issues in thermal gasification of biomass and its application to electricity and fuel production, Biomass Bioenergy, 32 (2008), pp. 573–581.
2. Probstein, Ronald F., and R. Edwin Hicks. Synthetic fuels. Courier Corporation, 2006.
3. Liu, C.; Shy, S.; Chiu, C.; Peng, M.; Chung, H. Int. J. Hydrogen Energy 2011, 36, 8595−8603.
4. Nikolaou, Zacharias Marinou, Jyh-Yuan Chen, and Nedunchezhian Swaminathan. "A 5-step reduced mechanism for combustion of CO/H 2/H 2 O/CH /CO2 mixtures with low hydrogen/methane and high H 2 O content." Combustion and flame 160, no. 1 (2013): 56-75.
5. Stylianidis, N., Azimov, U., Maheri, A., Tomita, E. and Kawahara, N., 2017. Chemical kinetics and CFD analysis of supercharged micro-pilot ignited dual-fuel engine combustion of syngas. Fuel, 203, pp.591-606.
6. Kéromnès, Alan, et al. "An experimental and detailed chemical kinetic modeling study of hydrogen and syngas mixture oxidation at elevated pressures." Combustion and Flame 160.6 (2013): 995-1011.
7. A. Cuoci, A. Frassoldati, T. Faravelli, E. Ranzi, Formation of soot and nitrogen oxides in unsteady counterflow diffusion flames, Combustion and Flame 156 (10), pp. 2010-2022 (2009), doi:10.1016/j.combustflame.2009.06.023.
8. http://creckmodeling.chem.polimi.it

9. Smith GP, Golden DM, Frenklach M, Moriarty NW, Eiteneer B, Goldenberg M, et al. GRI-Mech home page http://www.me. berkeley.edu/gri_mech/
10. Azimov, Ulugbek, Eiji Tomita, Nobuyuki Kawahara, and Yuji Harada. "Effect of syngas composition on combustion and exhaust emission characteristics in a pilot-ignited dual-fuel engine operated in PREMIER combustion mode." international journal of hydrogen energy 36, no. 18 (2011): 11985-11996.
11. Park, Okjoo, Peter S. Veloo, Ning Liu, and Fokion N. Egolfopoulos. "Combustion characteristics of alternative gaseous fuels." Proceedings of the Combustion Institute 33, no. 1 (2011): 887-894.
12. Mathieu, O.; Kopp, M. M.; Petersen, E. L.Proc. Combust. Inst. 2013, 34, 3211−3218.
13. Watson, Graeme MG, Jeffrey D. Munzar, and Jeffrey M. Bergthorson. "NO formation in model syngas and biogas blends." Fuel 124 (2014): 113-124.
14. Han, Zhiyu, and Rolf D. Reitz. "Turbulence modeling of internal combustion engines using RNG κ-ε models." Combustion science and technology 106, no. 4-6 (1995): 267-295.
15. Pan, Lun, Sage Kokjohn, and Zuohua Huang. "Development and validation of a reduced chemical kinetic model for dimethyl ether combustion." Fuel 160 (2015): 165-177.
16. Løvås, Terese. Model reduction techniques for chemical mechanisms. INTECH Open Access Publisher, 2012.
17. Konnov A. Remaining uncertainties in the kinetic mechanism of hydrogen combustion. Combust Flame 2008;152:507-528.
18. Li J, Zhao Zh, Kazakov A, Dryer FL. An updated comprehensive kinetic model of hydrogen combustion. Int J Chem Kinet 2004;36:566-575.
19. H. Hippler, H. Neunaber, J. Troe, J. Chem. Phys. 103 (1995) 3510 - 3516.
20. Hong, Zekai. An improved hydrogen/oxygen mechanism based on shock tube/laser absorption measurements. Diss. Stanford University, 2010.
21. Ellingson BA, Theis DP, Tishchenko O, Zheng J, Truhlar DG. Reactions of hydrogen atom with hydrogen peroxide. J. Phys. Chem. A 2007;111:13554–13566. 22. L.F. Keyser, J. Phys. Chem. 92 (1988) 1193–1200.
23. Li SC, Williams FA. A reduced reaction mechanism for predicting knock in dual-fuel engines. SAE paper 2000-01- 0957.
24. Fernandez-Galisteo D, del Alamo G, Sanchez AL, Linan A. Zeldovich analysis of hydrogen-air premixed flames. In: Proceedings of the European Combustion Meeting 2007.
25. Li SC, Williams FA. Reaction mechanisms for methane ignition. J Engr Gas Turb Power 2002;124:471-480.
26. Frassoldati A, Faravelli T, Ranzi E. The ignition, combustion and flame structure of carbon monoxide/hydrogen mixtures. Note 1: Detailed kinetic modeling of syngas combustion also in presence of nitrogen compounds. Int J Hydrogen Energy 2007;32:3471-3485.
27. Sutherland JW, Michael JV, Pirraglia AN, Nesbitt FL, Klemm RB. Rate constant for the reaction of O(3P) with H2 by the flash photolysis-shock tube and flash photolysis-resonance fluorescence techniques; 504K≤T≤2495K. Proc Combust Inst 1988;21:929–941.
28. Westbrook CK. Chemical kinetics of hydrocarbon ignition in practical combustion systems. Proc. Combust. Inst 2000;28:1563–1577.
29. Z. Han, R.D. Reitz Turbulence modeling of INTERNAL combustion engines using RNG k-e models Combust Sci Technol, 106 (1995), pp. 267–295.
30. Azimov, Ulugbek, et al. "Multidimensional CFD simulation of syngas combustion in a micro-pilot-ignited dual-fuel engine using a constructed chemical kinetics mechanism." international journal of hydrogen energy 36.21 (2011): 13793-13807.
31. Lee, Hsu Chew, Abdulmajeed A. Mohamad, and Lei-Yong Jiang. "Comprehensive comparison of chemical kinetics mechanisms for syngas/biogas mixtures." Energy & Fuels 29, no. 9 (2015): 6126-6145.

32. Kathrotia, Trupti, M. Fikri, M. Bozkurt, M. Hartmann, U. Riedel, and C. Schulz. "Study of the H+ O+ M reaction forming OH*: Kinetics of OH* chemiluminescence in hydrogen combustion systems." Combustion and Flame 157, no. 7 (2010): 1261-1273.
33. 2015 CD-Adapco Inc., DARS Basic 2.10.
34. 2017 CD-Adapco Inc. Methodology, Star-CD V4.6.
35. Donkerbroek AJ, van Vliet AP, Somers LMT, Frijters PJM, Klein-Douwel RJH, Dam NJ, Meerts WL, ter Meulen JJ. Time- and space-resolved quantitative LIF measurements of formaldehyde in a heavy-duty diesel engine. Combust Flame 2010;157:155-166.
36. Kee, R.J., Miller, J.A., Evans, G.H. and Dixon-Lewis, G. 1989. A computational model of the structure and extinction of strained, opposed flow, premixed methane–air flames. Symposium (International) on Combustion, 22:14791494Combust., [Proc.] 1989, 22, 1479–1494.
37. Roy MM, Tomita E, Kawahara N, Harada Y, Sakane A. Performance and emission comparison of a supercharged dual-fuel engine fueled by producer gases with varying hydrogen content. Int J Hydrogen Energy 2009;34:7811-22.
38. Reitz RD, Diwakar R. Effect of drop breakup on fuel sprays. SAE paper, 860469.
39. Reitz RD, Diwakar R. Structure of high-pressure fuel spray. SAE paper, 870598.

A simulation approach to meeting Euro VI emissions legislation with a natural gas lean burn heavy duty engine

A Auld, M Keenan, W Safi, A Gurr, R Pickett
Ricardo UK Ltd, UK

ABSTRACT

The energy specific carbon content of natural gas as a fuel is about 25% lower than that of diesel fuel. This, alongside the lower fuel cost and tax incentives, are some of the benefits of Natural Gas (NG) as an alternative fuel for long haul transport. Owing to the current stringent methane emissions limits in Europe, natural gas heavy-duty truck applications generally operate under stoichiometric conditions, which carries some efficiency penalties particularly in part load operation. To improve the fuel economy and reduce CO_2 emissions, operating a heavy-duty engine lean gives significant advantages over stoichiometric operation. As part of the co-funded collaborative European Union (EU) research project, HDGAS (Heavy-Duty Gas Engines integrated into Vehicles), Ricardo has developed the specification for a Euro VI capable lean burn long haul heavy-duty engine and state-of-the-art aftertreatment system using the Ricardo Integrated Model Based Development process (IMBD).

Both 1D and 3D analysis were used to identify appropriate engine components and calibration approaches. 3D simulation was also used to conduct detailed assessments around the geometry of a novel combustion system. The combustion process utilises an innovative direct injection and advanced ignition system. Drive-cycle simulation was performed to aid the design and specification of an aftertreatment system for the management of both regulated and unregulated pollutants within real-world operation. The results of the simulation highlight the CO_2 benefit of the lean burn variant compared to the baseline engine and the challenges surrounding emissions control.

The challenges and simulation results of developing a lean burn heavy-duty gas engine application, including aftertreatment system capable of meeting Euro VI emissions, are discussed and a technology route is proposed.

DEFINITIONS/ABBREVIATIONS

ASC - Ammonia Slip Catalyst
AT - Aftertreatment
BMEP - Brake Mean Effective Pressure
BSFC - Brake Specific Fuel Consumption
Cf - Flow Coefficient
CH_4 - Methane
CO - Carbon Monoxide
CO2 - Carbon Dioxide
CR - Compression Ratio

degCA - degrees crank angle
DI - Direct Injection
DoE - Design of Experiment
EGR - Exhaust Gas Recirculation
EHC - Electrically Heated Catalyst
EIVC - Early Intake Valve Closing
EQR - Equivalence Ratio
ETM - Exhaust Thermal Management
EU - European Union
Euro VI - Euro 6 HD Emissions Legislation
EVC - Exhaust Valve Closing
EVO - Exhaust Valve Opening
FL - Full Load
HC - Hydrocarbons
HD - Heavy-Duty
HDD - Heavy-Duty Diesel
HDGAS - Heavy-Duty Gas Engines integrated into Vehicles
HDV - Heavy-Duty Vehicle
IMBD - Integrated Model Based Development
KP - keypoint
LDV - Light-Duty Vehicle
MFB - Mass Fraction Burn
MOC - Methane Oxidation Catalyst
NG - Natural Gas
NH_3 - Ammonia
NMHC - Non-methane Hydrocarbons
NO_x - Nitrogen Oxides
OEM - Original Equipment Manufacturer
PL - Part Load
PV - Pressure:Volume
RDE - Real Driving Emissions
SCR - Selective Catalytic Reduction
TKE - Turbulent Kinetic Energy
VVT - Variable Valve Timing
WHTC - World Harmonised Transient Cycle

1 INTRODUCTION

Within the EU, Heavy-DutyVehicles (HDVs) are a major contributor to on-road CO_2 emissions, accounting for 30% thereof (1). Heavy-Duty Diesel (HDD) engines currently dominate the HDV sector due to their relatively high thermal efficiency and proven durability. In recent years, advances in engine thermal efficiency within the HDV sector have reduced the per vehicle CO2 emissions. However, given the relative increases seen in on-road freight traffic across the EU, CO2 emissions are still rising. Contribution of heavy-duty diesel engines to local pollution, particularly NO_2, is also a concern. As such, methods to reduce fuel consumption, CO_2 and other emissions within the HDV market are of primary concern to legislative bodies, Original Equipment Manufacturers (OEMs) and operators alike. Effective and practical CO_2 reduction technologies within the Heavy-Duty (HD) haulage sector are limited. Strategies now being introduced in the Light-Duty Vehicle (LDV) sector, such as electrification and hybridisation are ineffective in HDVs due to prohibitive on-cost, high associated mass and compromised range and load carrying capacity. Without a sector-specific carbon reduction technology, the relative contribution and importance of HDV carbon emissions will inevitably rise as other sectors adopt effective low carbon propulsion solutions.

To address the significant CO_2 production within the sector, the advancement of Direct Injection (DI) Gas technology, which is considered to be a viable alternative to diesel powered HDVs, could facilitate this market's need. At present, the European market for both Natural Gas and Dual Fuel Vehicles is modest in absolute vehicle numbers. However, a major opportunity exists to expand the deployment of Heavy-Duty Natural Gas vehicles within inter-city haulage applications and beyond. The main aim of Heavy-Duty Gas Engines integrated into Vehicles project, referred to as HDGAS herein, is to accelerate targeted development within the HDV sector. HDGAS aims to increase technology know-how, with a specific Ricardo focus on lean burn NG engine operation, and drive increased operator uptake of NG powered HDVs, ultimately leading to reduced vehicle CO_2 emissions for inter-city haulage applications.

HDGAS is a three-year Horizon 2020 (H2020) project to develop knowledge and know-how in several areas surrounding Natural Gas powertrains, including: engine design, fuel storage, supply systems, aftertreatment system specification, development and calibration. Within this project, Ricardo are developing a lean burn variant capable of achieving the project performance targets whilst also achieving emissions compliance. For the lean burn positive ignition direct injection NG engine, the targets are detailed below:

- Euro VI RDE emissions compliance (Table 1)
- 10% improvement in fuel economy compared to state-of-the-art stoichiometric 2013 natural gas engine
- 10% reduction in CO_2 emissions compared to state-of-the-art stoichiometric 2013 natural gas engine
- 10% improvement in peak torque and rated power compared to the baseline engine

Table 1: Euro VI on-highway positive ignition emissions limits for World Harmonised Transient Cycle (WHTC)

Pollutant Emissions	Limit
Carbon Monoxide (CO)	4,000 mg/kWh
Non-methane hydrocarbons (NMHC)	160 mg/kWh
Methane (CH_4)	500 mg/kWh
Nitrogen Oxides (NO_x)	460 mg/kWh
Ammonia (NH_3)	10 ppm (peak)
Particulate Mass (Pm)	10 mg/kWh
Particulate Number (Pn)	6.0×10^{11} #/kWh

2 APPROACH AND METHODOLOGY

The approach chosen was to use a sub-set of the Ricardo Integrated Model Based Development (IMBD) environment and simplified tools to provide a faster than real time evaluation, therefore enabling system optimisation over the large matrix of input conditions. The block diagram of the IMBD environment for the simulations is shown below in Figure 1.

With this method, the likely success or failure of a specific technical solution can be reviewed before committing to a costly development activity. This approach, frequently ultilised by Ricardo, has been shown to enable fast adaptation to varying inputs and system properties.

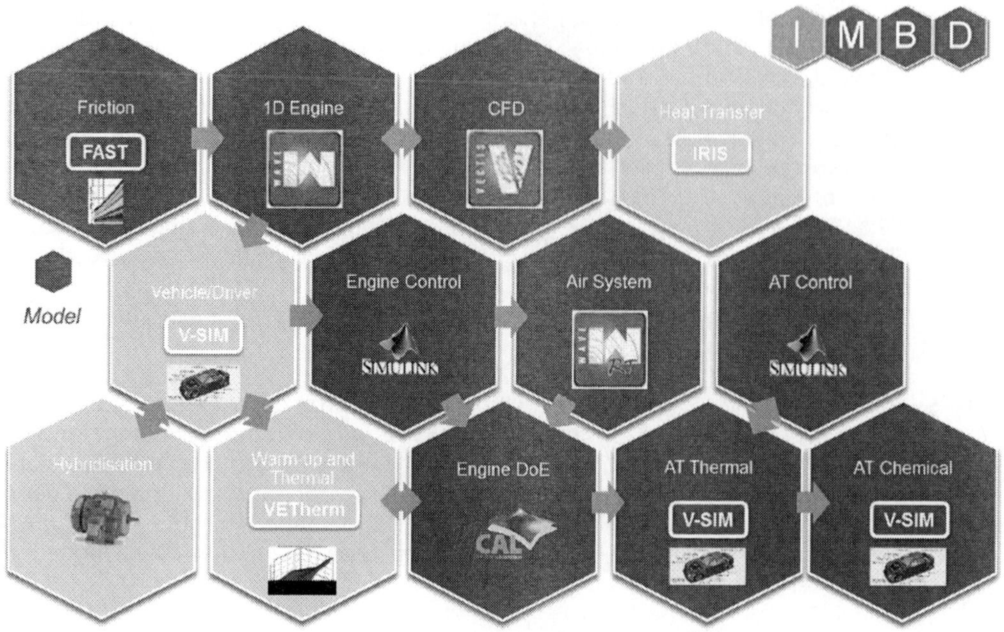

Figure 1: IMBD Environemnt (models ued in HDGAS are highlighted in blue)

To specify the lean burn engine variant, the baseline engine was simulated within 1D and 3D environments and the results validated against test data. Then additional technologies were evaluated with regard to CO_2 reduction, emissions control capability, performance improvement and any design limitations. The approach chosen was to use multiple simulation tools in conjunction with benchmark data from the baseline engine and, where appropriate, suitable engines known from other Ricardo activities.

For air management, turbocharging and target compliance, a Ricardo WAVE model (1D), validated to the baseline engine, was used. For the combustion system and intake port specification, 3D simulations were performed using Ricardo VECTIS software (3D). Since emissions compliance is key to the lean burn variant, due to challenges with methane control, the estimated tailpipe emissions were evaluated using a Ricardo VSIM model of the lean burn engine and aftertreatment system.

2.1 Input data

The input data for the specification process was supplied from a variety of sources including HDGAS partners, suppliers, and where appropriate, internal Ricardo data was used. The baseline engine is representative of current European natural gas commercial vehicle engines utilising stoichiometric port fuel delivery operation.

2.2 Assessment method

To create a specification of a lean operating long-haul truck engine capable of meeting the project targets, various additional technologies were assessed in simulation with regards to the WHTC CO_2 benefit, effect on engine emissions and aftertreatment performance, and structural limitations. The technologies or strategies considered were:

- Lean and stoichiometric operation
- Direct injection of fuel gas and high energy ignition system
- Optimised turbocharging for lean operation
- Variable valve timing and optimised valve profiles
- Highly turbulent port development

- EGR (cooled and uncooled)
- NO_X:CO_2 trade-off
- Combustion system geometry

Simulations were performed with the relevant system information and then, where appropriate, an optimisation process was completed for the calibration or design of the additional technology or strategy. This was conducted to achieve the lowest fuel consumption within emissions and design constraints. The estimated CO_2 benefit was then calculated using keypoints relevant to the WHTC cycle, as shown in Figure 2, and detailed in Table 2. The combustion system and port design specification process was concerned with achieving suitable in-cylinder air motion for charge mixture homogeneity and combustion performance. These were assessed against Ricardo guidelines.

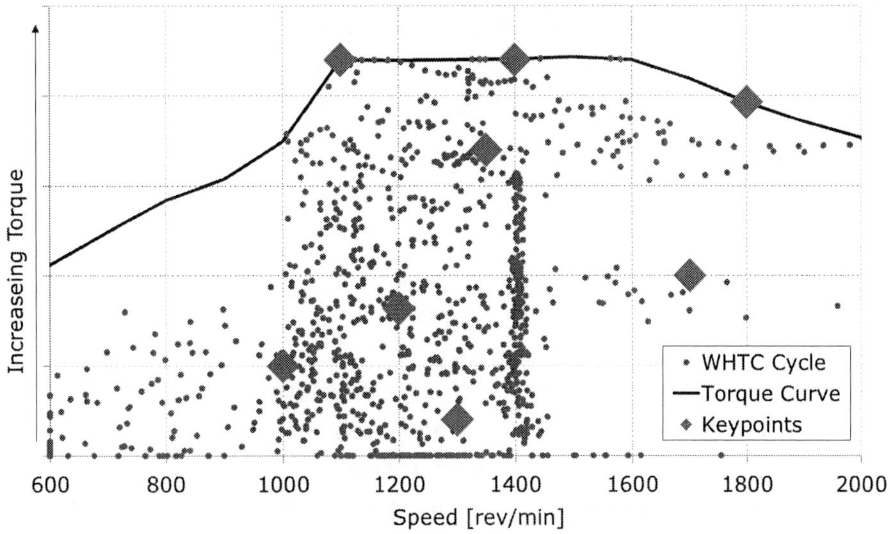

Figure 2: WHTC, torque curve and simulation keypoints

Table 2: Simulation keypoints – engine speed and load (Part and Full Load)

Keypoint	Speed [rev/min]	Load [%]
KP 1	1200	37
KP 2	1350	77
KP 3	1000	29
KP 4	1700	48
KP 5	1300	9
KP 1	1100	100
KP 2	1400	100
KP 3	1800	100

3 AIR MANAGEMENT

Air system and charge mixture preparation are key to the successful operation of a lean burn direct injection engine (2). To define the specification of the air system hardware, 1D simulation was used to assess the boost and air delivery systems, exhaust gas recirculation assessment and valve profile and timing optimisation. The output of these specifications was translated into 3D simulation to assess the impact on in-cylinder charge preparation.

The aim of the air system is to ensure delivery and preparation of the correct excess air, whilst minimising system losses and maintaining transient performance.

3.1 Turbocharging system

Turbocharging enables a lean burn engine to achieve the excess air level and the required power and torque. The increase in, not only pressure ratio, but also torque at low engine speeds requires careful attention to be paid to the compressor surge limits. Whereas the power demand at high speed requires a high efficiency to achieve sufficient air flow to limit high intake and combustion temperatures. A suitable turbocharger was chosen from a range of potential options. The plot below (Figure 3) shows a comparison between the selected and baseline (stoichiometric) turbochargers. It can be seen, from this plot that the baseline turbocharger struggles to meet the full load lambda target at high engine speed (achieved by the lean burn variant) - this would result in higher NO_x emissions and exhaust system temperatures beyond the range of capability of a lean aftertreatment system.

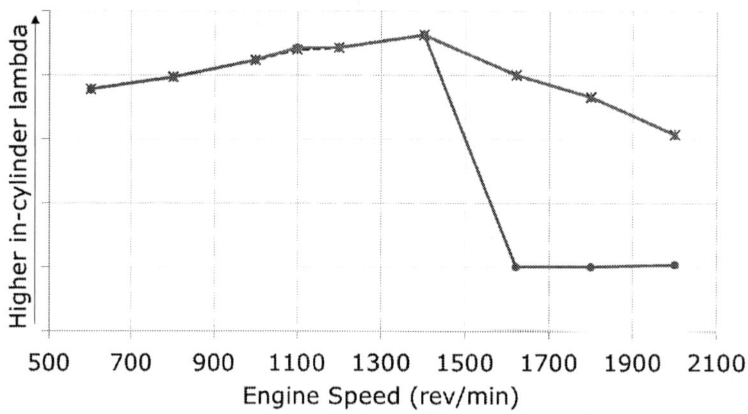

Figure 3: Lambda acheived at full load with stoichiometric and lean burn turbochargers

The increase air mass flow for lean operation at high loads, compared to that for a stoichiometric engine, dictates the need for an optimised turbocharging system However, as a consequence, there is a small impact on part load performance owing to the larger turbine size.

3.2 Valve timing and profile optimisation

Variable valve timing (VVT) has been shown to be an effective enabler of CO_2 reduction on gasoline engines by improving scavenging at low speed high load and reducing pumping losses within throttled regions. To fully assess VVT technology, both intake and exhaust VVT were considered, with a phasing range similar to current production systems.

The plots below (Figure 4) show the Brake Specific Fuel Consumption (BSFC) at various keypoints for a full range of intake and exhaust phasing within the structural limitation of the engine. The zero on each axis is the baseline phasing. At the throttled part load keypoint, it can be seen that an increased overlap, leading to higher residual content within the cylinder, lowers the pumping work, thereby improving fuel economy. Whereas at rated power, the optimum is for a slightly reduced overlap to improve engine breathing. At the low speed, full load point, a longer overlap period and positive engine pressure difference results in improved in-cylinder scavenging, hence BSFC.

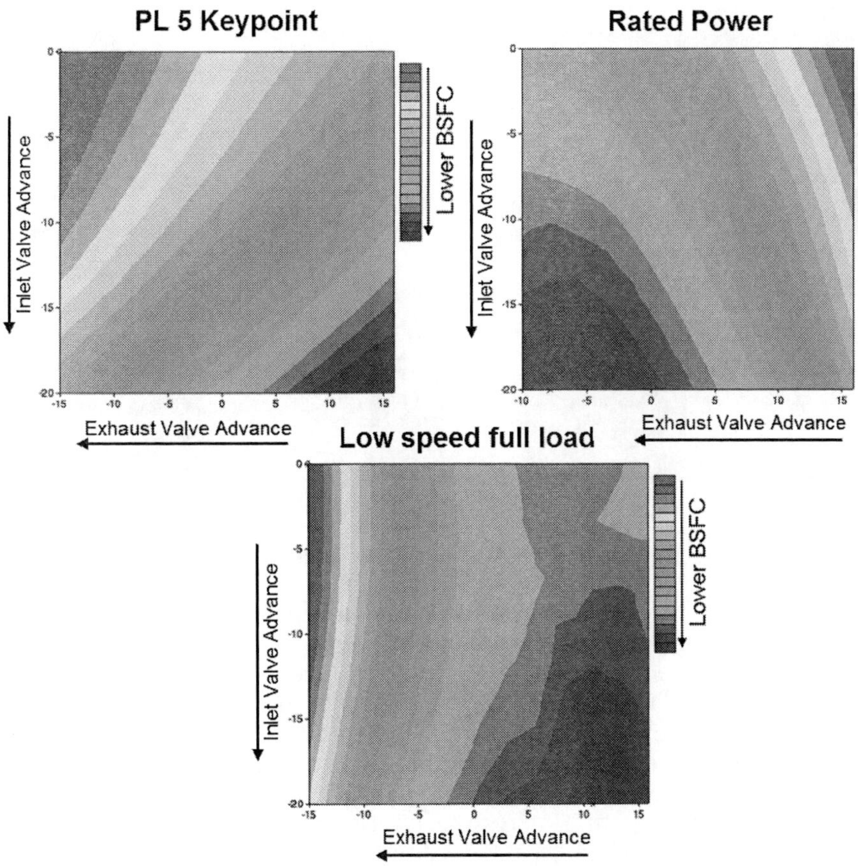

Figure 4: Intake and exhaust valve timing sweeps

In addition to the phasing of the valve events, the shape of the profiles themselves will affect the performance of the lean burn engine. To optimise the valve profiles, a design of experiments (DoE) approach was used, which targeted lowest fuel consumption. As a variable valve actuation system was not considered, the valve profiles defined would have to be suitable for the entire operation of the engine, therefore adhering to the pressure and temperature limits of full load, as well as the combustion stability requirements typical of part load operation.

The Ricardo DoE toolset enabled a model of engine operation related to the parameterised valve events to be created and, from this, an optimiser was used to define the valve profiles that result in the lowest cycle BSFC within system constraints. The baseline exhaust valve profile was maintained, whereas the optimisation process resulted in an intake valve profile with an earlier closing event (EIVC), referred to as 'Aggressive Miller'. The main reason for the BSFC benefit compared to the baseline profile is the reduced pumping loss at throttled conditions. This is shown in Figure 5, which highlights the effect of EIVC on the part load throttled keypoints. For the high load boosted conditions, the lower effective compression ratio (from the additional expansion during the intake stroke) reduces combustion temperature allowing combustion advance within knock constraints and a reduction in effective compression work.

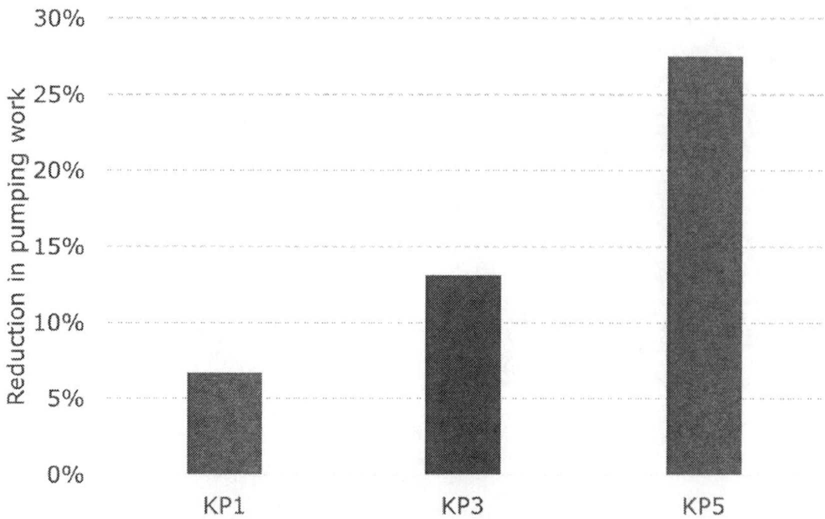

Figure 5: Effect of EIVC on throttled keypoints

The effect of closing the intake valve before the piston has reached bottom dead centre (BDC) on the induction stroke is shown below in the two pressure:volume (PV) plots for part load and full load (Figure 6). Noticeably, the pumping loop for both keypoints is reduced compared to the baseline condition. For the part load keypoint (left) this is a reflection of the reduced throttling with EIVC, whereas for the full load keypoint (right) the lower compression work with EIVC, owing to the lower effective compression ratio, has been matched by an increase in work done by the compressor. For the full load keypoint (in Figure 6) to match the same excess air ratio, there is a 15% increase in compressor power for the EIVC case.

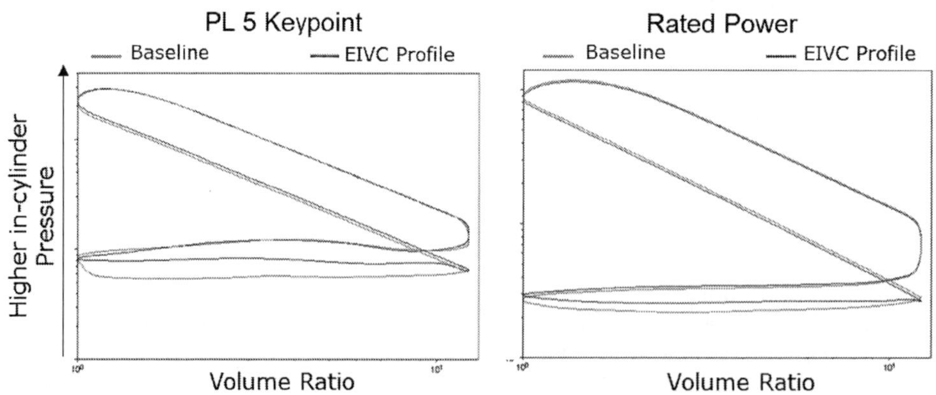

Figure 6: Log pressure volume ratio diagram for two KPs with EIVC

Another benefit of the EIVC profile for the high load points is that, for the same lambda target, there is additional expansion of the charge mass before compression stroke, resulting in lower pre-ignition temperature. This would help mitigate the onset of knock and/or allow combustion phasing advance for BSFC reduction. This approach has been used by large gas engines and gasoline engines to achieve high specific power (3).

The potential effect of the EIVC profile on in-cylinder air motion, charge mixing and flame propagation is not captured within the 1D simulation. 3D simulation, as detailed later, indicates that the reduced lift and duration of the optimised 'Aggressive Miller' profile would have a detrimental effect on the in-cylinder air motion, the level of turbulent kinetic energy and, potentially, combustion. To improve the in-cylinder air motion, but maintain the benefits of an EIVC profile on BSFC and knock mitigation, a second intake profile ('Soft Miller') was devised with a higher peak lift than the 'Aggressive Miller' profile, thereby allowing better development of the air motion within the cylinder. The three intake valve profiles devised are shown below in Figure 7.

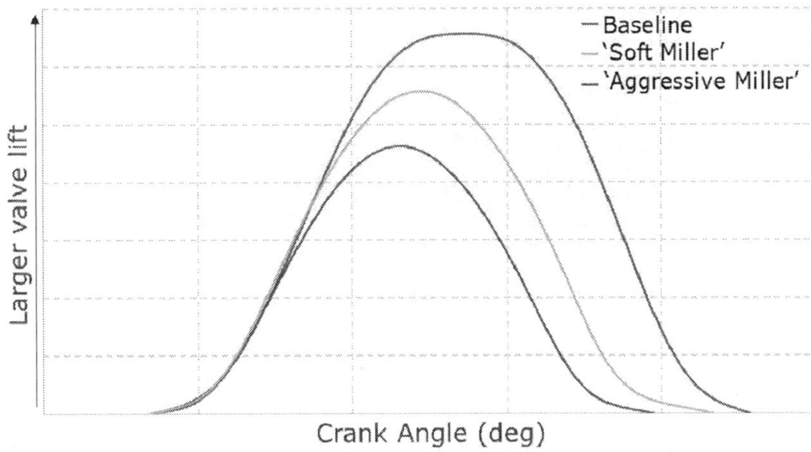

Figure 7: Intake valve profiles simulated

The highest BSFC improvement from the altered valve profiles is realised at the lowest load keypoint. There is a small benefit seen at the boosted keypoints. The effect of increasing the peak lift and duration with the 'Soft Miller' profile is to reduce the BSFC benefit compared to the 'Aggressive Miller' profile.

3.3 Exhaust gas recirculation

Exhaust gas recirculation (EGR) has been used successfully on stoichiometric gas engines to aid NO_X control, mitigate the onset of knocking at full load, reduce pumping losses when throttling at part load and lower in-cylinder heat loss (4). Since NO_X control is achieved during lean operation chiefly through increased excess air, the need for EGR to lower combustion temperature and, therefore, NO_X production is limited. However, the other benefits for stoichiometric engines can still be translated to lean burn operation (5).

To evaluate the benefits of EGR on the lean burn engine, an EGR circuit was added to the 1D model and a sweep of EGR rate and lambda target was performed at a part load keypoint, as shown in Figure 8. During the lambda optimisation process, a lean stability limit was adhered to, reflecting assumed combustion stability and misfire limits. However, maintaining this limit and adding EGR would likely result in combustion instability or misfire. Therefore, a charge mass flow limit (highlighted by the dotted white lines) was used based on the lambda limit with no EGR mass flow. It can be seen below, in Figure 8, that a low level of EGR provides some BSFC benefit whilst maintaining the same charge mass flow as the baseline condition – this is mainly due to reduced pumping work.

EGR was also assessed at full load, with the aim of improving knock mitigation - the inert gases within the externally cooled EGR lower the pre-ignition temperatures thereby delaying the onset of knock. It was seen that a small quantity of EGR reduces the pumping work, as the fresh air target is reduced accordingly, and lowers the NO_X emissions. A reduction in predicted knock intensity was also seen and, therefore, allowed for more optimal combustion phasing at high load; consequently there would be a benefit for exhaust system temperatures.

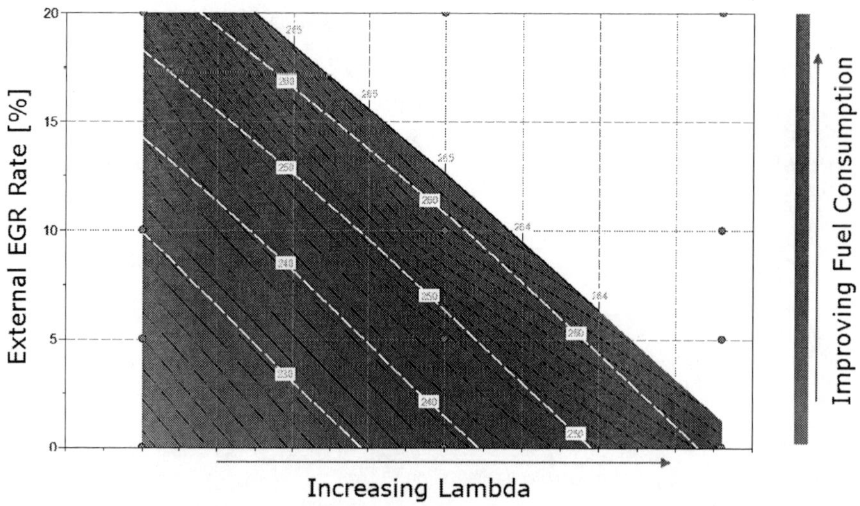

Figure 8: External EGR sweep at a part load KP

3.4 In-cylinder charge management

Previous experience (6) and literature (7) (8) (9) has shown the importance of the charge conditions preceding ignition of the charge. These relate to mixture velocity and kinetic energy within the cylinder, globally and locally around the point of ignition. To achieve levels that have been shown to achieve fast and stable combustion, the intake ports, valve profiles and timing and combustion chamber geometry were assessed using 3D simulation.

Fast analysis of multiple intake port geometries was performed to provide suitable characteristics for 1D simulation and define a range of port geometries for further screening within 3D simulation. The aim of port development was to increase the desired air motion whilst maintaining a suitable flow coefficient (Cf). The development centred on removing air flow restrictions (to improve Cf) within the ports whilst directing airflow to target tumble motion within the cylinder. A summary of the results of the iterative process are shown below, in Figure 9. Literature and previous experience has shown that tumble ratio is key for in-cylinder turbulence development and mixture homogeneity: the 1D simulations indicate that fuel consumption is fairly insensitive to the small reduction in Cf detailed below (between iteration 2 and 3).

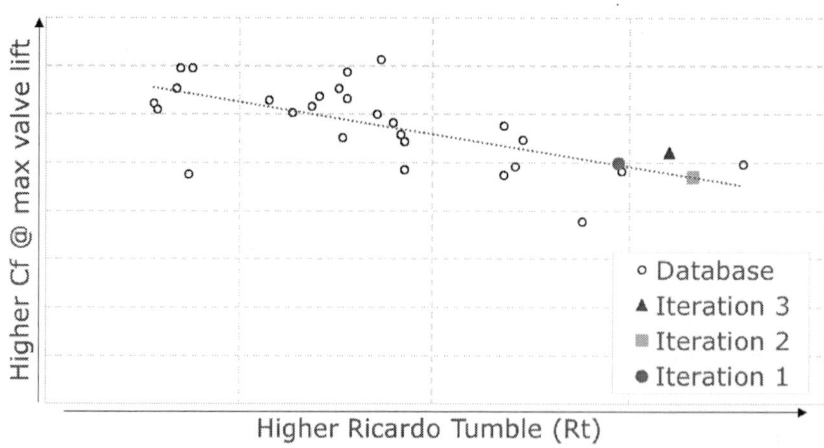

Figure 9: Intake port air flow development (database of relevant engines)

Transient gas exchange simulation was used to further refine the port geometry and aid development of the combustion chamber. The gas exchange process was simulated from exhaust valve opening through to ignition to ensure the conditions within the cylinder were accurate. Both exhaust and inlet valve events were modelled, and the boundary conditions to the simulation were defined by the 1D simulation.

The range of valve profiles developed during the 1D optimisation process were simulated and the cycle in-cylinder tumble motion results are shown below. It can be seen that the effect of a lower valve lift with earlier closing, as exhibited by the "Aggressive Miller" profile is detrimental to the desired air motion. Figure 10 below shows a reduction in tumble motion with the 'Aggressive Miller' intake valve profile, also the turbulent kinetic energy (TKE) was seen to be much lower than the baseline. Higher turbulent kinetic energy is preferred at the time of ignition and globally, within the cylinder, to ensure ignition and flame propagation and thus stable and more efficient combustion. The 'Aggressive Miller' profile simulations were also run with the optimised valve phasing derived from the 1D simulation, and it was seen that high valve overlap can influence the initial development of the in-cylinder tumble motion. The 'Soft Miller' profile was developed to provide similar, yet muted fuel consumption benefit whilst still enabling high levels of tumble consistent with the baseline valve profile.

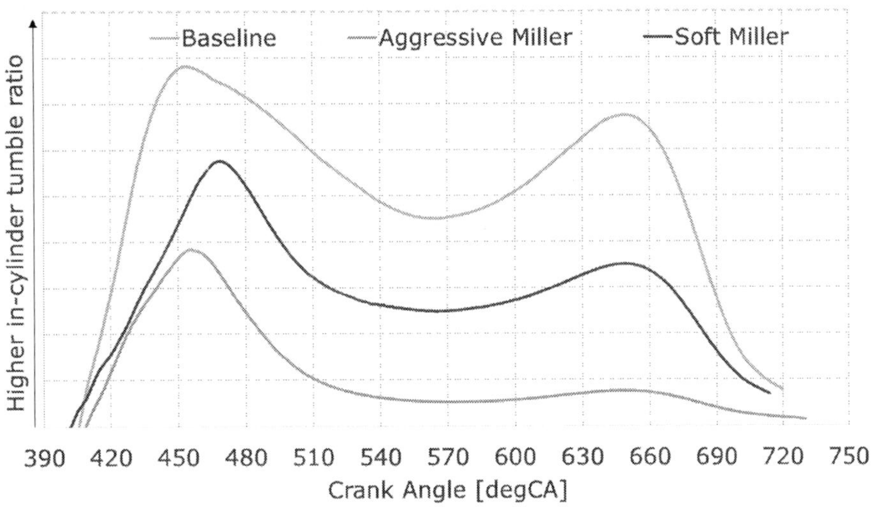

Figure 10: In-cylinder tumble at KP5 with different intake valve profiles (ignition ~710 degCA)

4 FUEL DELIVERY

4.1 Lean operation and direct injection

One of the main benefits of lean burn operation compared to stoichiometric operation is the ability to vary the lambda according to the speed, load and engine conditions. At low load throttled conditions, allowing higher excess air reduces the pumping losses incurred from throttling, although there is a trade-off with longer combustion burn duration. At higher loads, where boosting is required, the increased excess air reduces the peak combustion temperature thereby reducing heat transfer - again this is offset against the increase in combustion burn duration.

For the lean burn engine, the full load lambda target is controlled by the boosting capability of the turbocharging system and the structure and temperature limits of the intake and exhaust valves. At part load, when throttled, temperature and pressure limits are not a concern and, therefore, the main limitation on lambda is combustion stability, which is a function of combustion system design, charge mixture homogeneity and fuel gas composition. For the initial developments, a lean stability limit representative of recent production engines was set.

For appropriate simulation of the lean burn engine, an understanding of the combustion speed is required. Although the 1D model was validated to test data, the change in engine specification and intake conditions will affect the combustion (10) and (11). To fully capture these effects, a combustion model was used that defines the duration between 10% mass of fuel burned and 90% mass of fuel burned (MFB) to be a function of:

- In-cylinder lambda
- Mean piston speed
- In-cylinder temperature and pressure before combustion
- Combustion chamber geometry
- Flammability limits (lean and rich)

This model was correlated to test data and then tuned to match the baseline combustion duration at stoichiometric conditions. The effect of increased lambda operation on the burn duration is shown below, in Figure 11; as expected the higher level of excess air results in a longer combustion burn duration. Ignition timing was optimised at both full load and part load keypoints to achieve lowest BSFC within knock and NO_X limits.

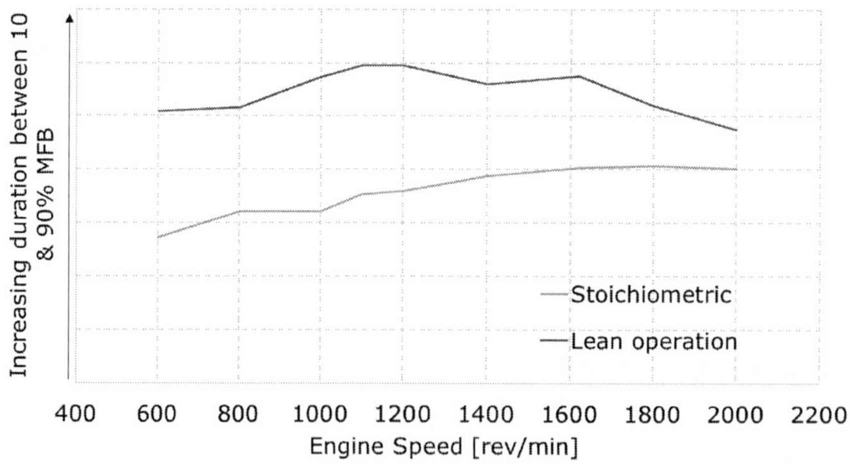

Figure 11: Combustion duration - stoichiometric vs. lean operation

As expected, the main benefit of lean burn operation is apparent at part load, where the pumping losses are greatly reduced at throttled keypoints; this is shown below in Figure 12 and in pressure versus volume traces given in Figure 13. The higher compression and, as a result, higher expansion ratio of the lean operation can be clearly seen at both keypoints. The higher load keypoint, owing to increased boosting, has a higher peak cylinder pressure for lean operation. However, this is well within structural limitations of the engine. The difference in the pumping work can clearly be seen for the lower load keypoint.

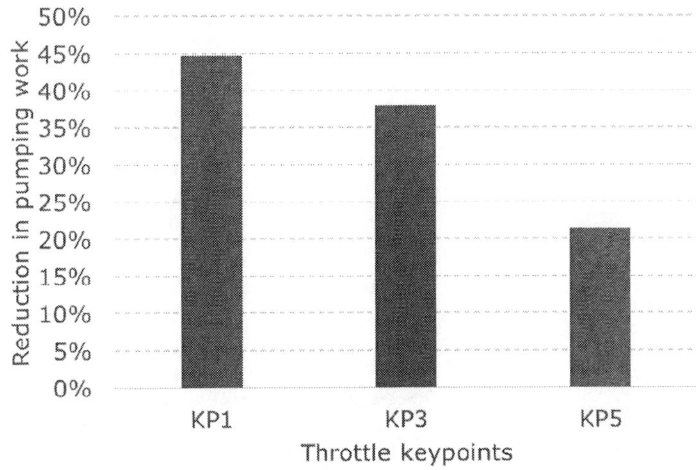

Figure 12: Reduction in part load pumping work from lean operation

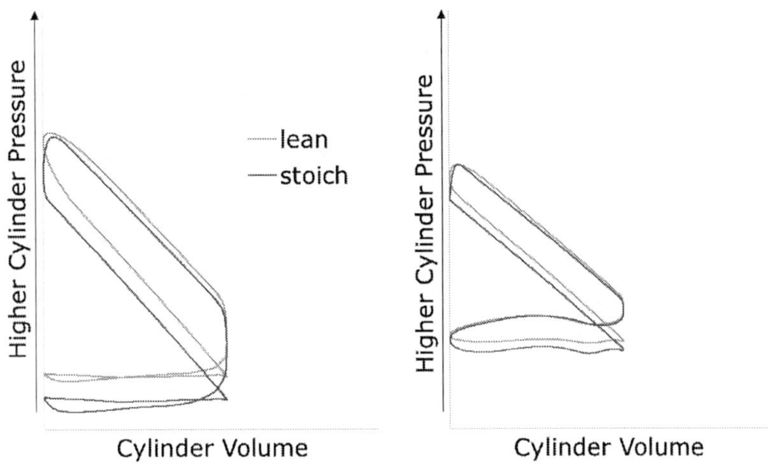

Figure 13: PV comparison high load boosted (left), low load throttled (right)

This reduction in fuel consumption does lead to a corresponding reduction in exhaust gas temperature, for the three throttled keypoints. However, the increase in mass flow, owing to the higher excess air, leads to a similar exhaust enthalpy between stoichiometric and lean burn operation.

The fuel injection period at high loads results in overlap with the induction stroke owing to the pressure level of the direct injection gas system and the requirement to allow time for in-cylinder mixing. The optimisation of the injection timing and duration is a trade-off between volumetric efficiency and potential stratification from to combustion instability from late injection and poor mixing (12) (13). It was not possible to fully evaluate this within a 1D simulation environment. Therefore, to define the injection parameters, a fixed end of injection was set, based on previous experience, and evaluated during subsequent 3D simulations. The earliest start of injection was also defined as exhaust valve closing (EVC) to prevent any fuel gas slip to the exhaust system. The injection window for the full load curve is shown below, in Figure 14. It can be seen that to provide sufficient margin between EVC and ignition, the gas injection for 1600-2000 rev/min occurs during the intake stroke. Only between 600 and 800 rev/min does the gas injection occur fully within the compression stroke.

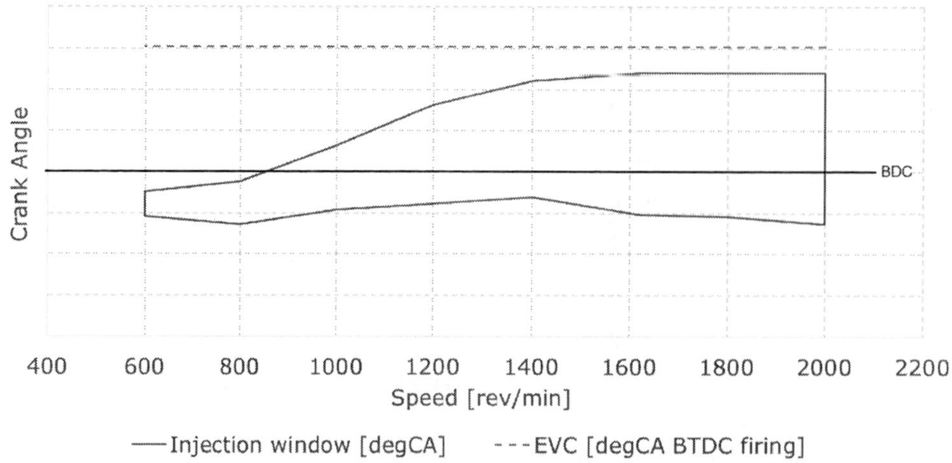

Figure 14: Gas injection duration at full load

4.2 Charge mixture preparation

Key to successful operation of a direct injection NG engine is the charge mixture preparation. Typically, port or manifold fuelled engines will have sufficient mixing length and time to ensure mixture homogeneity at the point of ignition, whereas a direct injection approach, although beneficial, carries risk, especially for long injection duration periods and highly transient operation.

The following results highlight the effect of both calibration and hardware choice on achievable mixture preparation. The aim of the mixing simulation was to assess the homogeneity of the charge around the ignition region and globally, within the cylinder, at the point of ignition. The effect of the gas spray injection on turbulence was also reviewed.

For the rated power point, it can be seen in Figure 15 that the gas jet has little effect on the local and global turbulence. The benefit of an early injection is highlighted by the reduction in equivalence ratio distribution above the lean stability limit. The local lambda around the ignition region is close to the target and well below any lean misfire limit for both early and late timing.

The benefit of advanced injection timing is most apparent at the low load keypoint, as detailed in Figure 16, where the global equivalence ratio band (EQR between min and max probability) has decreased by ~75%. The earlier timing results in slightly reduced global turbulent kinetic energy, as the fuel spray interacts with the initial tumble vortex formation when the intake valves are opened.

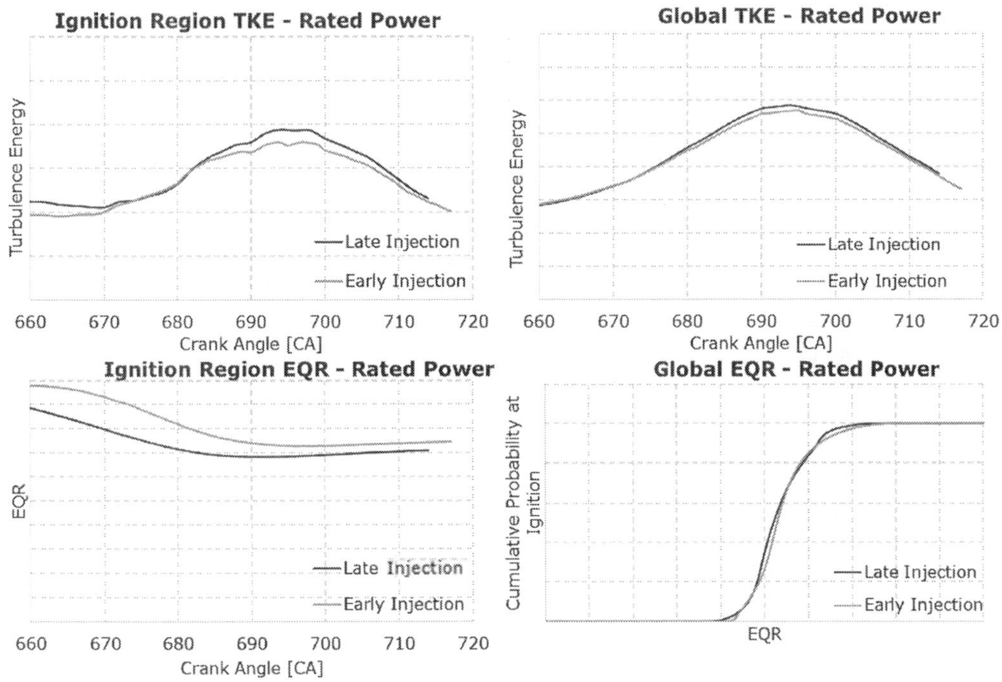

Figure 15: Rated power injection timing assessment

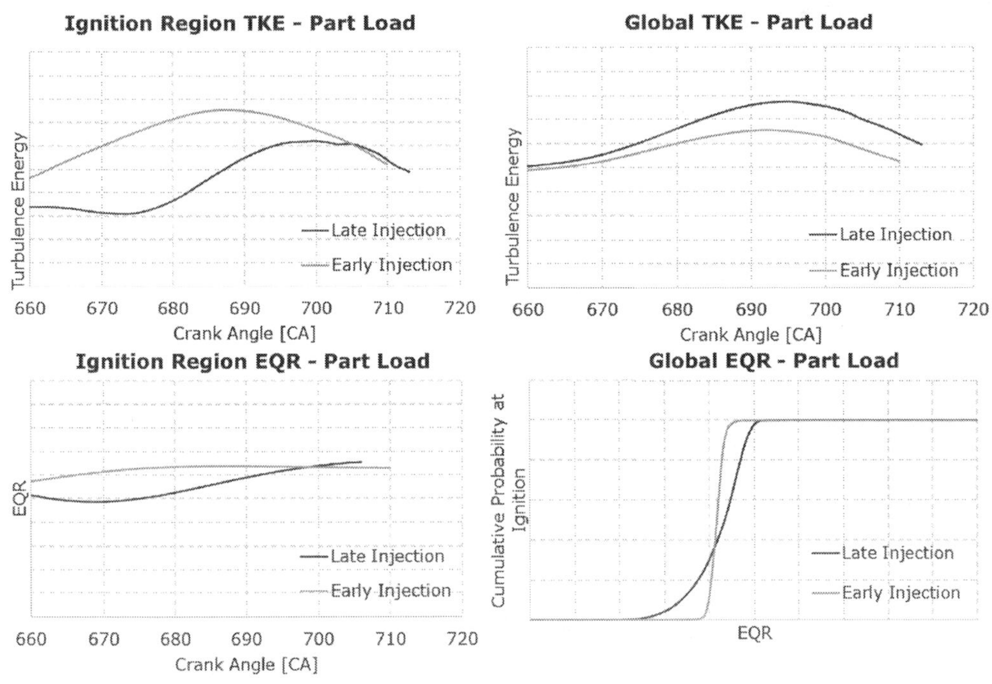

Figure 16: Part load injection timing assessment

It is clear that injection timing has a strong impact on fuel-air mixing and, therefore mixture homogeneity globally as well as locally around the ignition source. Another area of interest is the fuel-gas injection system with regard to the position of the injector and the spray pattern. The results shown below, in Figure 17 highlight the effect of fuel spray cone angle on the equivalence ratio distribution at the point of ignition, the cell equivalence ratio probability metric indicates the likelihood that a cell within the 3D simulation will have a specific EQR at ignition timing. The optimisation of the spray pattern is greatly related to the air induction motion and the combustion chamber geometry. For this engine, it can be seen that a 'mid' cone angle is optimal.

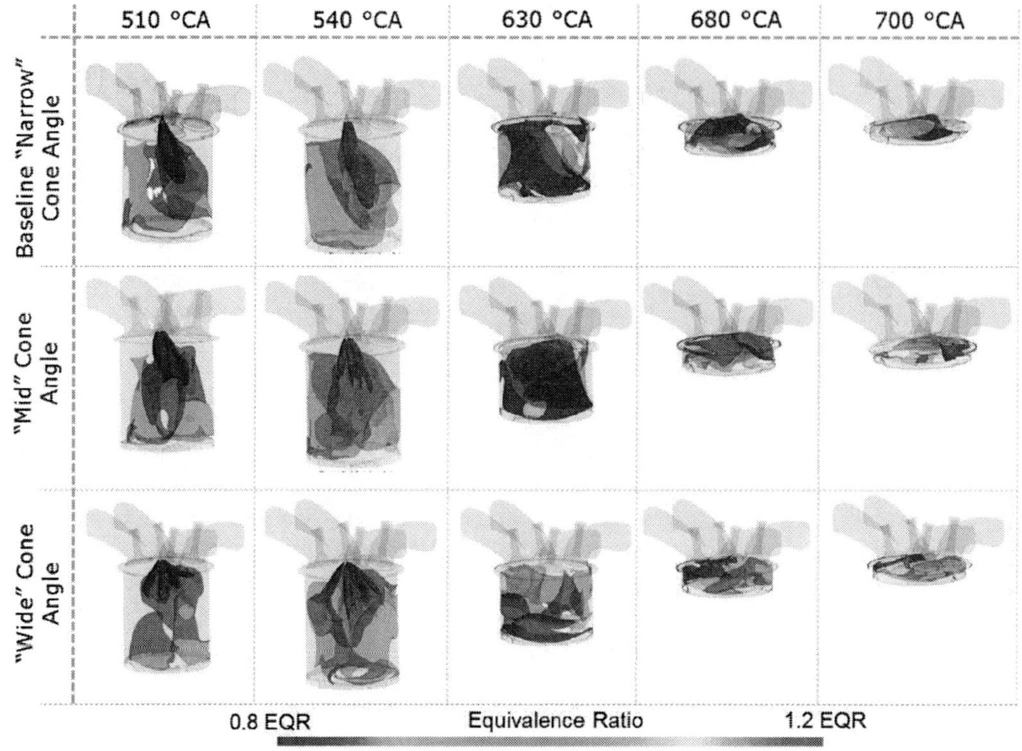

Figure 17: In-cylinder equivalence ratio planes at various crank angles and nozzle cone angles

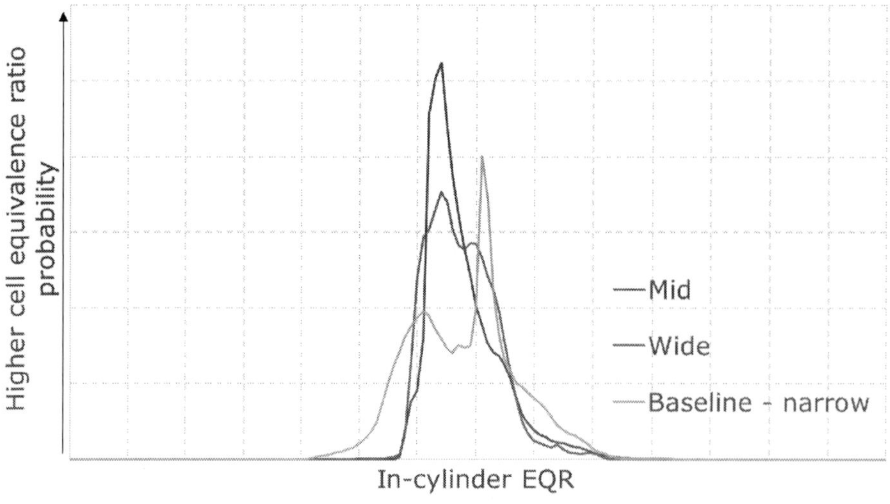

Figure 18: EQR assessment at SOC for various cone angles

5 EMISSIONS CONTROL

Natural gas emissions control via exhaust catalysis is a significant challenge due to the stability of methane. Methane does not start to oxidise over a highly loaded precious metal catalyst until above 400°C. Traditionally, for HDD engines hydrocarbon (HC) control is not an issue, moving to methane as the fuel significantly increases the challenge of exhaust hydrocarbon control. The added challenge, when operating the engine lean, is controlling NO_X emissions, which require the use of a dedicated SCR catalyst and urea injection system. To evaluate the ability of the lean burn engine to meet to the emissions limits (Euro VI), an approach was chosen, as detailed below in Figure 19, to develop a Simulink based model of a developed aftertreatment system, within the IMBD architecture, using the Ricardo vehicle simulation framework (VSIM). This software has been frequently used by Ricardo and reported in other work (14) (15).

The VSIM model uses the output of the 1D calculations and steady-state emissions maps to simulate the behaviour of the aftertreatment system. This modular simulation approach allowed for detailed review of each sub-system and the rapid addition and subtraction of various characteristics, thereby showing the sensitivity of the final result. Additionally, the output of the 1D steady-state map for exhaust gas temperature was manipulated to create a more realistic transient exhaust gas temperature profile, which is key to accurately predicting aftertreatment behaviour.

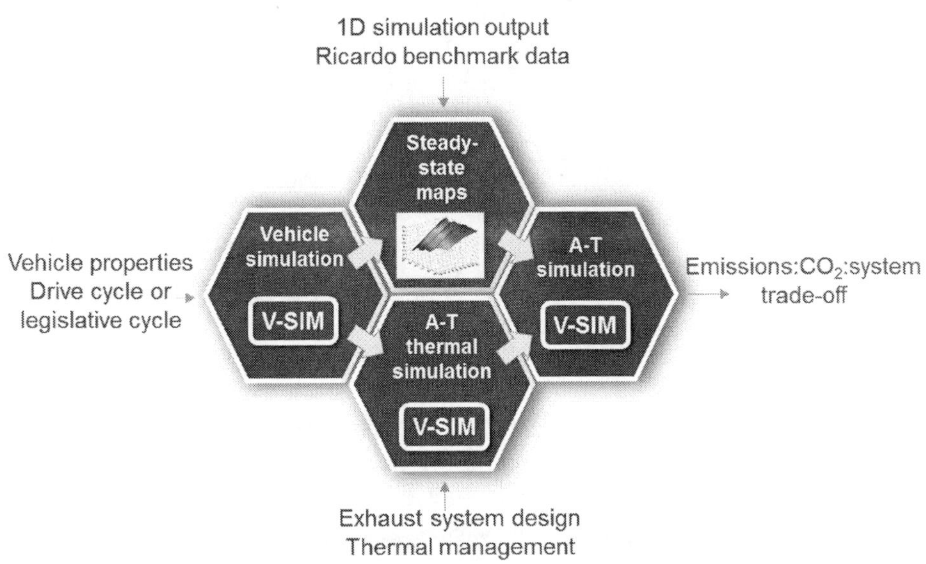

Figure 19: IMBD setup for aftertreatment evaluation

5.1 Aftertreatment model input data

The emissions-catalyst interactions within the aftertreatment system were modelled based on data for the various reactions associated with each catalyst – this included both reduction and oxidation reactions, including exothermic reactions. For these simulations, the data was accessed through lookup maps based on characteristic data of the latest generation of catalysts that was derived from synthetic gas reactor and engine bench testing. The characteristics always reflected the catalysts in an aged condition.

Engine out emissions for NO_X, CO and THC were derived from the 1D simulation and benchmark data. The target combined cycle aftertreatment conversion efficiencies to meet the project targets were ~90% for both NO_X and CH_4, whereas CO required only ~30% reduction.

A control strategy, representative of current production systems, was used to govern the timing of urea injection, this ultilised various model parameters. The different settings were evaluated to ensure appropriate levels were set for the range of conditions investigated during the simulation.

5.2 Thermal model tuning

An accurate transient temperature profile of the aftertreatment system is key to predicting emissions conversion performance. Specific temperature profiles for the wide matrix of systems, calibration settings and drive cycles were not available. Therefore, an approach was developed using steady-state test data and system properties, to predict the exhaust feed-gas temperature profile.

To convert steady state exhaust temperature data, used as a model input, into a transient drive cycle temperature profile, an additional thermal system was developed. The effect of this approach is shown in Figure 20, where the predicted and measured temperatures have a close match over the WHTC. The steady-state exhaust temperature map was also corrected for cold coolant temperatures, to better simulate engine warm-up. This was found to be highly influential for relatively short cycles where the warm-up period is dominant.

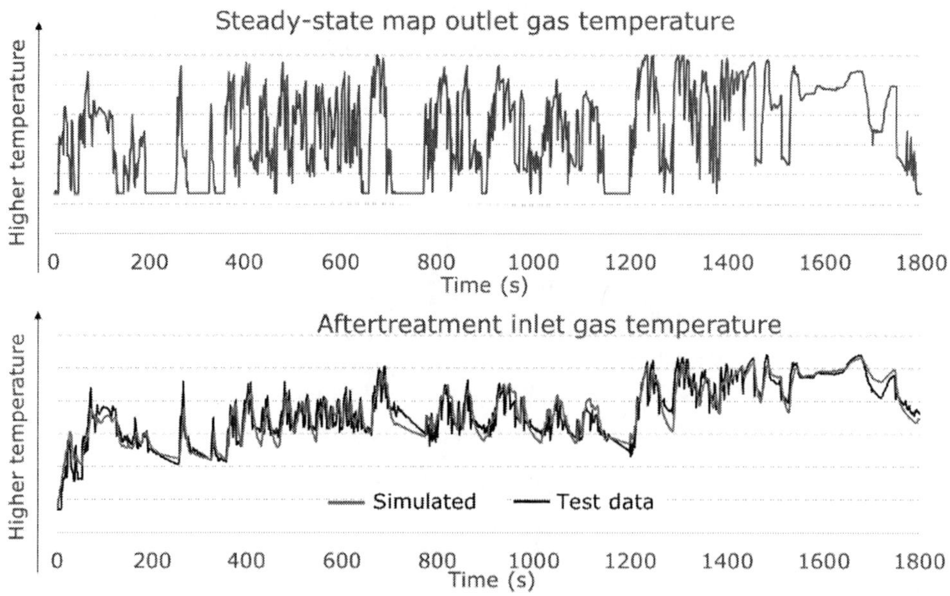

Figure 20: Thermal model effect

5.3 Emissions simulation results

The aftertreatment simulations were performed for CO, CH_4 and NO_X emissions. It was assumed that the non-methane hydrocarbons (NMHC) emissions target, predominately sourced from the fuel gas, would be achieved if the CH_4 limit was met. Both cold and hot WHTCs were simulated. The initialisation temperature conditions for the hot WHTC were based on the end of cold cycle with the addition of a cooling period. Owing to the typically lower exhaust gas temperature from lean combustion, compared to stoichiometric operation, as a result of the increased excess air, the control of methane slip (CH_4) is a significant challenge. Figure 21, below shows the difference in temperature between the two gas combustion modes and that using diesel fuel. To mitigate this risk, an assessment of thermal management strategies, with the purpose of increasing the methane oxidation catalyst (MOC) brick temperature, was also completed. An electrical heater applied to the exhaust gas pre MOC and stoichiometric operation of the engine, at the start of the cycle, were the strategies assessed.

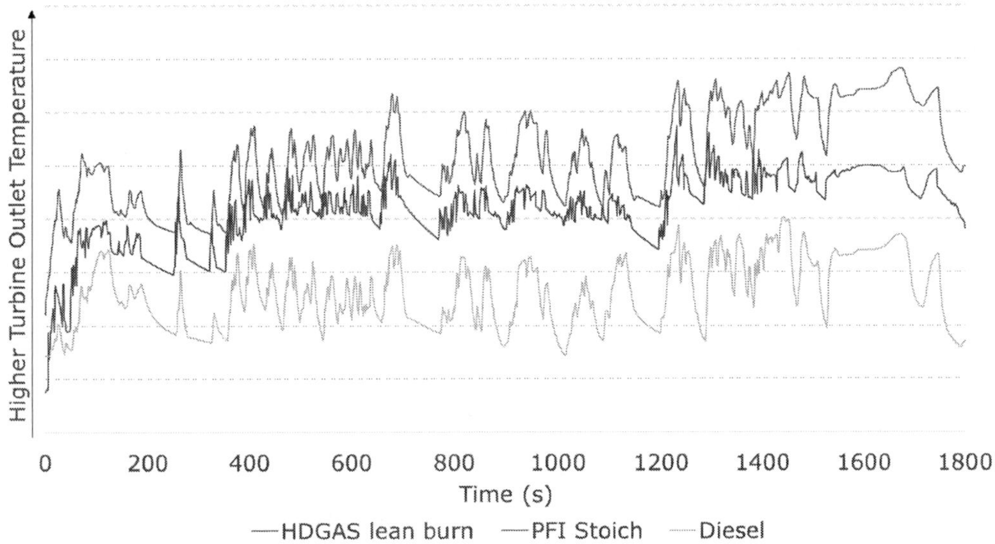

Figure 21: Comparison of turbine oulet temperature for diesel, lean burn and stoichiometric over the WHTC

Due to the excess air present within the exhaust gas, the typical aftertreatment approach of a three-way catalyst (TWC) for NG emission control is not appropriate and, therefore, multiple catalysts are required to control both Methane and NO_X. The baseline aftertreatment system devised is shown below in Figure 22.

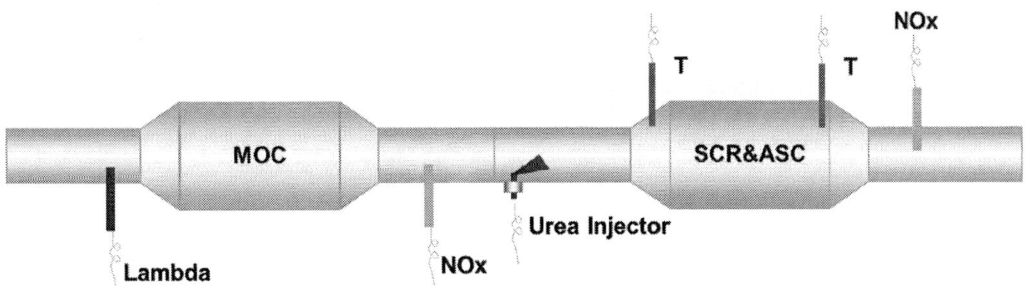

Figure 22: Baseline aftertreatment system layout with sensors (T=Temperature)

The cumulative emissions results highlight the importance of the first 400 seconds for both the cold and hot WHTC for CH_4 emissions control, as shown in Figure 23. This is a reflection of the light-off temperature of a MOC, which is typically around 400°C formulation dependant.

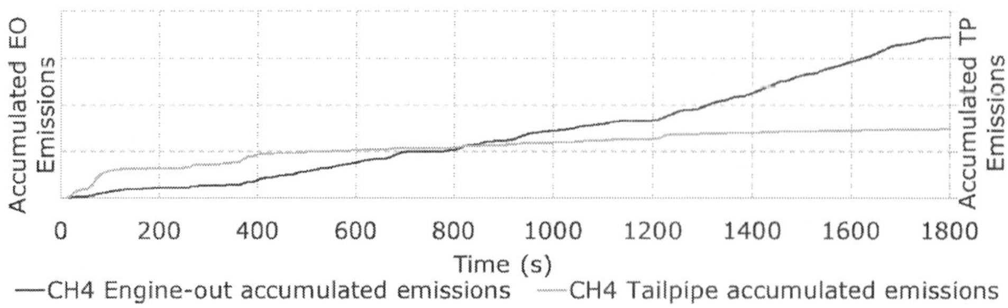

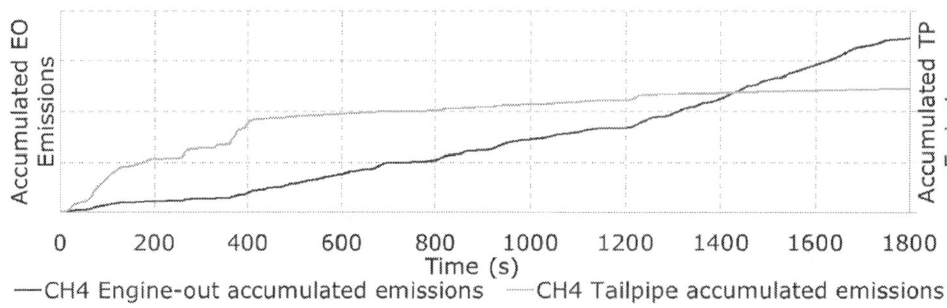

Figure 23: CH_4 emissions at engine out (EO) and tailpipe (TP) over both cold (bottom) and hot (top) WHTC

Thermal management is not necessary for NO_X control via the SCR, as the start of its peak efficiency window is lower than that of the MOC. As can be seen in Figure 24, during the cold WHTC, with no thermal management, the majority of the NO_X slip occurs during the first 100s, when the brick temperature is still below the urea dosing temperature: beyond this temperature the high conversion efficiency of the SCR minimises any tailpipe NO_X slip.

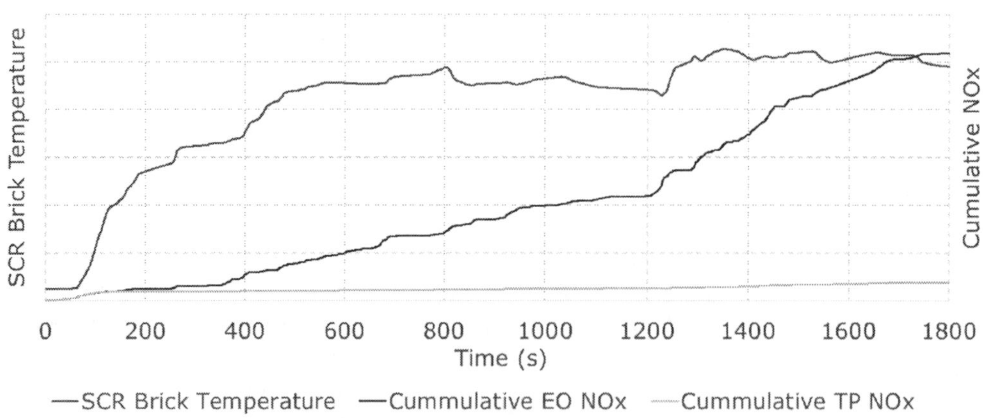

Figure 24: NO_X emissions and SCR temperature over WHTC cold – EO=Engine out, TP = tailpipe

Results of the aftertreatment simulation, in Figure 25, show that without thermal management, NO_X and CO emissions targets can be achieved, whereas CH_4 is above the limit unless thermal management is applied either through an electrically heated catalyst (EHC) or through lowering the in-cylinder lambda target. All the thermal management strategies investigated carry a CO_2 penalty. Owing to the lower temperature requirement for NO_X and CO emissions control by an SCR and oxidation catalyst respectively, and the low weighting of the cold WHTC to combine result, both NO_X and CO are fairly insensitive to exhaust thermal management.

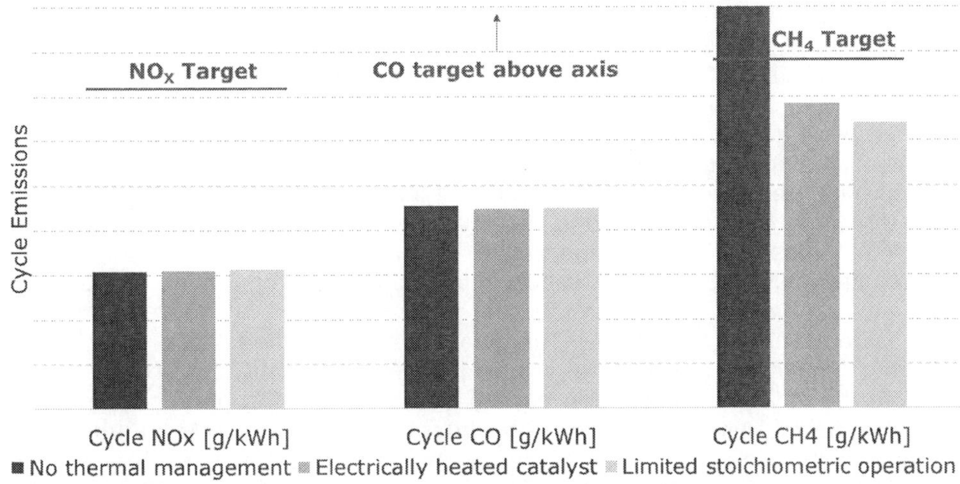

Figure 25: Combined cold and hot cycle results with and without thermal management

6 DISCUSSION AND CONCLUSIONS

Heavy-duty natural gas engines that operate under lean conditions are not currently in production at Euro VI. However, if the right combustion characteristics can be achieved, a lean burn application could be viable within Euro VI emissions legislation when used in conjunction with a state of the art aftertreatment system.

To enable the specification of the lean burn variant, simulation tasks using the Ricardo IMBD approach have been completed to identify appropriate components and calibration settings. The required combustion characteristics were defined, and an engine specification has been developed, supported by 1D analysis, appropriate literature and relevant benchmarks. 3D simulation was then used to conduct detailed assessments around the developed geometry of the combustion chamber. The results seen within the simulation tasks highlight the CO_2 benefit of the lean burn variant compared to the baseline engine. A full assessment of the relative CO_2 benefits will be completed during the engine testing stage of the project.

The main benefit for lean burn operation and the subsequent freedom to vary lambda was lower pumping losses at low load throttled conditions. This allowed higher excess air and reduced the throttling requirement, although there is was trade-off with longer combustion burn duration. At higher loads, where boosting is required, the increased excess air reduces the peak combustion temperature thereby reducing heat transfer losses - again this is offset against the increase in combustion burn duration. Further investigations highlighted the benefits of optimising the intake valve profiles and both intake and exhaust valve timing to lower pumping losses at low load throttled conditions. Simulations have also indicated that EGR, at both part load and full load, would be beneficial for both CO_2 and NO_X emissions. 3D simulations were performed to aid the combustion system development, targeting high levels of turbulent kinetic energy and fuel-air mixture uniformity. The simulation results indicate the sensitivity of fuel-air mixing to injection timing, and the benefit of injecting as early as possible to increase the available time to achieve homogeneity has been shown.

The evaluation of the aftertreatment system has shown the challenge of achieving the hydrocarbon limits for a lean burn engine, owing to lower exhaust temperature compared to stoichiometric operation. Simulations indicate that the emissions targets can be met with exhaust thermal management employed.

The relative CO_2 benefit of the specified lean burn engine is shown below as a technology route, in Figure 26. It can be seen that the 10% target improvement over a baseline 2013 NG engine is realised with the various additional technologies. A further 2% CO_2 improvement is achieved over a comparable diesel fuelled engine. The majority of CO_2 improvement is realised through lean operation and direct injection of the gaseous fuel - this highlights the relatively high pumping and increased heat transfer penalty from stoichiometric operation. Further incremental benefits are seen from incorporating Miller intake valve profile to further lower part load pumping losses; the CO_2 benefit from optimised variable valve timing is also included within this step. The simulation indicated an additional benefit of the Miller intake valve profile was to reduce the charge temperature and, thereby, lower the knocking propensity of the engine. This enabled a compression ratio (CR) increase for further efficiency gains. The reduced fuel consumption from the various technologies and lower combustion temperatures from lean operation combine to delay oxidation catalyst light-off and subsequent hydrocarbon control during warm-up. Consequently there is a fuel penalty due to exhaust thermal management penalty in order to meet the emissions targets, however this is carefully managed through limited stoichiometric operation to minimise the CO_2 increase.

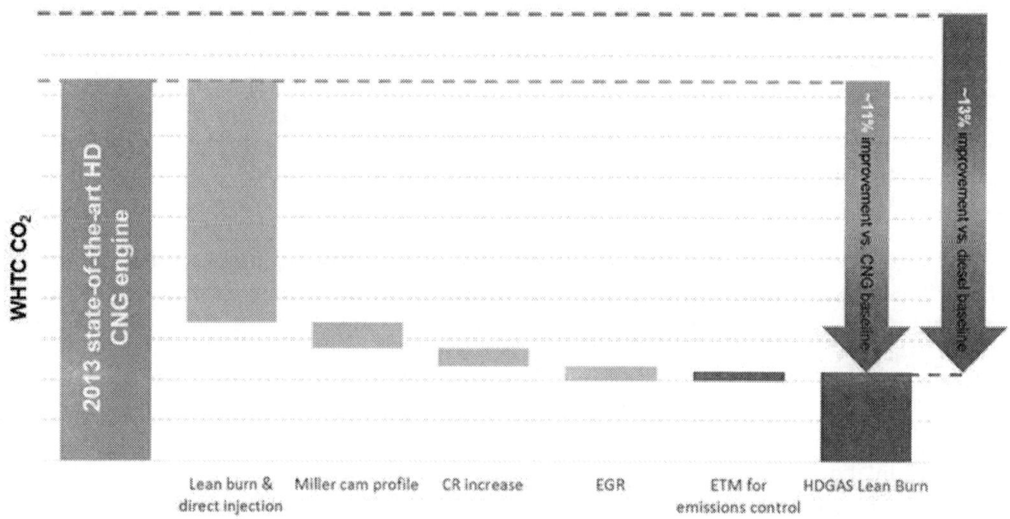

Figure 26: Overall CO_2 walk for HDGAS lean burn engine

7 ACKNOWLEDGEMENTS

The authors would like to thank the Ricardo team members Joshua Dalby, Andy Noble and Tiago Carvalho for their contributions to the work reported here. Thanks also go to the partners within the HDGAS project consortium for their assistance and permission to publish this paper.

The work reported here received funding from the European Union's Horizon 2020 research and innovation programme, under grant agreement No 653391.

8 REFERENCES

(1) Muncrief, R. and Sharpe, B., "Overview of the heavy-duty vehicle market and CO_2 emissions in the European Union", ICCT Working Paper 2015-16

(2) Clarke, D., Such, C., Overington, M. and Das, P., "A Lean Burn Turbocharged, Natural Gas Engine for the US Medium Duty Automotive Market", SAE Technical Paper 921552, 1992, doi:10.4271/921552

(3) Osborne, R. et al., "A Miller Cycle Engine without Compromise - The Magma Concept", SAE Int. J. Engines 10(3):2017

(4) Zelenka et al., "The Potential of Exhaust Gas Recirculation in Large Gas Engines", CIMAC Congress Presentation, 2013

(5) Bhargava , S., Clark, N. and Hildebrand, M., "Exhaust Gas Recirculation in a Lean-Burn Natural Gas Engine", SAE Technical Paper 981395

(6) Kyriakides, S. and Glover, A. "A study of the correlation between in-cylinder air motion and combustion in gasoline engines", I.Mech.E C55/88, 1988

(7) He et al., "Impact of Tumble on Combustion in SI Engines: Correlation between Flow and Engine Experiments", SAE Technical Paper 2007-01-4003, 2007

(8) Millo et al., "Effects of Different Geometries of the Cylinder Head on the Combustion Characteristics of a VVA Gasoline Engine", SAE Technical Paper 2013-24-0057, 2013

(9) Johansson et al., "Combustion Chambers for Natural Gas SI Engines Part 2: Combustion and Emissions", SAE Technical Paper 950517

(10) Ayala, F. and Heywood, J., "Lean SI Engines: the Role of Combustion Variability in Defining Lean Limits", SAE Technical Paper 2007-24-0030, 2007

(11) Ayala, F. et al, "Effects of Combustion Phasing, Relative Air-fuel Ratio, Compression Ratio, and Load on SI Engine Efficiency", SAE Technical Paper 2006-01-0229, 2006

(12) Zoldak, P. and Naber, J., "Spark Ignited Direct Injection Natural Gas Combustion in a Heavy Duty Single Cylinder Test Engine - Start of Injection and Spark Timing Effects", SAE Technical Paper 2015-01-2813, 2015

(13) Thipse et al., "Injection Strategies, Optimisation and Simulation Techniques on DI CNG Technology", SAE Technical Paper 2015-26-0046, 2015

(14) Downes, T. et al, "The Potential of Cost Effective LNT-Based Diesel NO_X After-treatment for Real Driving Emissions", Paper presented at the SIA Powertrain Conference and Exhibition, Rouen, 1-2nd June, 2016

(15) Auld, A., Ward, A., Mustafa, K., and Hansen, B., "Assessment of Light Duty Diesel After-Treatment Technology Targeting Beyond Euro 6d Emissions Levels", SAE Int. J. Engines 10(4):2017

POWERTRAIN DESIGN

The development of the Szorenyi four-chamber rotary engine

P King
Rotary Engine Development Agency, Australia

ABSTRACT

A four-chamber Otto cycle rotary engine, the Szorenyi Rotary Engine, has been invented and developed by the Rotary Engine Development Agency (REDA) in Melbourne, Australia. The engine concept has been awarded a U.S. Patent (Number 6,718,938 B2). A prototype engine has been constructed and a successful proof-of-concept engine test was achieved in 2008.

The stator of the Szorenyi engine is a similar shape to a Wankel engine. However, the geometric shape of the engine rotor is a rhombus, which deforms as it rotates inside the contour of the mathematically defined stator. This geometry translates to a rotary engine with four combustion chambers. Each revolution of the crankshaft produces one revolution of the rotor, a complete engine cycle in each of the four chambers, and therefore four power strokes. In contrast, the Wankel engine produces one power stroke per crankshaft revolution. Additionally, the Wankel engine is rev limited, typically to 3,000 rpm rotor speed, due to the excessive crankshaft bending caused by the centrifugal forces of the eccentric rotor. The Szorenyi engine has a balanced rotor and so is not rev limited in this regard. However, the actual rev limit is yet to be established by modelling and testing.

The advantages of the Szorenyi engine over the Wankel engine is higher power density due to the potentially higher revving rotor; a larger space inside the rotor for internal cooling of the rotor; four power strokes per crankshaft revolution against the Wankel's single power stroke; no need for a balancing flywheel in a single rotor engine; no need for a reduction gear in aircraft and UAVs with large propellers; and, suitability for multi-rotor engines. Therefore Szorenyi engine could be used in any applications where the reciprocating and Wankel engines are currently used and provide all of the above advantages.

RMIT University has conducted ideal mathematical modelling of the engine geometry and fuel burn. The model analysed the Szorenyi engine, the Wankel, and a reciprocating engine of the same displacement. This modelling has shown that the Szorenyi engine thermal efficiency is 0.46% greater than the reciprocating engine and 0.38% greater than the Wankel engine. Additional modelling has shown that the Szorenyi engine has approximately the same power density as the Wankel engine when operated up to the rev limit of the Wankel engine rotor, and much greater power density above that speed.

The prototype engine used in the proof-of-concept test experienced an internal failure, but now has redesigned rotor hinges. That engine is awaiting a program of further testing to assess engine performance. Also, the RMIT University mathematical modelling of the Szorenyi engine, while providing good results, is ideal and so complex modelling is required to more accurately predict performance.

DEFINITIONS AND ABBREVIATIONS

ABDC	After Bottom Dead Centre
ATDC	After Bottom Dead Centre
BBDC	Before Bottom Dead Centre
BDC	Bottom Dead Centre
BTDC	Before Top Dead Centre
CAD	Computer Aided Design
CFD	Computational Fluid Dynamics
NSU	Neckarsulm Motorenwerke AG
P-V	The pressure and associated volume in the combustion chamber
REDA	Rotary Engine Development Agency
RMIT	Royal Melbourne Instite of Technology
TDC	Top Dead Centre
UAV	Unmanned Aerial Vehicles

INTRODUCTION

A new configuration of a rotary engine has been developed by the Rotary Engine Development Agency (REDA), based in Melbourne. The special features of the engine are that it is a rotary engine with four chambers and the engine stator profile that contains the four chambers is mathematically defined. Named after its inventor, Peter Szorenyi, the Szorenyi Rotary Engine promises to be smaller than an equivalent Wankel engine; deliver higher power by revving higher than a Wankel; and have much greater power density than a reciprocating engine. The invention has been awarded a US patent (No 6,718,938 B2). A prototype has passed the proof-of-concept stage of development, and initial mathematical modelling has produced favourable results.

THE INVENTION

The discovery of the Szorenyi curve was made in 2000. REDA was formed shortly after the discovery and was awarded an ACT Research and Development Grant of $63,000 in 2001. The US patent was awarded in 2004.

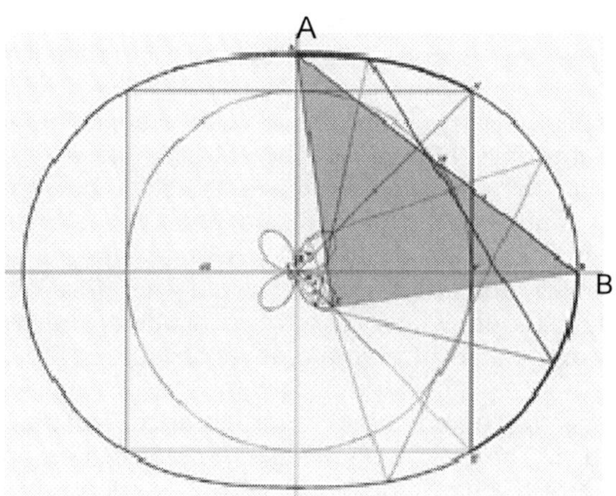

Figue 1. The Geometry of the Stator Profile

The invention is the geometric shape that can contain a four-sided rhombus with the corners of the rhombus in any orientation. The contour of what is used as the stator profile of the engine has been patented as 'The Szorenyi Curve'. The geometry of the stator profile is shown in Figure 1. The discovery of the curve was made while examining the geometry of the stator profile of a Wankel engine. Using a right-isosceles triangle with the hypotenuse (the line AB in Figure 1) representing the face of a four-segment rotor, it was noticed that as the triangle was rotated, the apex of the triangle traced out the rough shape of a four-leaf clover around the centerline of the stator. It was then a simple matter to mathematically describe the profile of the stator by utilising the mathematical shape similar to four-leaf clover – the sin2θ function.

In the engine design the mathematically described curve represents the profile of the engine stator; the hypotenuse of the triangle represents the face of each of the four hinged segments of the engine rotor; and, the gap between the stator and the segment represents the chamber of the rotary engine.

The U.S. Patent patent describes the geometric derivation of the stator - the Szorenyi Curve. The mathematical expression for the stator contour can readily be derived[1]. The power of the discovery of the curve and the mathematical expression is that a wide range of stator profiles can be explored to determine the optimum shape of the stator and combustion chamber for the particular application of an engine.

DESIGN AND DEVELOPMENT

After the invention was patented, the Rotary Engine Development Agency was formed and work began translating the geometry of the invention into a working rotary engine. Some of the early designs of the engine components can be seen Figure 2. A CAD depiction of a later complete design is shown in Figure 3.

Figure 2. Partly Assembled Prototype Engine

The stator appears similar to the Wankel rotary engine but the shape of the Szorenyi engine stator is able to contain a four-segment hinged rotor. The apexes at the hinged joints of the segments of the rotor remain in physical contact with the stator profile as the rotor rotates. The gap between the engine stator and the rotor segments creates the four chambers. As the engine rotates each chamber completes the induction, compression, ignition and exhaust phases of the Otto cycle.

[1] Although the geometry of the stator appears simple visually, the mathematical expression is somewhat cumbersome. The x and y co-ordinates of the shape contain the angle in the sin2θ expression for the four-leaf clover. The most usable form of expression for the contour of the stator profile is:

$x = \{ \sin 2\theta + \sqrt{(w^2 - \sin^2 2\theta)} \} \{ \sin \theta + \cos \theta \}/2$

$y = \{ \sin 2\theta + \sqrt{(w^2 - \sin^2 2\theta)} \} \{ \sin \theta - \cos \theta \}/2$

where θ = the generating angle of the four-leaf clover (sin2θ)
and w = rotor face width compared to the four-leaf clover radius (unity).

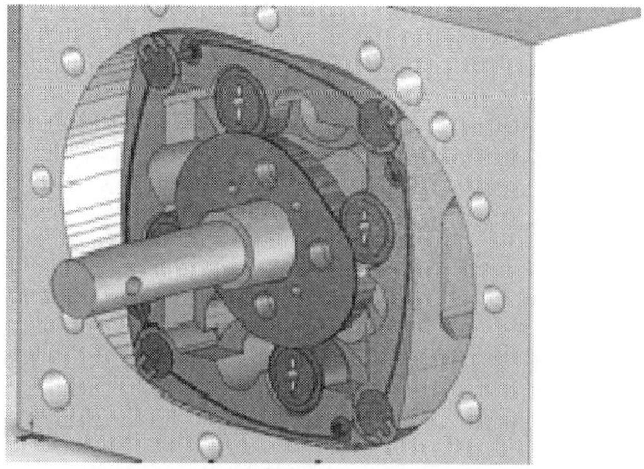

Figure 3. CAD depiction of the engine internals

Each of the four chambers in turn draws the air-fuel mixture in through a peripheral port (indicated by the left arrow in Figure 2). As the rotor rotates (in a clockwise direction in Figure 2) the trailing apex seal of the chamber passes over the intake port thus trapping the air-fuel mixture in front of it. The mixture is then compressed and ignited. (The spark plug can be seen in Figure 2. It is positioned slightly after top dead centre.) As the rotor rotates, the high-pressure gases of combustion generate a force that acts in a line eccentric to the drive-shaft centerline thus spinning the rotor and producing usable power. Further rotation of the rotor allows the gases to expand until the leading apex seal passes a peripheral exhaust port and allows the gases to be expelled. (The exhaust port is indicated by the right arrow in Figure 2.) Because the rotor and stator form four chambers, one rotation of the rotor produces a complete engine cycle in each chamber. Therefore there are four power strokes per revolution of the crankshaft. A single rotor Szorenyi engine is therefore equivalent to an eight cylinder reciprocating engine. Also, the Szorenyi engine does not need the cumbersome valve-train of the reciprocating engine. The result is a simpler engine with exceptional power density.

The prototype engines were produced with stator internal dimensions of 220 x 160 x 70 mm. A compression ratio of 9.5 was used and this resulted in a swept chamber volume of 287cc. So the swept volume of the four-chamber Szorenyi engine prototype was 1148cc. (Because the Szorenyi engine produces four power strokes per crankshaft revolution, the prototype was equivalent to a 2.3 litre eight cylinder reciprocating engine.)

Various design solutions were considered in the lay-out of the engine rotor. The major challenge was to design the hinge joints of the four rotor segements. The initial prototype design of the hinge-pin was as per the patented design. It is labelled as item 15 in Figure 4.

This design did not provide insufficient sealing. The gas leaked from the high pressure combustion chamber through the hinge and into the centre void of the engine. This was revealed by a bench test of the

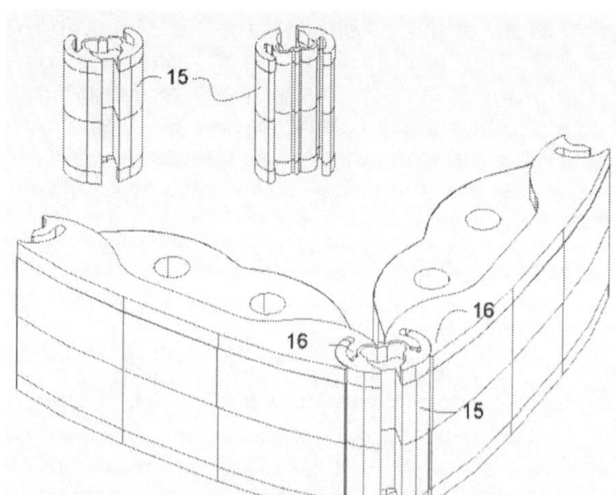

Figure 4. Hinged Rotor Segment Design

prototype which produced good pressure in some chambers, but one chamber could only achieve 70 psi. An attempt to run the engine was made but was unsuccessful.

The hinge seal design subsequently went through a number of iterations. The solution was to incorporate an additional seal which acted in a radial direction on the hinge-pin. This can be seen in Figure 3. The redesigned hinged segments were incorporated into the prototype and the engine was bench tested. The test demonstrated that a minimum of 120 psi could be maintained in each chamber which was considered sufficient for combustion to occur.

The design of other components of the rotor pack was relatively straightforward. These components can be seen in Figure 5 which depicts the rotor pack sitting outside the engine side plate. (The additional hinge seals are not visible in this picture.) The four sides of the deforming rhombus can be seen as a four-segment rotor. Roller-bearing wheels were mounted on either side of the rotor segment. The wheels roll around a cam whose profile ensures that the apex seals remain in contact with the engine stator throughout a complete revolution of the engine rotor. The racetrack shaped cam is mounted around the output shaft. The cam-plates are fixed to the engine side-plates in order to correctly locate the rotor pack inside the engine stator.

Since the stator profile is mathematically defined, the profile of the cam can also be determined mathematically. If the stator and cam profile are perfectly manufactured then, as the rotor rotates, the apex seal will stay in contact with the stator surface without moving in a radial direction. Manufacture of the cam-plates to a sufficient accuracy proved difficult and expensive. Ultimately 11 cam-plates were produced before two matching plates were accepted as suitable for their purpose.

The prototype engines did not incorporate any cooling or lubrication systems. Figure 2 shows the first stator design which did include some air-cooling fins. However, the fins added significantly to the cost of producing the stator. Also, this first stator did not have a sufficiently accurate profile and had to be remade. The later stators were made without the cooling gills as that was an unnecessary expense.

As can be seen in Figure 3 and 5, the output shaft passes through a hole at the center of the cam-plates and is supported by roller bearings mounted in the side-plates. As the engine rotates, each rotor segment bears on a lobe that is attached to the output shaft. The force produced by the high pressure combustion gases bears on the face of the rotor flank and is transferred to the output shaft via the lobe attached to the output shaft. The force is eccentric to the output shaft and creates a torque which rotates the shaft and provides output power.

Figure 5. Exposed Rotor Pack

The rotor assembly has side-seals which run along the length of each of the rotor segments and bear on the engine side-plates. Together with the hinge seals and apex seals, they create the four airtight chambers. The design of the seals in the hinges of the rotor segments in the first two prototypes was inadequate. Redesign of the seal around the hinge-pin, as described above, was successful and in a test on 26 February 2008 the engine achieved a sustained idle of 700 rpm.

Unfortunately, the engine would not accelerate away from idle revs. After the test, the engine was stripped and it was found that the hinge-pin joining two of the rotor segments had broken. The side plate of the engine was severely scored and so the lack of acceleration from idle revs was attributed to this additional friction of the rotor. The cause of the broken hinge-pin was attributed to the inaccuracy of the manufacture of the cam-plates and the small diameter of the hinge pin. The combination of poor tolerances of the cam-plates allowed the rotor segments to wobble as they rotated inducing sufficient loads to break the hinge-pin.

The hinge-pins were redesigned and manufactured, and more cam-plates were manufactured to a tighter tolerance. A bench test of the re-assembled prototype indicted that each chamber could achieve 120 psi. This was considered sufficient for combustion to occur. One test of this engine was attempted but the ignition timing equipment was problematic and the engine did not start. Before any further testing of this prototype could be attempted the inventor, Peter Szorenyi, sadly died. No further testing of the prototype has been attempted since that time. Instead, mathematical modelling has been undertaken by REDA and RMIT University.

MODELLING

The Royal Melbourne Institute of Technology (RMIT) University has conducted ideal mathematical modelling to determine the performance of a Szorenyi, a reciprocating, and a rotary engine [1]. The model assumes that the ideal Otto cycle is achieved in each of the three engines[2]. The model simulated the three engines using a nominal total volume of 125 cc and a compression ratio of 10:1. The modelling was a numerical analysis performed using an Excel Spreadsheet. Firstly, the geometry of the combustion chamber of each engine was established and the volume calculated for small increments in crankshaft angle. The fuel burn characteristics were modelled using a Wiebe function. The same efficiency factor and exponent were used in the Wiebe equations. Throughout the power stroke, the fraction of fuel burned was then calculated and, together with combustion chamber volume, the chamber pressure was determined. An area calculation of the P-V diagram provided the net work output. The ignition point was then varied and iterative calculations performend to determine the optimum timing for each engine. Because the model did not include the effects of friction, heat transfer or pumping loss it only predicted the ideal performance of the engines.

The modelling undertaken by RMIT University as described above indicated that, in the Szorenyi engine, the combustion chamber volume remained closer to its minimum value for longer than is the case for the reciprocating or Wankel engines. This characteristic improves the fuel burn in the engine and results in higher pressures in the chamber throughout the power stroke. The volume history of the three engines is shown in Figure 6.

---

[2] The Wankel engine has apexes 120^0 apart. So, in order that a chamber does not have the intake and exhaust ports open at the same time, the intake port should close no earlier than 60^0 BTDC and the exhaust port should open no later than 60^0 ATDC. If this were the case then the maximum chamber volume of the engine, which occurs at 90^0 before and after TDC, would not be utilised. In the ideal modelling conducted by RMIT, the ports of the Wankel engine are assumed to close and open at the point of maximum chamber volume (90^0 before and after TDC). In an actual engine this would mean that both ports would be open in the same chamber for 15^0 of rotation.The assumption therefore translates to a actual engine where judicious port location overcomes the adverse effect of the two ports being open concurrently and 100% volumetric efficiency is achieved. The assumed port opening and closing positions are 'ideal' but may favour the Wankel engine because they are not necessary in modelling the reciprocating and Szorenyi engines. More advanced modelling, such as computational fluid dynamics (CFD), is necessary in order to model the Wankel engine without making these assumptions.

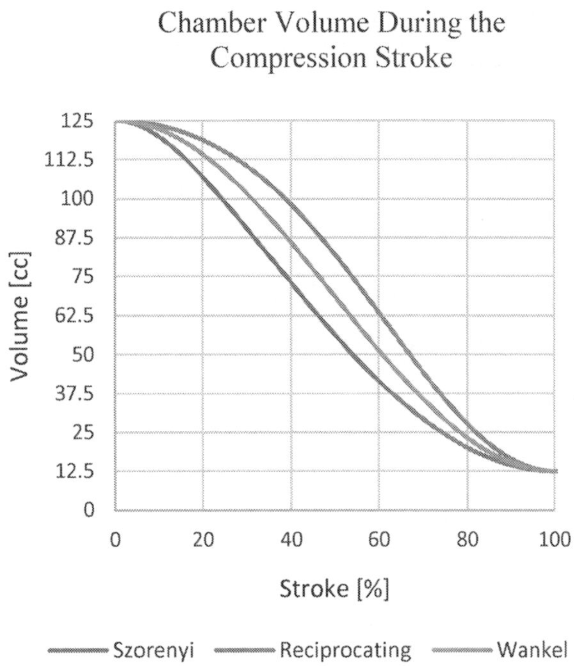

Figure 6. Chamber Volume during the Compression Stroke

The modelling also included the predicted fuel burn using a Wiebe function. The fuel fraction burnt is shown in Figure 7. Note that the crank angle and chamber position is different for each engine. The power stroke of the Szorenyi engine occurs in 90^0 of crankshaft rotation; the Wankel takes 90^0 rotation of the rotor which is 270^0 of the crankshaft rotation; and the reciprocating engine takes 180^0 of crankshaft rotation.

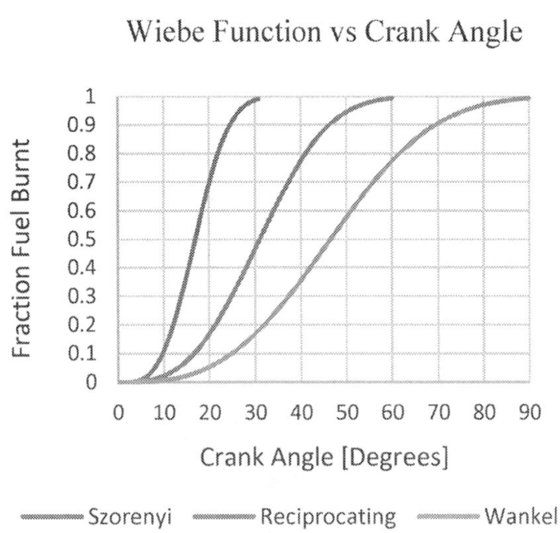

Figure 7. Wiebe Function vs Crankshaft Angle

By combining the fuel burn with the chamber volume, the net work done was then optimised for each engine by altering the fuel ignition timing. The result of this procedure is shown in Figure 8.

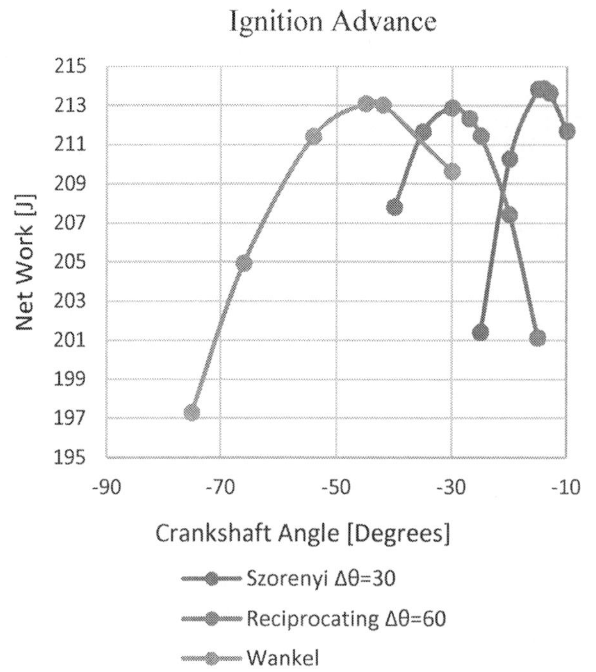

Figure 8. Ignition Advance of Crankshaft Angle

The resultant variation of pressure with volume in the combustion chamber is shown in Figure 9. The higher peak pressure of the Szorenyi engine results in more net work, higher specific fuel consumption and higher thermal efficiency.

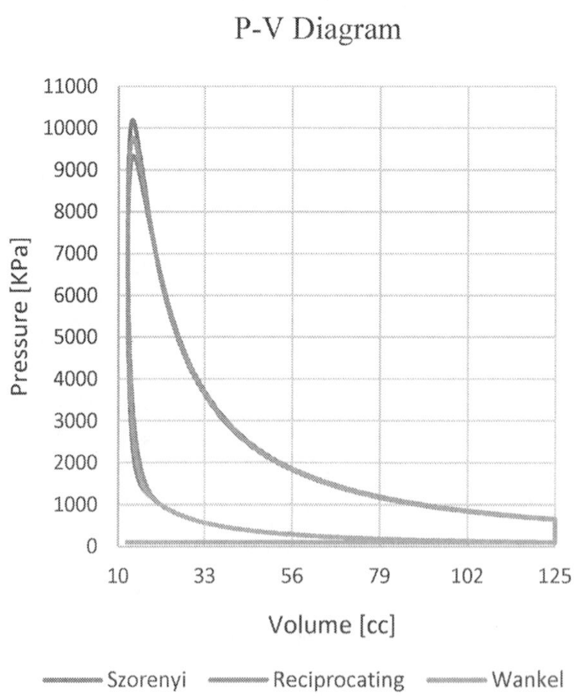

Figure 9. P-V Diagram

The numerical results of the RMIT University modelling are shown in Table 1. This modelling determined that the net work and thermal efficiency of the Szorenyi engine is 0.46% greater than the reciprocating engine and 0.38% greater than the Wankel.

Table 1. Comparative Engine Performance

Engine	**Net Work per Power Stroke (Joules)**	**Thermal Efficiency (η_f)**
Szorenyi	213.8593	0.474707
Wankel	213.0435	0.472900
Reciprocating	212.8758	0.472524

More sophisticated modelling is required to include better representation of the combustion process as well as kinematic effects such as friction, heat transfer and pumping losses. However, the ideal modelling performed by RMIT University indicates that the Szorenyi engine exceeds the power of the reciprocating and Wankel rotary engines.

ROTARY CHARACTERISTICS

Rotary engines are much more compact than reciprocating engine but have suffered in the past from poor fuel economy and more exhaust gas pollutants than reciprocating engines. The Wankel engine was initially developed and sold by Neckarsulm Motorenwerke AG (NSU). The engine was later developed by a variety of companies. General Motors, Mercedes-Benz and Citroen all produced Wankel engine powered cars in prototype or production form. However, the high fuel usage ended these endeavours. The Mazda Motor Corporation has produced the only commercially available automotive Wankel engine. The last design of the Wankel engine by Mazda was for their RX-8 sportscar. This RENESIS[3] engine incorporated a number of innovations which improved fuel economy and enabled the engine to meet the USA LEV-II (LEV) exhaust emissions standard.[2]

The Wankel engine has also been used in a range of light-aircraft and unmanned aerial vehicles (UAVs). Madza twin-rotor engines have been used for these applications as well as a range of other manufactures' single-rotor and twin-rotor engines. The Wankel engine is particularly suitable for light-aircraft and UAVs due to its high power density and low vibration.

Some of the significant advantages and disadvantages of the Wankel, Szorenyi and reciprocating engines are as follows:

a. Rotary engines benefit from not requiring the valves as in reciprocating engines to control the intake and exhaust phases of the gases. These valves restrict the flow and have dynamic limitations. The use of ports in rotary engines results in engines that are less complex and cheaper to manufacture.

---

[3] The Mazda RENESIS engine is a Wankel twin-rotor engine with a chamber displacement of 654 cc and a compression ration of 10:1. It has the intake and exhaust ports located in the side-plates of the engine. The earlier version of the engine (the 13B-REW) had a side intake port, but the exhaust port was 'peripheral', that is, it was located in the rotor housing.

b. The shape of the combustion chamber of the Szorenyi engine is very similar to that of the Wankel engine. Fully compressed it resembles a cresent-moon and fully expanded a half-moon shape. The shape of the combustion chamber is not conducive to good combustion with a homogenous charge. However, the rotary engine is well suited to improve combustion by using stratified-charge techniques [3]. In this technique, fuel is injected at the base of the spark-plug and the motion of the chamber passing the stationary injector (located in the stator housing) spreads the ignition point within the chamber. This enhances fuel-air mixing and improves combustion. The method is suitable for using heavy fuels in rotary engines and therefore makes the rotary engine more attractive for use in aircraft and UAVs[4].

c. The Szorenyi (and Wankel) engine can also be run on gasoline, liquid-petroleum gas, or hydrogen. The use of hydrogen is possible because the inlet and exhaust ports are well separated, and the intake port is located in the coolest area of the engine. Thus pre-heating of the hydrogen, which can lead to back-firing, can be avoided. Diesel fuel can also be used. However, the geometry of the Szorenyi (and Wankel) engines precludes compression ratios above about 15:1, and so a pre-compression phase would be required.

d. Modern Wankel engine designs incorporate a rotor cooling system. However, there is very little space inside the engine stator available to do this. The geometry of the Szorenyi engine provides a significant volume of vacant space in which to incorporate a rotor cooling system. A variety of rotor designs have been developed for the Szorenyi engine and some designs reveal that more than a quarter of the stator internal volume is available to be used for this purpose.

e. Both the Wankel and Szorenyi engine have an array of seals, friction surfaces and roller bearings. In addition to apex seals and side-seals, the Szorenyi requires hinge seals. However, the Szorenyi engine does not require the cut-off seals as used by Mazda in their RENESIS engine.[2] These cut-off seals are necessary with the intake and exhaust ports located in the side-plates of the engine. The seals are to prevent leakage of the gases between the intake and exhaust ports which is detrimental to exhaust gas composition and volumetric efficiency of the engine.

f. The power density of the two engines have been modelled using the net power estimates obtained from the RMIT modelling and an estimate of the package size of the Szorenyi and Wankel engines. For equivalent power output, the overall engine package size of a single rotor Wankel engine is 2% smaller than the Szorenyi engine[5]. The stator of the Wankel is about 11% smaller than the Szorenyi stator.

---

[4] Heavy fuels are the the kerosene-type fuels such as Jet-A, Jet-A1, JP-5, JP-8. These heavy fuels are used for aviation engines. JP-5 and JP-8 are used by the military. The ability of rotary engines to use heavy fuels in aircraft and UAVs is an advantage in the logistics of deployed military operations.

[5] The assessment of package size made here is based on the assumption in the RMIT ideal modelling that 100% volumetric efficiency is achieved at the maximum chamber volume (90^0 BTDC for the intake stroke). This assumption is valid for the Mazda RENESIS engine because the side-ports of that engine ensures that there is no overlap of the ports. However, earlier version of Mazda's Wankel engine had side-port intake and peripheral exhaust ports which resulted in considerable port overlap. For a Wankel engine with peripheral ports to achieve no overlap, those ports would need to be positioned nominally at 60^0 from TDC. The maximum chamber volume occurs at 90^0 from TDC, so this would result in a smaller combustion chamber volume. To restore the chamber volume, the engine dimensions would need to be scaled up. When this is done, the package size of a single-rotor peripheral-port Wankel engine becomes about 30% larger than an equivalent Szorenyi engine. Then the power density of a Szorenyi engine is about 30% greater than the single-rotor peripheral-port Wankel engine for all rotor speeds up to the Wankel's rotor rev limit of 3,000 rpm, and much greater above that.

This is due to the assumed 100% volumetric efficiency of the Wankel and the eccentric motion of its rotor. However, a single-rotor Wankel engine requires a balancing rotor. This is attached to the crankshaft outside the engine, and the additional space occupied by the balancing rotor is about 13% of the stator package size. Due to the similar power output of the two rotary engines, the power density is therefore essentially the same. Of course, the power density of both rotary engines greatly exceed that of any reciprocating engine.

g. A major disadvantage of the Wankel engine is the limited rotational speed of the rotor. The revs are limited because the centroid of the rotor remains at a constant radius from the centreline of the crankshaft and causes excessive bending loads on the crankshaft. At high crankshaft speeds, the bending of the crankshaft mis-aligns the rotor and can cause the side-seal or apex-seal to bottom out and make solid contact with the side-plate or stator. Ultimately, this can result in catastrophic failure of the engine. The crankshaft bending cannot be alievated by increasing its diameter as this also increases the size of the rotor as well as its radius of rotation. Because of the crankshaft bending, commercially produced Wankel engines are typically rev limited to 9,000 rpm of the crankshaft (3,000 rpm of the rotor). The Szorenyi engine does not have a simlar limitation because its rotor is symmetrical and so there is no eccentric load on the crankshaft caused by the rotor. If the Szorenyi engine rotor can rev much higher than 3,000 rpm, its power density will greatly exceed that of the Wankel engine. It is therefore an objective in the development of the Szorenyi engine to establish the revving potential of the engine by mathematical modelling and testing, and prove its power density advantage over the Wankel engine.

h. The Wankel engine is becoming more popular in light aircraft and UAVs mainly due to its high power density and low vibration, despite its slightly higher fuel economy than reciprocating engines. Early Wankel engine designs utilised peripheral ports and a mixture of peripheral and side ports, but these configuration resulted in over-lapping of the intake and exhaust port timing and consequently compromised volumetric efficiency, fuel economy and exhaust gas emissions. The most recent Wankel design from Mazda, the RENESIS engine, has the exhaust port located in the side-plate.[2] As a result there is no over-lapping of the ports[6]. However, the use of side-ports means that Wankel engines with more than two rotors require thicker side-plates to accommodate the port ducts. This compromises the power density of the Wankel engine for applications in general aviation. The Szorenyi engine can utilise peripheral ports without having over-lapping ports[7]. Consequently, Szorenyi engines with three or more rotors will have higher power density than similar Wankel engines. The Szorenyi engine also has an added advantage over the high-revving Wankel in aircraft and UAV applications because it does not require a gear to reduce the crankshaft revs to achieve the lower propeller revs necessary for efficient operation. This further improves the Szorenyi engines power density for aircraft and UAV applications. Doing away with the reduction gear results in a more compact and lighter engine installation, and also avoids mechanical complexity and a potential source of failure.

---

[6] The triangular shape of the Wankel rotor means that, with peripheral ports, to avoid port over-lap the ports should be positioned 120^0 apart. But this means that the exhaust port would open at 60^0 ATDC, that is, two-thirds of the way through the power stroke. Positioning the exhaust port in the side-plate avoids port over-lap, allows the opening time of the port to be retarded, and results in better volumetric efficiency.

[7] The rotor has four equal-size segments and so the apex seals are 90^0 of rotation apart. Since each of the four engine strokes take 90^0 of rotation, the ports can be located 90^0 apart with no overlap.

APPLICATIONS

The Szorenyi Rotary Engine is suitable to be used for all applications where a reciprocating or Wankel engine are currently used. Small or large versions of the engine could be made.

The Szorenyi engine is particularly suited to aircraft and UAV applications. The high power density of the Szorenyi engine translates to larger payloads or greater range or endurance than is possible with current Wankel and reciprocating engines. Multi-rotor Szorenyi engines can be more easily created than multi-rotor Wankel engines because they avoid the complex side port arrangement. They also benefit from not needing the reduction gear used for the Wankel engine where high power is required and large diameter propellers.

FUTURE DEVELOPMENT

REDA is confident that the Szorenyi Rotary Engine has the potential to replace the reciprocating and the Wankel engine wherever they are currently used. However, further prototype testing and mathematical modelling is required to support these claims.

The current prototype engine is ready for limited testing. However, the prototype engine does not have a cooling or lubrication system so any testing will not provide indicative performance of the Szorenyi engine. Therefore more sophisticated mathematical modelling of the engine will be undertaken. This modelling needs to include the dynamics of the combustion process, and the effects of friction and heat transfer. The modelling also needs to determine the rev capability of the rotor and hence demonstrate the superior power and power density of the Szorenyi engine.

SUMMARY AND CONCLUSIONS

The Rotary Engine Development Agency (REDA) in Melbourne, Australia has developed a new configuration internal combustion rotary engine. Known as the Szorenyi Rotary Engine, it has four combustion chambers and offers a significant advantage over the Wankel rotary engine and reciprocating engine in a range of applications. The engine concept has been awarded a U.S. Patent (Number 6,718,938 B2). A prototype engine has been designed and produced by REDA and passed a proof-of-concept test. Modelling of the engine has been conducted by RMIT University indicates that the Szorenyi engine is more thermally efficient than the reciprocating and Wankel engine. Additional modelling has shown that the Szorenyi engine has approximately the same power density as the Wankel engine when operated up to the rev limit of the Wankel engine rotor (typically 3,000 rpm) and much greater power density above that speed. Modelling and testing is required to determine the rev limit of the Szorenyi engine and hence its potential power density.

Although very similar to the Wankel engine, the Szorenyi engine has advantages over the Wankel in regard to higher power density due to the higher revving rotor; a larger space inside the rotor for additional internal cooling of the rotor; four power strokes per crankshaft revolution against the Wankel's single power stroke; no need for a balancing flywheel in a single rotor engine; no need for a reduction gear in aircraft and UAVs with large propellers; and, suitability for multi-rotor engines. The Szorenyi engine could therefore be used in any application where the reciprocating and Wankel engines are currently used but provide these aditional benefits.

REFERENCES

1. L. F. Espinosa MSc and Prof P. Lappas PhD, *Mathematical Modelling Comparison of a Reciprocating, a Szorenyi Rotary and a Wankel Rotary Engine,* 2017 (son to be published), School of Engineering, RMIT University, Melbourne, Australia.
2. Masaki Ohkubo, Seiji Tashima, Titsuharu Shimuzu, Suguru Fuse and Hiroshi Ebino, *Developed Technologies of the New Rotary Engine (RENESIS)*, 2004, Mazda Motor Corporation, SAE Paper 2004-01-1790.
3. Chol-Bum M. Kweon, *A Review of Rotary Engine Combustion Technologies,* 2011, Army Technology Directorate, Army Research Laboratory, Aberdeen Proving Ground, MD, ARL-TR-5546.

GASOLINE ENGINES

Advanced gasoline engine technology for minimising fuel consumption

P Freeland, G Jones, M Feasey
Cosworth Limited, UK

S Chen, W Huang
Haitec / CEC, Taiwan

R Kaiser, M El-Kassam
Pierburg Rheinmetall Automotive, Germany

ABSTRACT

The paper outlines the fuel consumption benefits available from combining the latest state-of-the-art low-friction and de-throttling technologies in a concept demonstrator engine. Most notably this work explored the possibility to combine variable valve lift and cylinder deactivation for the first time in a production feasible engine concept, outlining the benefits available by doing so.

The paper demonstrates the potential of a production-feasible 2.0 litre Central GDI engine to deliver in excess of 105 kW/litre, and yet to achieve less than 300 g/kW.h BSFC at 2000/2.0 bar BMEP with λ=1.0 Fuelling. In this way, the authors have achieved an efficiency value competitive with that of diesel engines, whilst retaining the cleaner combustion and simpler aftertreatment processes inherent with positive ignition gasoline engines, utilizing pre-mixed flame stoichiometric combustion.

DEFINITIONS/ABBREVIATIONS

BEV	Battery electric vehicle
BMEP	Brake mean effective pressure
BSFC	Brake specific fuel consumption
CAN	Communications area network
CDA	Cylinder deactivation
CVVL	Continuously variable valve lift
ECU	Electronic control unit
EIVC	Early inlet valve closing
EMS	Engine management system
EU	European union
GDI	Gasoline direct injection
GPF	Gasoline particulate filter
HLA	Hydraulic lash-adjuster
LNV	Lowest normalized value
NEDC	New European drive cycle
NOx	Oxides of Nitrogen
NVH	Noise, vibration and harshness
PHEV	Plug-in hybrid electric vehicle
PMEP	Pumping mean effective pressure
RDE	Real-world driving emissions
RFF	Roller finger follower

SD	Standard deviation
TDC-E	Top dead centre - gas exchange
VCU	Valve control unit
V_{max}	Maximum possible vehicle speed
VVT	Variable valve timing
λ	Lambda

1. INTRODUCTION

In terms of impact on human civilization and culture, the internal combustion engine has been one of the most influential and beneficial machines to have been developed. The mobility it has afforded has facilitated levels of trade and communication that were unimaginable even 100 years ago, providing a significant contribution to the improvements in the standard of living now enjoyed across the world.

This success has led to extremely wide-spread use, which has, in turn, brought impacts on air quality from particulates and NO_X emissions in densely populated areas, and increased the reliance on fossil fuels. These aspects need to be managed responsibly to ensure that this quality of life can be maintained with minimal impact on human health and the environment. To this end, tremendous improvements have been made in system efficiency, the cleanliness of combustion processes and the effectiveness of exhaust aftertreatment systems. However, ever more detailed understanding and optimisation of powertrain systems is essential to keep pace with global targets on air quality, greenhouse gas effects and the management of natural resources.

Legislation has driven a lot of this development, but regional differences in legislative focus have forced diverse and sometimes conflicting objectives. For example, a strong focus on CO_2 emissions in Europe led for a while to an increased reliance on diesel engines to help meet fleet average targets. However, more recent concerns over air quality, and the cost and efficiency implications that RDE legislation measures will impose for these vehicles mean that the contribution offered by diesel engines towards lower fleet-average CO_2 values will be reduced from now on. This fact alone means that the basic efficiency of gasoline-based propulsion units will need to be improved to match or exceed the CO_2 levels previously offered only by diesel engines.

Several approaches are being explored to achieve this. A strong favourite in the EU and Chinese markets is the electrification of vehicle powertrains, which offers direct benefits from energy recovery and torque smoothing. Further potential benefits with PHEV and BEV technology are available for short journeys, providing that a carbon neutral grid supply is available [1]. However, such technologies are not without significant environmental impact of their own, and on a practical basis the benefits available are restricted by the weight penalty, efficiency and cost implications of electric energy systems. These constraints become amplified if significant electric-only range is sought.

Downsizing has been used to very good effect over the past ten years to reduce fuel consumption via load-point shifting. Up to 24% improvements in vehicle fuel consumption have been reported [2]. This trend is seen to continue further as boosting systems develop, and hybrid systems are developed to augment torque delivery and transient response. However, high load fuel consumption is usually compromised by the need for high levels of boosting to retain engine power, requiring additional technologies to maintain driving flexibility [3].

In practice, to achieve the best vehicle fuel consumption and emissions characteristics under a variety of driving situations, the efficiency of the powertrain needs to be optimised over the whole operating range. To achieve this without restricting the flexibility of use, aspects of all the approaches above need to be combined synergistically, such that the limitations of each approach are minimised. In this respect, an approach which offers a good degree of downsizing without excessive boosting, combined with de-throttling [4] and low-friction technologies was considered by the authors to offer the best solution. An investigation was undertaken to demonstrate the potential of such an approach and as part of this study, a detailed understanding of de-throttling methods and potential was sought, and the work presented here was conducted.

2. MOTIVATION

The studies to which this paper relates were conducted as part of a concept engine programme constructed to evaluate appropriate technologies for meeting vehicle certification requirements for the Chinese passenger car market into and beyond 2020.

The predominant aim was to prove a powertrain which, in conjunction with other vehicle technologies could achieve better than 5.5 l/100km fuel consumption for a medium-sized passenger vehicle, over an NEDC drive cycle.

The authors and their colleagues worked closely together to specify, design, procure and validate bespoke engine hardware to support this aim. This paper focuses on the de-throttling technologies evaluated, and the resulting specific fuel consumption achieved.

3. ENGINE DETAILS

The concept engine was designed as a four cylinder, sixteen valve gasoline engine with central direct injection. The primary objective for this engine design was to achieve the lowest fuel consumption possible over the broadest range of speeds and loads, whilst achieving state-of-the-art specific performance levels and 50% downsizing potential (ie, specific power output greater than 105 kW/litre).

The basic dimensions and specifications of this engine are given in Table 1.

Table 1. Fundamental Dimensions and specifications selected for the concept engine.

Cylinder configuration	In-Line 4
Stroke	92.0 mm
Bore	83.0 mm
Swept volume	2.0 litres
Compression ratio	10:1
Fuelling	Central GDI
Target mixture	λ=1.0, up to hardware temp limit
Injector	Bosch HDEV5 multi-hole
Number of valves per cylinder	4
Inlet valve lift	0.0-10.0 mm
Inlet cam phaser range	70° crankshaft
Exhaust valve lift	10.0 mm
Exhaust cam phaser range	50° crankshaft
Emissions aftertreatment	3-way catalyst (+protection for GPF)

Several approaches to reduce frictional losses were incorporated in the engine design, including managed oil-circuit demand in conjunction with a variable displacement oil pump and optimised crankshaft and cylinder block geometry to reduce the side-thrust forces of the pistons under firing loads. In addition, plasma-sprayed cylinder bore technology from Oerlikon Metco was specified, to provide improved heat-transfer characteristics to enable the highest compression ratio to be selected. This bore technology provides a microscopically porous structure with excellent inherent oil retention properties, which enable a much smoother bore-surface to be used thereby also reducing piston-ring frictional losses.

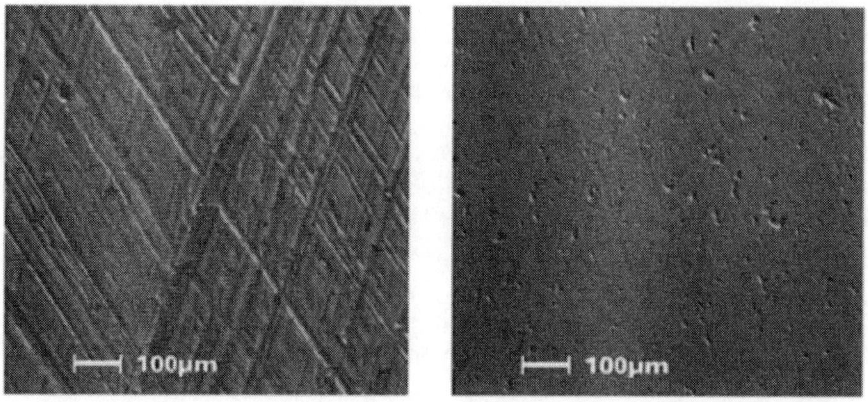

Figure 1. Comparison of the bore surface texture permissible between a conventional honed cylinder bore (left) and a super-finished plasma-sprayed bore (right) [picture courtesy of Oerlikon Metco].

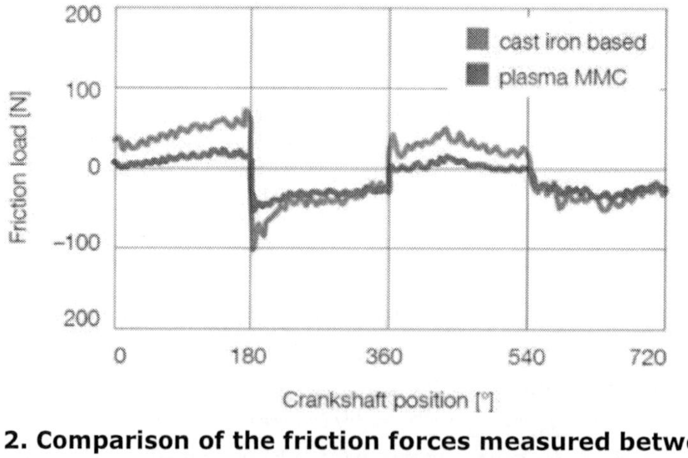

Figure 2. Comparison of the friction forces measured between a conventional honed cylinder bore and a super-finished plasma-sprayed bore [picture courtesy of Oerlikon Metco].

The mechanism used to provide control of the inlet valve lift and duration was provided by Pierburg [5]. The mechanical design and operation of this system is described in more detail below. In summary, this system allowed peak valve lift to be varied from zero mm to full lift according to the angular position of a control shaft which governed the pivoting motion of an intermediate rocker.

For this engine, the valve control mechanism was designed such that the valve lift for cylinders 2 & 3 could be reduced to zero, to enable cylinder deactivation for these cylinders. The profiles of peak valve lift vs control shaft angle for each cylinder are shown in figure 3.

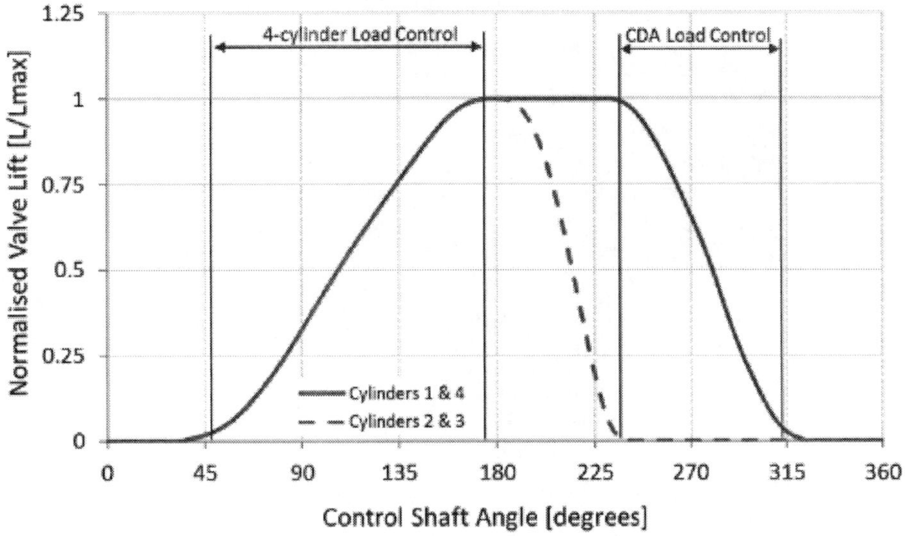

Figure 3. Pierburg UpValve valve lift vs control shaft angle, showing the regions used for load control in both 4-cylinder mode and CDA made.

The control shaft cam profile was effectively divided into two halves, the first 180° providing valve lift control for all cylinders, whilst the second 180° of shaft rotation duplicated the valve lift profile for two cylinders (1 and 4) but reduced peak lift for the remaining cylinders (2 and 3) to zero. Thus, cylinder deactivation (CDA) or "half engine" operation could be provided and employed in conjunction with variable intake valve lift and duration. To this end, cylinders 2 and 3 were also fitted with collapsible HLAs to disable the exhaust valves, switched via oil pressure acting on a locking pin.

This ability to provide both early inlet valve closing (EIVC) de-throttling and cylinder deactivation from the same hardware made this system an excellent development tool for evaluating the relative merits and synergies of these two approaches concurrently on the same engine. Reciprocally, because this system was fundamentally implementation ready at the time of testing, this demonstrator engine provided the ideal tool to validate design options for mass-production [6].

4. FULLY VARIABLE VALVETRAIN DETAILS

The main components of the variable intake valve lift system are shown in figure 4:

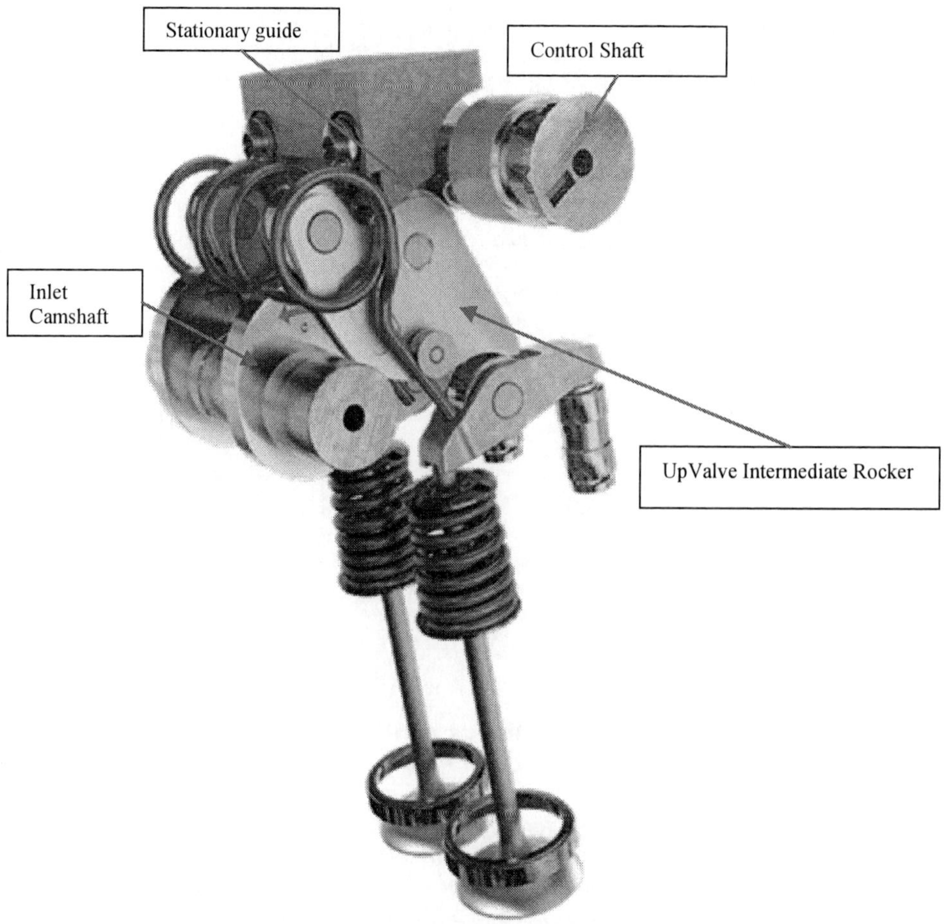

Figure 4. The Pierburg UpValve system.

Design and kinematics

The UpValve system selected enabled infinitely variable control of valve lift from zero to full lift. The valve opening was performed by a low-friction roller finger follower (RFF) with hydraulic lash adjuster (HLA) in a very conventional manner. However, to achieve the variations in valve lift and duration, an intermediate rocker was interposed between the camshaft and the RFF, the pivot point for which could be varied by sliding around a profiled stationary guide, controlled by the profile of an eccentric cam profile ground on the control shaft.

The variation in valve lift achieved is illustrated in figure 5.

In (a) the inlet cam contact is shown on the base circle, and the inlet valve is closed. The intermediate UpValve rocker acts on the conventional RFF via a profiled surface (which comprises a non-lifting section concentric with the pivot roller, and a lifting section which is not concentric to the roller and therefore provides deflection into the RFF).

In (b) the control shaft <S> has been rotated to the maximum lift position, moving the UpValve roller <RS> to the right of its lateral range. In this position, as the Inlet camshaft <N> rotates and deflects roller <RN>, the pivoting motion of the UpValve rocker <H> is sufficient to engage the lifting portion of the UpValve rocker profiled face <A> which depresses the RFF roller <R> thereby opening the inlet valve.

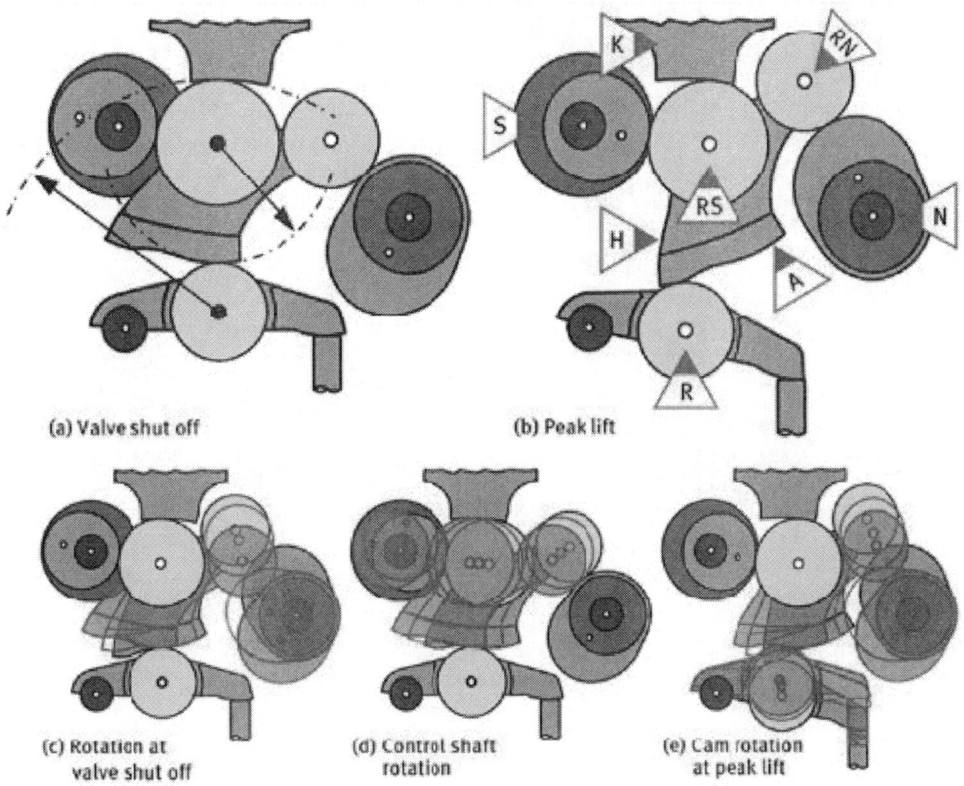

Figure 5. Layout and function of the UpValve system.

To achieve differing valve lifts and durations, the control shaft <S> is rotated to displace the UpValve roller <RS> into a different position against the stationary guide <K>. This alters the amount that the UpValve rocker pivots about <RS> and degree to which the lift portion of the surface <A> depresses the RFF roller <R>. In the zero-lift position (c) the full cam deflection only pivots the UpValve rocker within the zero-lift range of the contoured lift surface <A>. As the control shaft is rotated (d) the UpValve rocker is displaced to the right, the contact point between the UpValve roller <RS> and the stationary guide <K> is shifted around the circular profile of <K> and the roller <RN> is brought further within the range of the cam lobe profile, such that the pivoting action of the UpValve rocker now engages the lift portion of the profiled surface <A> against the RFF roller <R>. The design of the control shaft profile and the concentricity of the stationary guide surface to the RFF roller centre mean that the valve lift is infinitely variable and can be adjusted continuously during operation.

In figure 6, curve <H> depicts the required full lift. The lift of the cam follower is depicted <S>. The opening parts of <S> and <H> determine the profiled surface of the rocker lobe. Therefore, the closing flank of <H> provides the second half of the cam shape, as the rocker lobe profile is traversed again in rearward direction.

Because each valve has its own and independent UpValve rocker, individual cam profiles, and control shaft and working lobe were possible. Thus, this system allowed options for individual valve opening requirements to be exploited - for example facilitating varying charge motion characteristics at low-load through differential lift of the valves per cylinder.

This individuality of valve control between different cylinders also gave the possibility to provide cylinder deactivation. This was achieved by utilising half of the control shaft operating angle to provide valve lift load-control for all cylinders together in four-cylinder mode, whilst for the other half of the control shaft operating angle the control profiles gave zero lift for two cylinders (2 & 3) and varying valve lift load-control for the remaining two only. This study is the first in which both variable valve lift and cylinder deactivation are employed in a production feasible engine.

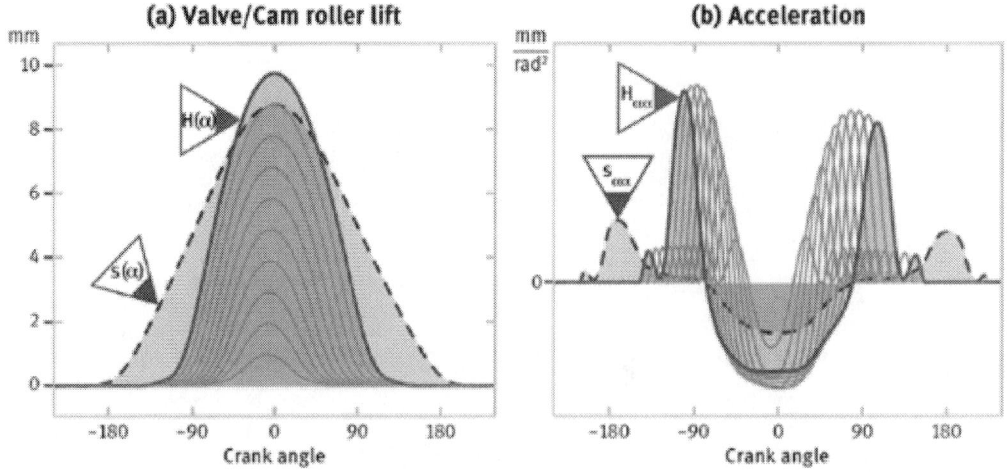

Figure 6. Valve kinematics.

5. RESULTS

Engine installation on dynamometer testbed

Several sets of hardware were procured and built into operating engines. These were fitted to a steady-state dynamometer test cell capable of measuring engine speed and torque, fuel flow rate, key temperatures and pressures.

The engine installation in one of Cosworth's dynamometer testcells in its Northampton engineering facility is shown in figure 7.

Figure 7. The concept engine installed in the Cosworth Northampton test facility.

Dynamic cylinder pressure data were monitored and recorded to provide indicated mean-effective pressure values, heat-release and combustion quality information, and peak cylinder pressure and knock monitoring. The engine control systems were operated with a development MED17 EMS provided by Bosch, capable of controlling to a torque model and providing steady-state control functions for all the engine systems employed.

Once the basic mechanical functionality had been validated, a programme of performance and fuel economy optimisation was conducted, in which the generic trends of valve timing and inlet valve lift, together with injection timing and rail pressure were evaluated within different regions of the speed / load map. With this basic information and understanding, at each speed and load point tested each variable was swept locally around the generic optima to fine-tune the engine operation, together with spark timing. Using this approach, the results shown below were obtained.

Engine Performance

The engine was tested with a selection of turbochargers to find the configuration which gave the best torque curve shape. The final assembly chosen was based on an MHI TD04 assembly, with modified housings and featuring a 56mm compressor and a 47mm turbine wheel.

With this configuration, Peak power potential exceeded the target value of 210 kW, and the target Torque value of 360 Nm could be achieved between 1500 rev/min to 4500 rev/min.

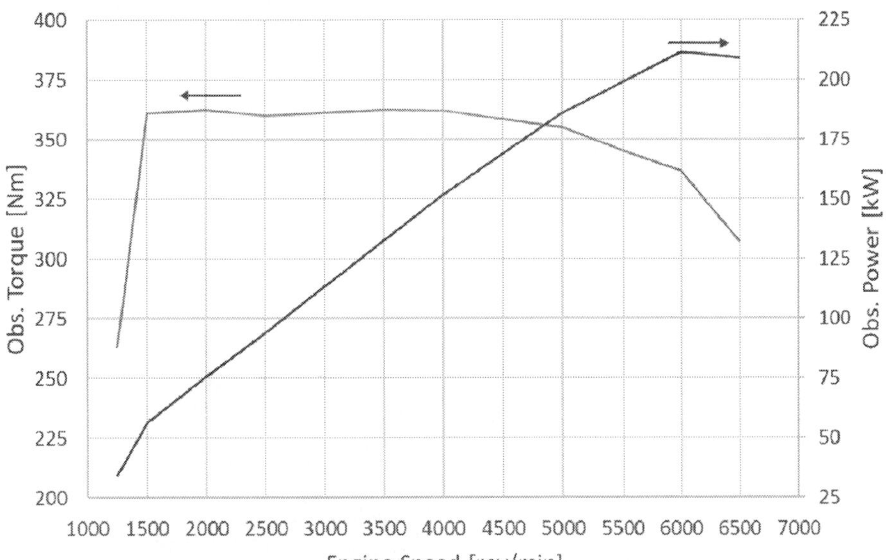

Figure 8. Performance of the concept engine.

With this hardware and performance established, attention was then focused on the part load fuel consumption study.

Valve lift and timing investigations

To augment the inherent de-throttling afforded by this engine's downsizing potential, the fully variable inlet valve control coupled with the variable phasing on both inlet and exhaust camshafts gave the ability to reduce throttling losses via both trapped residual gas fraction (through valve overlap) and via fresh charge trapping efficiency (through reduced valve duration) simultaneously.

Figure 9 illustrates the range of valve opening and timing facilitated by this system.

Although shown as a series of discreet curves in the fully advanced position for illustration purposes, the inlet valve lift (and hence duration) was infinitely variable between zero and full lift throughout the cam phaser range.

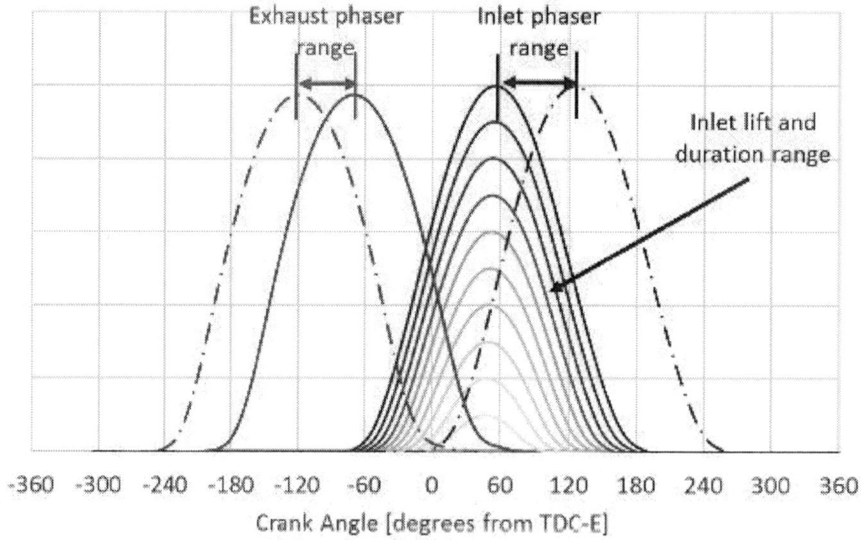

Figure 9. Valve lift and timing provided by the Pierburg UpValve system and cam phasers.

A comprehensive study was conducted to investigate the limits associated with these effects, the best reduction in throttling losses, and the minimum BSFC that could be achieved at different speed / load conditions. An example of this optimisation is illustrated in figure 10, where the BSFC response to inlet valve lift and valve overlap is shown for the 1000 rev/min, 1.0 bar BMEP operating point.

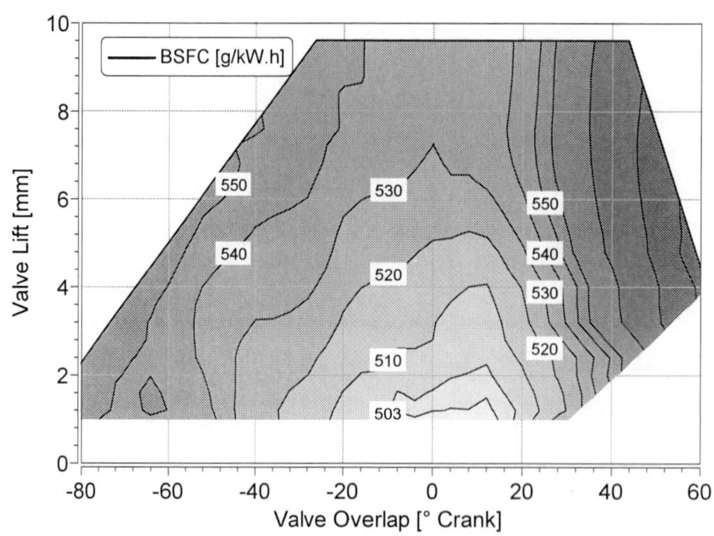

Figure 10. Engine BSFC as a function of inlet valve lift and valve overlap at 1000 rev/min, 1.0 bar BMEP (operating in four-cylinder mode).

With very low valve duration settings, although plenum throttling could be practically eliminated, at very low valve lifts there were still significant irreversible flow losses across the inlet valve as the charge air entered the cylinder. Thus, intake-induced pumping work could not be entirely eliminated and under such situations, increased residual gas fractions could be used to good effect to minimise gas exchange losses up to the point at which combustion quality became compromised - an unacceptable limit being denoted from statistics, ie, where the lowest normalized value (LNV) of IMEP fell below approximately 65%, or the standard deviation (SD) of IMEP became greater than 0.15 bar. In this way, the limiting factor was generally the combustion stability achievable with high residual gas fractions.

The specific fuel consumption values measured with this approach are shown in the BSFC map in figure 11. Superimposed on this plot are contours of constant power, to indicate the relevance of the de-throttling benefits under various driving conditions.

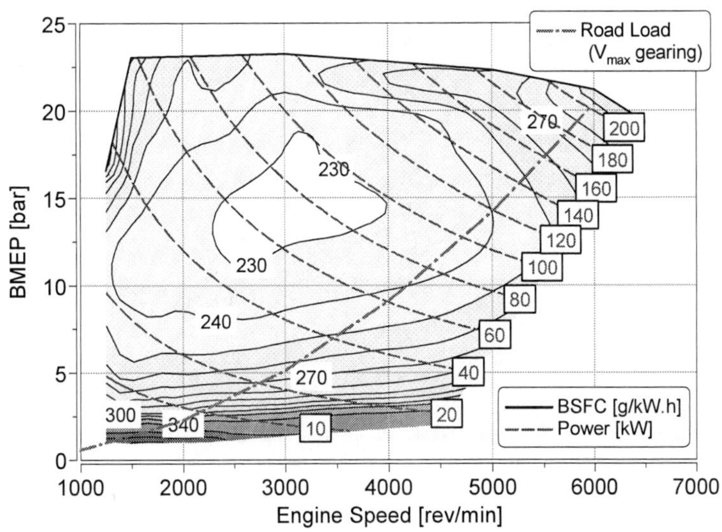

Figure 11. BSFC contours measured over the engine speed-load operating range, with contours of constant power overlaid (operating in four-cylinder mode).

The UpValve system enabled a very broad area of low BSFC to be maintained. 90% of the operating area had a BSFC value less than 300 g/kW/h.

With gearing to enable Vmax to be reached (ie, the road-load denoted by the pink dotted line) the engine could operate with less than 270 g/kW.h (31% brake thermal efficiency) from approximately 20 kW (at 2700 rev/min) up to 180 kW (at 5500 rev/min).

With overdrive gearing under higher-load conditions, significantly better than 240 g/kW.h (35% brake thermal efficiency) could be achieved within a significant range of engine speeds.

Cylinder deactivation investigations

At the time of testing, the engine management functions to switch into and out of CDA mode had not been fully developed for this particular approach. Functionality existed to switch the hydraulic circuit for the exhaust valve deactivation via the collapsible HLAs, and to rotate the UpValve control shaft for inlet valve deactivation independently, but coordinated deactivation was not possible. Therefore, a detailed study of the effects of cylinder trapping with exhaust first, or inlet first deactivation strategies were not possible at this stage. However, the overall effects on pumping losses and fuel consumption from disabling two cylinders and invoking "half engine" mode were assessed from steady-state dynamometer testing. Complete engine mapping with cylinder deactivation has not been conducted to date, and more development is required to establish the precise boundaries to where this system can be operated. Fuel consumption data was recorded for six part-load points to give an indication of the overall benefits available.

A comparison of engine BSFC at these points, between the 4-cylinder CVVL case and two-cylinder CDA operation was made, and the results of this is shown in figures 12:

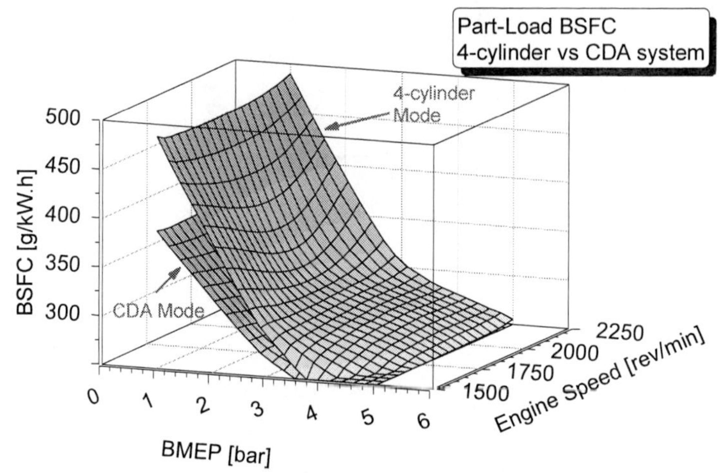

Figure 12. BSFC contours derived from part-load mini-map measurements, with optimized CVVL settings in 4-cylinder mode and CDA mode.

The precise range of engine speeds and loads over which cylinder deactivation could be employed had not been established at the time of testing, but the area covered by the mini-map points tested was considered representative of the typical usable area, and served to illustrate well the degree of benefit that CDA could offer over that given by the variable valve lift alone.

The percentage difference in BSFC between CDA operation and CVVL operation, over the range of speed / load cases measured was calculated from the contours derived, and is shown in figure 13.

Over the operating range evaluated, CDA gave between 3% and 20% better BSFC than was achieved with the CVVL system alone. The magnitude of the benefit was greatest at the lowest loads, reducing significantly as engine load increased. This reflected the degree to which pumping losses were incurred still with the CVVL system at low loads due to the irreversible flow losses across the inlet valves.

Because of this, the actual fuel consumption / CO_2 benefit given in a vehicle application would strongly depend on the gearing strategy, the vehicle weight, and the driving style being adopted.

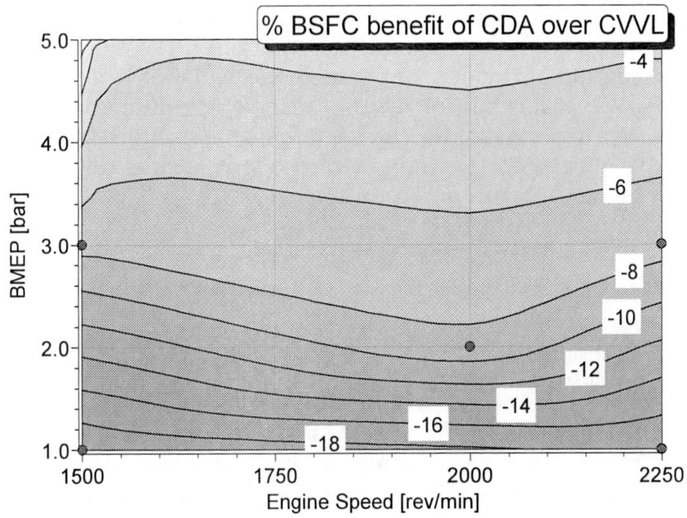

Figure 13. The percentage difference in BSFC offered by CDA operation, over optimum 4-cylinder mode operation.

6. LEGISLATED EMISSIONS

At the time of writing, injector spray-targetting work for was still underway, and detailed emissions optimisation testing had not been completed. However, in both four-cylinder mode and in CDA mode, the engine's basic combustion was good, and HC and NO_X concentrations in the raw feedgas were well within expected limits.

With the deployment of CDA, at any given engine speed and torque, the cylinder charge for the operating cylinders was approximately twice that of operation in four cylinder mode. Therefore there was a slight reduction in feedgas HC concentrations, and a slight increase in NO_X concentrations at the lighter-load operating points. This effect may have some minor impact on the ideal catalyst loading and constitution, but the differences seen were well within the dynamic range of normal driving styles and it is not expected that there would be any significant impact on tailpipe emissions during vehicle operation.

Once the CDA switching strategy has been developed, further work is planned to investigate the emissions penalty of switching into and out of CDA mode, and the maximum time that CDA mode could be used before an unacceptable emissions penalty was incurred. It is hoped that these results will be included in future publications.

7. SUMMARY/CONCLUSIONS

A concept demonstrator engine has been developed that is capable of delivering a specific power output exceeding 105kW/litre, and giving in excess of 23 bar BMEP from 1500 rev/min up to 4500 rev/min.

In addition to the downsizing capability offered by this specific power output, this engine demonstrated significant specific fuel consumption benefits from several low friction and de-throttling approaches.

The combination of approaches evaluated are all production feasible in high volume and when combined together enable specific fuel consumption values that rival that of diesel engines, whilst maintaining the inherently cleaner combustion and emissions control processes of premixed stoichiometric combustion processes. This technology combination provides an ideal combination of characteristics for efficient, flexible powertrains to deliver the best vehicle fuel consumption, CO_2 and emissions levels. In such ways, the full benefits of the internal combustion engine can continue to be utilized to the benefit of modern society with minimal negative impact to human health and the environment.

8. REFERENCES

1. J. Meisel, An analytical Foundation for the Toyota Prius THS-II Powertrain with a Comparison to a Strong Parallel Hybrid-Electric Powertrain. *SAE 2006-01-0666, Advanced Hybrid Powetrains 2006*
2. V. Korte; J. Hall; J Taylor; T. Raatz; et al. Anforderungen an den Ladungswechchsel und die Audladung bei extremen Downsizing, *4. MTZ-Fachtagung "Ladungswechsel im Verbrennungsmotor" 2011*
3. P. Freeland; J. Taylor; D. Oudenijeweme, P. Stansfield; et al. Advanced technologies for Downsized direct injection gasoline engines. *Thiesel 2012 Conference on Thermo- and Fluid Dynamic Processes in Direct Injection Engines*
4. Habermann, K. et al.: Gemischt-sequenzielle Aufladung am Motor mit getrennten Auslassfluten und vollvariablem Ventiltrieb *9. MTZ- Fachtagung Ladungswechsel im Verbrennungsmotor, Stuttgart 2016*
5. Breuer, M.; Grimm, K.; Moormann, S.; Schmitt, S.: Weiterentwicklung des vollvariablen Ventiltriebs UniValve *6. VDI- Fachtagung Ventiltrieb und Zylinderkopf, Würzburg 2015*
6. Breuer, M.; Grimm, K.; Schmitt, S.; Hunkel, D.: Developmnt of a fully variable valve train to production readiness. *MTZ worldwide 78(2017)2, 2017*

Further development of the Magma concept – a high-efficiency Miller cycle engine with uncompromised performance

R J Osborne, K J Pendlebury, T A Downes, S O' Brien
Ricardo UK Ltd, UK

ABSTRACT

The Magma engine concept is characterised by a high compression ratio combustion system employed in a downsized direct-injection gasoline engine. An advanced boosting system and Miller cycle intake-valve closing strategies are used to control combustion knock while maintaining specific performance. A key feature of the concept is the use of high CR without compromise to mainstream performance levels.

This paper focuses on further development of the Magma combustion system using a single-cylinder engine, including valve event, air motion and injection strategies. Key findings are that early intake-valve closing is effective both in mitigating knock and improving fuel consumption. A load equivalent to 23.6 bar brake mean effective pressure for a multi-cylinder engine has been achieved with a geometric compression ratio of 13:1.

DEFINITIONS/ABBREVIATIONS

ACAC	Air charge air cooler
BDC	Bottom dead centre (piston)
BMEP	Brake mean effective pressure
BSFC	Brake specific fuel consumption
CPS	Cam profile switching
CVVL	Continuously variable valve lift
EGR	Exhaust gas recirculation
EIVC	Early inlet valve closing
EVC	Exhaust valve closing
EVO	Exhaust valve opening
GIMEP	Gross indicated mean effective pressure
IMEP	Indicated mean effective pressure
ISFC	Indicated specific fuel consumption
IVC	Inlet valve closing
IVO	Inlet valve opening
LIVC	Late inlet valve closing
MFB	Mass fraction burned
NIMEP	Net indicated mean effective pressure
NISFC	Net indicated specific fuel consumption
PMEP	Pumping mean effective pressure
R_t	Tumble ratio
TDC	Top dead centre (piston)
V_c	Cylinder volume displaced during the effective compression stroke
V_e	Cylinder volume displaced during the expansion stroke
WCAC	Water charge air cooler

1 INTRODUCTION

Improvements in gasoline engine fuel economy are a key element in achieving fleet average vehicle CO_2 requirements. Many approaches for improved efficiency have been proposed, with varying implications for engine performance, packaging, cost and aftertreatment requirement. Two main approaches to reducing gasoline engine CO_2 have evolved in recent years [1]:

- Engine downsizing, with boosted direct-injection engines operating at higher load factors to reduce gas exchange losses and friction contribution
- Increased compression ratio for naturally-aspirated engines, with modified inlet valve closing strategies and often cooled exhaust gas recirculation

If a high compression ratio and advanced valve-timing strategy could be combined with downsizing and boosting, it may be possible to obtain the efficiency benefits of both approaches.

There are however significant challenges for this combination, most notably combustion knock which increases with increasing compression ratio [2]. High CR would typically dictate reduced load and/or retarded combustion phasing, leading to poor specific performance and thermodynamic efficiency. The solution to this problem originally proposed by Ralph Miller is to mitigate knock by using valve timing to reduce in-cylinder effective compression ratio, whilst maintaining the efficiency benefits of a high expansion ratio [3].

The history of Atkinson and Miller cycle engines is reported by the authors in a previous publication [2] and in comprehensive review papers [4,5]. Contemporary application of the Miller cycle can be seen in a number of recent production engines [6-13]. Some of these engines employ a conventional geometric CR and use late intake-valve closing (LIVC) in order to reduce pumping losses under part-load conditions [6-8]. LIVC is achieved using hydraulic cam phasers with an intermediate lock position that allows moderate retard angles. The boosting systems employed are conventional, consisting of a single waste-gated turbocharger.

In other current engines the geometric compression ratio is increased significantly, to 11.7, 12.5 or even 13:1 [9-13]. These engines use either short intake-valve opening durations to achieve early intake-valve closing (EIVC), in some cases with cam-profile switching, or LIVC again through retarded intake cam phasing. Thermodynamic efficiency improvements from an increased compression ratio are reported, but are generally accompanied by reduced specific performance.

The Magma concept seeks to deliver fuel efficiency benefits while maintaining full-load specific performance. This paper presents results achieved using a single-cylinder research engine over a wide operating range, and concludes with drive-cycle fuel consumption analysis and a multi-cylinder engine specification.

2 ENGINE CONCEPT

The Magma engine is the subject of previous publications [1,2,14], where development of the concept using simulation tools and initial engine testing is reported. Key elements of the concept are:

- A central direct-injection combustion system
- High compression ratio
- Miller cycle intake-valve closing strategies
- Increased maximum boost pressure

Magma could be described as a deeper application of existing Miller cycle concepts, using high boost pressure to further enable increased compression ratio and more extreme valve events at higher load. The compression ratio of 13:1 was chosen to deliver maximum benefit from increased expansion work, before the negative effects of increased surface area to volume ratio begin to dominate.

With this high compression ratio, knock mitigation is essential in order to maintain specific load. In this case, knock is controlled via early closing of the inlet valve to reduce the effective compression ratio in the cylinder.

In order to maintain the required airflow with a shorter inlet cam duration charge pressure is increased substantially. Two charge cooling stages are used to counteract the temperature rise associated with a high overall pressure ratio. It is important that this cooling occurs outside the cylinder. Some of the extra piston compression work associated with the higher CR is transferred to the boost system, thereby allowing cooling to occur before the air enters the cylinder. The net result is a charge temperature at the end of compression that is lower than would be the case if all the compression work had been carried out by the piston.

At low-load operating points, short inlet-valve opening duration is used to regulate airflow with a reduced requirement for throttling. This reduction in pumping work further increases the fuel consumption benefit. The overall effect of the combined elements is an engine with significantly reduced fuel consumption across the speed-load range, while maintaining peak BMEP.

3 EXPERIMENTAL SET UP

Experiments were conducted using a Ricardo Hydra single-cylinder research engine with direct injection and dual variable valve-timing (Table 1). The cylinder head, piston crown and cam profiles were developed specifically for the Magma engine.

Table 1 – Key parameters of the test engine

Displacement volume / no. cylinders	0.5 litres / 1 cylinder
Bore	84 mm
Stroke	90 mm
Number of valves	4
Compression ratio	13:1
Tumble ratio, R_t	0.8 - 2.8

Valve timing is illustrated in Fig. 1, showing inlet and exhaust cam in both parked and fully phased positions. The engine (Fig. 2) was installed in a research test facility with an external high-pressure air supply.

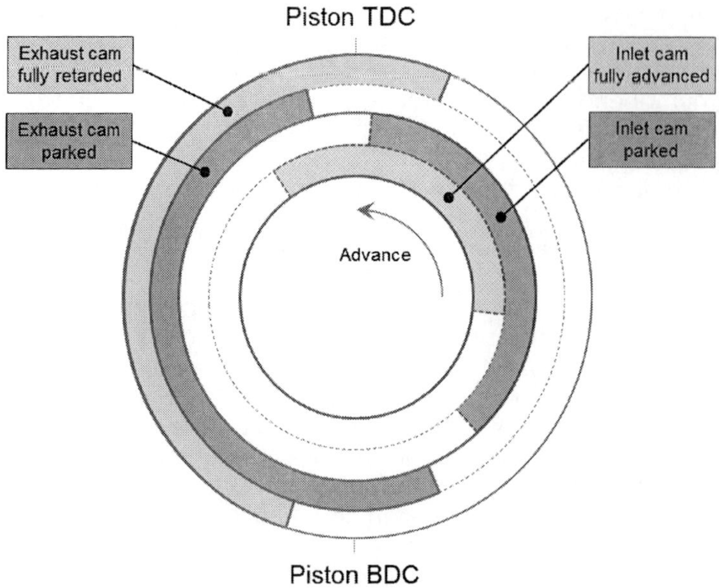

Figure 1 – Cam timing diagram including phasing range; opening and closing points are shown for 1 mm valve lift

Figure 2 – Hydra single-cylinder research engine installed in the test facility

4 METHODOLOGY

4.1 Testing Approach

A range of steady-state tests were carried out at key operating points in order to characterise the combustion system. At each point the following parameters were optimised for best net indicated specific fuel consumption at part load, or highest achievable IMEP at full load:

- Exhaust and inlet cam phasing
- Injection timing
- Ignition timing

All optimisations were carried out in adherence to standard Ricardo guidelines for parameters such as knock and combustion stability. For full-load operation at 2000 rev/min a peak knock amplitude of 2 bar was applied (measured over 300 cycles), and at 4000 rev/min the limit was 4 bar. A limit of 3% for coefficient of variation (CoV) of NIMEP was applied in all cases.

4.2 Key-Point Selection

Using the Ricardo vehicle simulation tool V-SIM, a D-segment vehicle model was developed and simulated over the WLTC and FTP-75 drive cycles. The output from this simulation was an engine speed vs. load trace. A common set of key-points was then generated based on residency time and cycle fuel consumption contribution (Table 2). For each cycle, a set of key-point weighting factors were also derived. In addition to the six part-load key points, a further three full-load points were also chosen.

Table 2 - Key points selected for testing and comparisons with baseline

	Engine speed [rev/min]	**BMEP** [bar]	**FMEP** [bar]
Part load	1500	4	0.63
	1500	10	0.67
	2000	2	0.71
	2000	8	0.74
	2000	14	0.77
	3000	8	0.93
Full load	1500	20	0.90
	2000	25	1.07
	4000	25	1.25

The part-load points were used to predict cycle fuel consumption over the WLTC and FTP-75 drive cycles for the D-segment vehicle, using the weighted key-point method. For single-cylinder engine testing, IMEP operating conditions equivalent to the appropriate BMEP levels for a multi-cylinder engine were set.

A friction model for the Magma engine was developed using the Ricardo internal toolset for friction analysis known as FAST. The FAST methodology uses a semi-empirical equation approach to build a friction model for engine architectures in a time efficient way. FAST builds on the methods of Sandoval, Heywood et al. [15, 16], but with the following enhancements:

- Correlated to measured data from modern engines
- Ability to calculate friction under fired conditions as well as motored
- Ability to deal with systems not otherwise considered (e.g. balancer shafts, vacuum pump)
- Physics-based treatment of piston and rings (enables user to examine effects of ring tension, connecting rod length etc.)

4.2.1 Baseline comparison

A comparison was made with conventional engine technology by replacing the Magma fuel consumption data with that of a baseline engine from the Ricardo benchmark database. Trend analysis shows that the mainstream (i.e. highest manufactured volume) 2017 rating for such a product in a D-segment vehicle is ~135 kW, while in 2022 it will be ~140 kW.

The baseline selected is thus a representative downsized, boosted direct-injection gasoline engine, of 2.0 litres and 140 kW. It has dual variable valve-timing and a compression ratio of 10.5:1. Comparisons between Magma and the baseline were made at a number of key operating points, and WLTC and FTP-75 cycle prediction comparisons were made.

5 RESULTS AND DISCUSSION

5.1 Air Motion

One of the key challenges arising from the application of EIVC is the impact on air motion. The combination of EIVC and high geometric compression ratio impact air motion as follows:

- Short intake-valve opening duration - reduced period for tumble formation
- Early intake-valve closing - increased period for tumble decay
- Reduced peak lift - tumble formation is reduced at lower lifts
- Flatter combustion chamber - affects the transition from solid body rotation to turbulence, increasing collapse of air motion leading to lower turbulent kinetic energy at ignition timing

Each of these factors needs to be addressed to avoid problems associated with insufficient air motion, such as extended ignition delay, slower combustion and greater cycle-to-cycle variability. Consequences manifest themselves as unstable combustion, particularly at low loads, and increased peak knock at higher loads.

To assess this impact of air motion in two port designs were developed, as shown in Figs. 3 and 4.

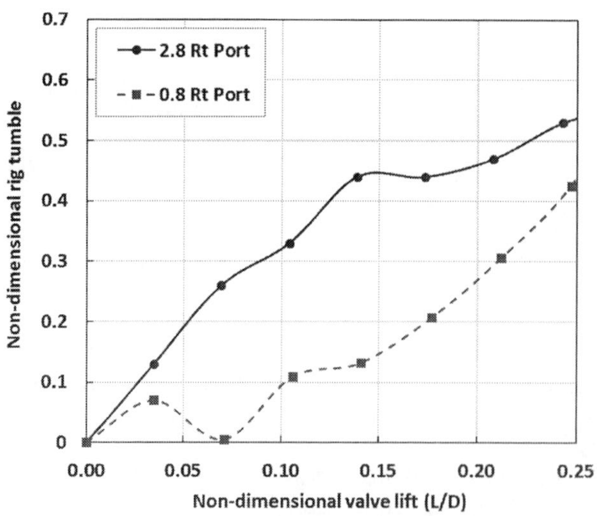

Figure 3 – Intake port tumble ratio (steady-state flow rig) vs non-dimensional valve lift

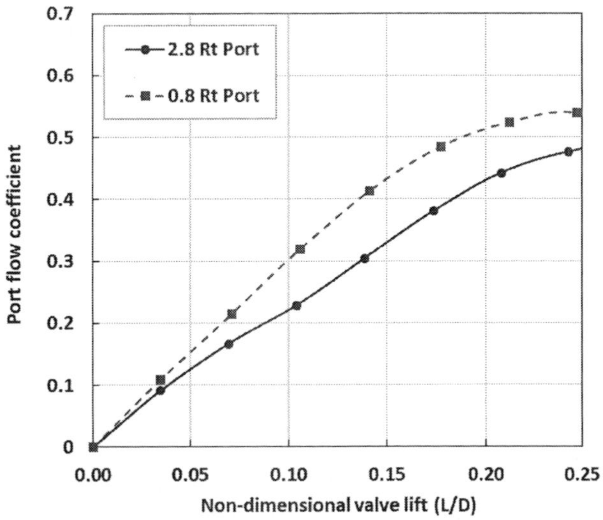

Figure 4 – Intake port flow coefficient (steady-state flow rig) vs non-dimensional valve lift

The lower tumble port represents a conventional design for a central-injector combustion system. However, as shown in Fig. 3, the tumble generation at low lift is significantly reduced compared to that at high lift. For this port, the tumble ratio when integrated over the Magma valve lift event was 0.8 R_t.

The higher tumble port was developed specifically to increase tumble at low valve-lift. This maximises the air motion generated with a short duration EIVC cam, such as the one tested. The result is a tumble ratio of 2.8 R_t when integrated over the Magma intake lift event. The modifications to increase tumble do have an impact on port flow coefficient (Fig. 4), but this impact can be tolerated.

When tested on the research engine, the impact on combustion at a low load operating point is significant (Fig. 5). As the intake cam phasing is advanced, the combustion with either port shows signs of degradation. The ignition delay increases, and the 10-90% mass fraction burned (MFB) duration also extends. However, at the same IVC timing, the higher tumble port clearly shows faster combustion and reduced ignition delay.

With the lower tumble port, combustion stability becomes the limiting factor precluding further inlet cam advance, as evidenced by the sharp turn up in CoV of IMEP in Fig. 6. The higher tumble port however, maintains stable combustion throughout the intake cam phasing range. This is a result of increased turbulent kinetic energy at ignition.

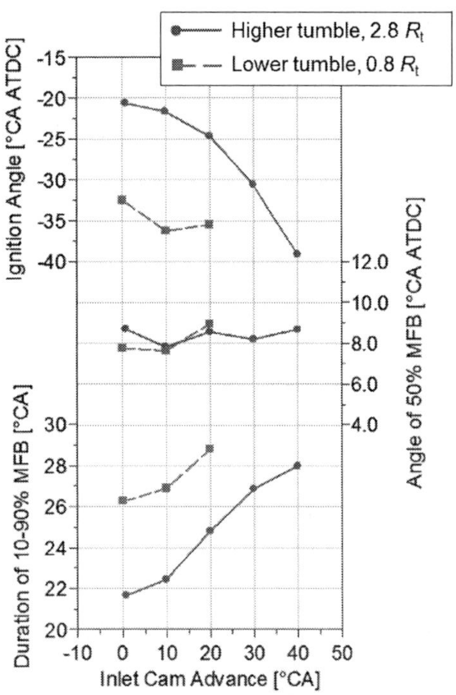

Figure 5 – Comparison of combustion characteristics resulting from increased tumble at 2000 rev/min, 2 bar BMEP equivalent

Advanced intake-valve closing can then be applied, thereby realising the associated benefits such as de-throttling at low load or advanced combustion phasing at higher load. The net result is improved fuel consumption (net indicated SFC), despite any penalty associated with the slightly more restrictive intake port.

Fig. 6 illustrates the gas exchange efficiency improving as the inlet cam is advanced, due to de-throttling. The optimum cam timing indicated in the figure results in minimum net ISFC with acceptable combustion stability, using the higher tumble port. The lower tumble port however, cannot exploit this cam timing because combustion stability degrades to an unacceptable level.

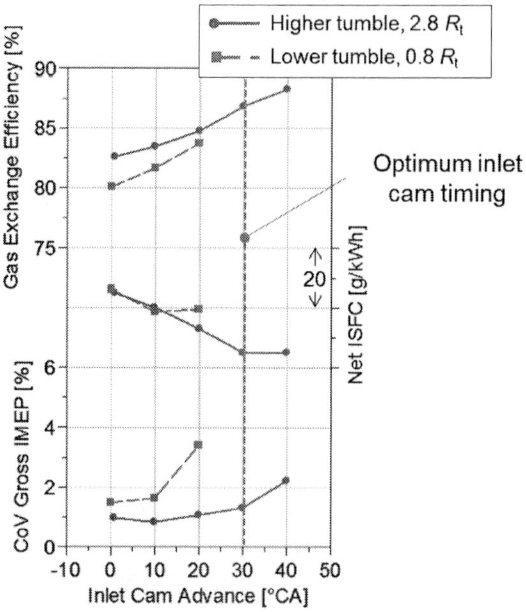

Figure 6 – Demonstration of improved combustion on ability to apply increased inlet cam advance and effect on NISFC (2000 rev/min, 2 bar BMEP equivalent)

5.2 Combustion Zones Identified

An overall picture of the combustion characteristics observed across the operating range is shown in Fig. 7.

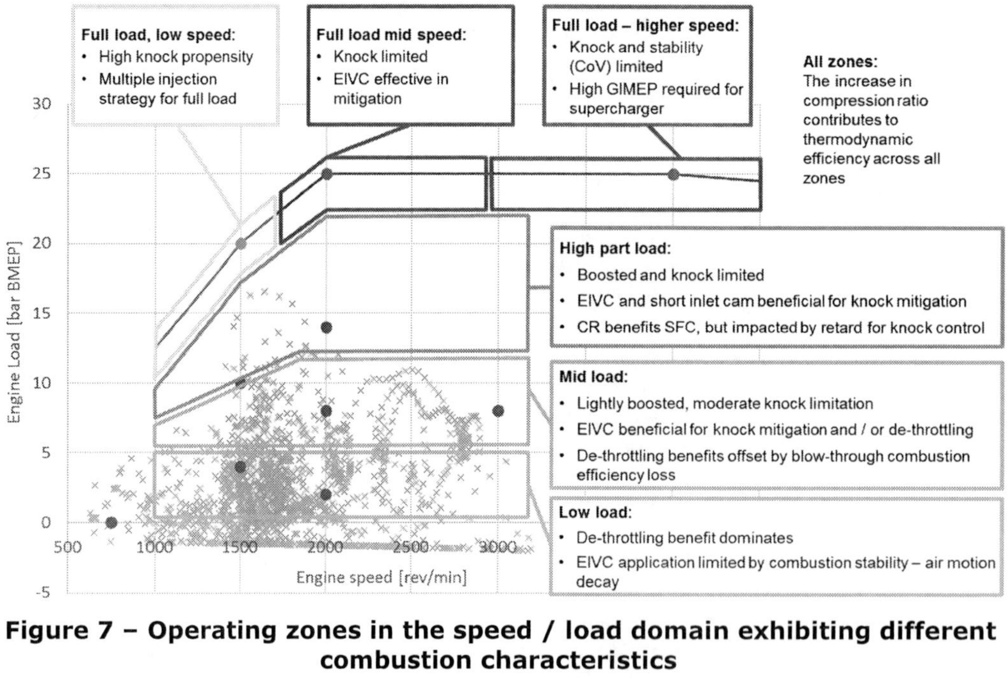

Figure 7 – Operating zones in the speed / load domain exhibiting different combustion characteristics

An explanation of the findings in each zone follows, split into full-load and part-load regions.

5.3 Full-Load Performance

5.3.1 Peak BMEP at 2000 rev/min

At 2000 rev/min the engine was successfully able to run at an NIMEP level equivalent to 23.6 bar BMEP on a multi-cylinder engine with stoichiometric air fuel ratio. Fig. 8 shows the differences between the conventional baseline build and the Magma build on a cylinder pressure vs. volume basis. Each stroke of the diagram is discussed here.

Inlet pressure is visibly higher in the Magma case, to compensate for reduced inlet valve opening period (NIMEP is actually also higher in this case, requiring increased airflow). Towards the end of the inlet stroke, following IVC, the charge is expanded causing the observed reduction in cylinder pressure.

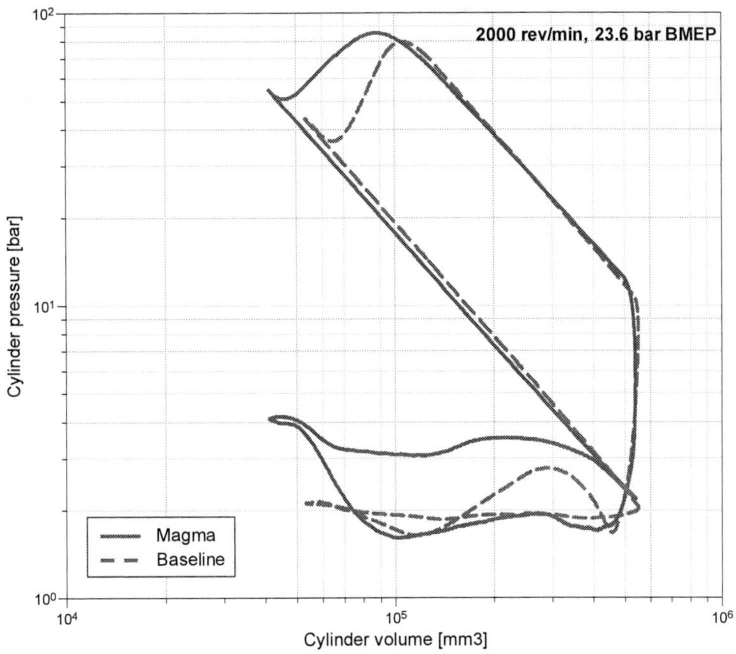

Figure 8 – Cylinder pressure vs volume at 2000 rev/min, 23.6 bar BMEP equivalent for Magma, 20.9 bar BMEP for the baseline engine

At the start of the compression stroke, the pressure can be seen to recover back to a similar level to that seen at IVC. This is the point at which 'effective' in-cylinder compression begins, and after this compression follows the normal trend.

Ignition occurs at 1 °CA ATDC(F). The ignition timing, delay, and burn rate are conventional (as shown in Table 3), indicating that combustion does not need to be retarded to mitigate knock. The expansion stroke follows the conventional shape, matching closely to the baseline case.

The exhaust stroke also resembles that of the baseline, with the exception of a pressure increase towards TDC, where the cylinder is exposed to high inlet manifold pressure during valve overlap. The considerable difference between net IMEP and BMEP in the Magma case is due to the supercharger mechanical input power.

Table 3 – Comparison of key parameters at 2000 rev/min full load

	Baseline	Magma
Gross IMEP [bar]	22.4	24.7
Net IMEP [bar]	22.0	25.9
Equivalent BMEP [bar]	20.9	23.6
Ignition timing [°CA ATDC(F)]	6.5	1.1
IVC (1 mm lift) [°CA ATDC(NF)]	186	110
MFB 50% [°CA ATDC(F)]	27.9	26.3
Ignition to 10% MFB [°CA]	13.8	14.8
MFB 10-90% duration [°CA]	14.6	18.4
Inlet manifold pressure (abs) [kPa]	205	345
Exhaust O_2 concentration [%]	0.14	1.36

The equivalent BMEP achieved at 2000 rev/min is compared to a benchmark database in Fig. 9, showing that the Magma engine does not lie on the normal trend line for geometric compression ratio vs. peak BMEP. This illustrates the challenge for Magma and other high compression ratio Miller cycle engines, and the resulting need to reduce the in-cylinder effective compression ratio.

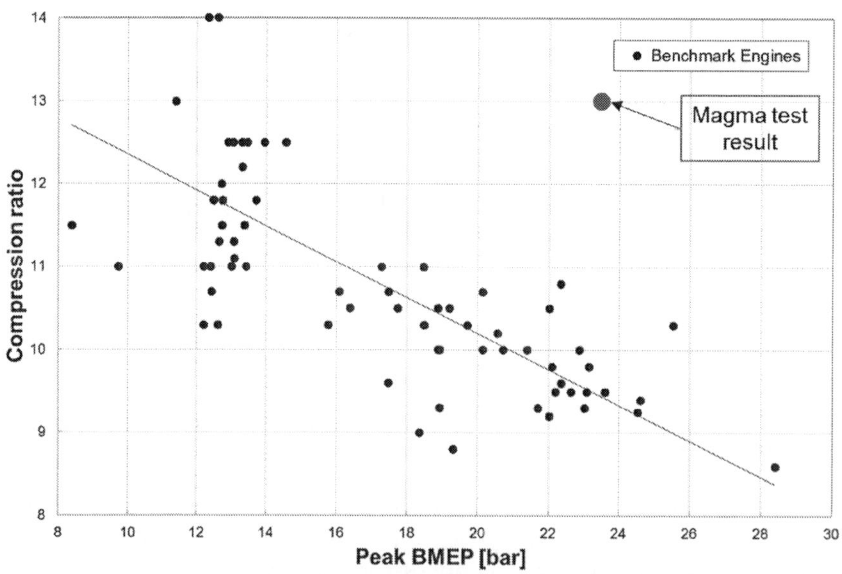

Figure 9 – Magma test result at 2000 rev/min full load compared with benchmark dataset of production engines on the basis of compression ratio vs. peak BMEP

5.3.2 Peak BMEP at 4000 rev/min

At higher engine speed, the pV diagram in Fig. 10 exhibits similar characteristics to the 2000 rev/min case; the maximum achieved gross IMEP is higher at 27 bar. Increased boost pressure is required at this speed, necessitating higher input power to the supercharger (as part of the two-stage boost system). This means a higher GIMEP is required to achieve the same BMEP as at 2000 rev/min, as shown in Table 4.

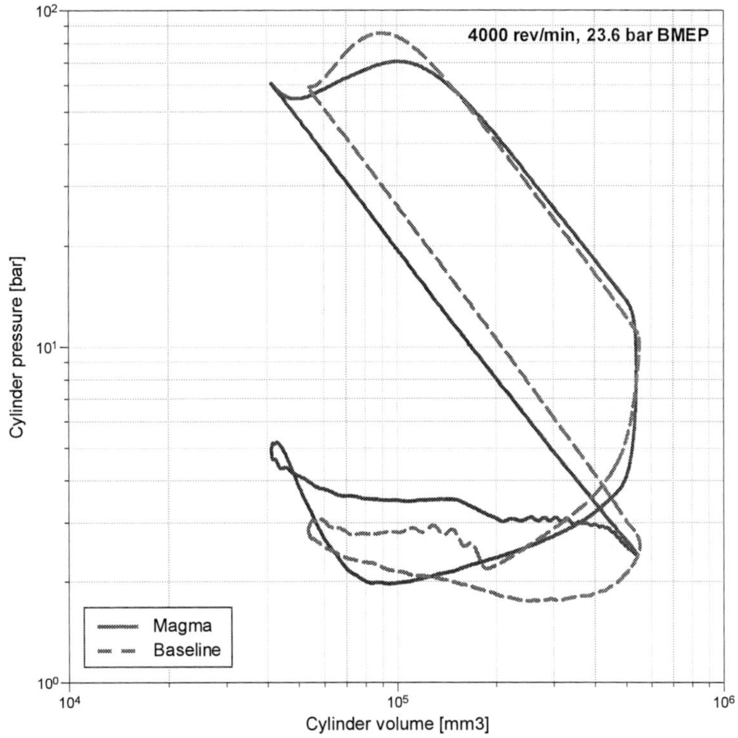

Figure 10 – Plot of cylinder pressure vs cylinder volume at 4000 rev/min 23.6 bar BMEP equivalent for Magma, 20.6 bar BMEP for the baseline engine

As seen at 2000 rev/min, knock is the limiting factor here; EIVC is shown to mitigate knock as observed at other key points. The combustion system is shown to operate successfully at this speed and load, indicating that the concept is applicable at higher speeds.

Table 4 – Comparison of key parameters at 4000 rev/min full load

	Baseline	Magma
GIMEP [bar]	23.4	26.9
NIMEP [bar]	21.9	26.6
BMEP [bar]	20.6	23.6
Ignition timing [°CA ATDC(F)]	-8.6	-2.9
IVC (1 mm lift) [°CA ATDC(NF)]	206	110
MFB 50% [°CA ATDC(F)]	21.9	25.2
Ignition to 10% MFB [°CA]	20.3	17.3
MFB 10-90% duration [°CA]	21.2	18.5
Inlet manifold pressure (abs) [kPa]	260	447
Exhaust O_2 concentration [%]	0.09	0.42

5.4 Part-Load Fuel Consumption

5.4.1 Thermodynamic Efficiency and Compression Ratio

An increased expansion ratio improves thermodynamic efficiency as it allows the engine to convert more released heat into useful work. By adjusting IVC timing the effective CR and ER can be disassociated to some extent. Thus the in-cylinder effective CR can be kept at a level for acceptable knock whilst the geometric ER is increased for improved thermodynamic efficiency [17]. This is illustrated using cylinder pressure measurements from the Magma engine in Fig. 11.

IVC occurs when the piston is significantly above bottom dead centre (BDC). Immediately after this point, the charge is expanded and then compressed back to the same condition as observed at IVC. From the pV diagram this can be seen to be a broadly reversible process. The compression stroke then effectively starts, leading to the effective in-cylinder compression ratio indicated by V_c, which is measurably smaller than the expansion volume denoted by V_e. As the geometric expansion ratio is high for all speed-load operating points, the thermodynamic efficiency benefits are also evident at all conditions.

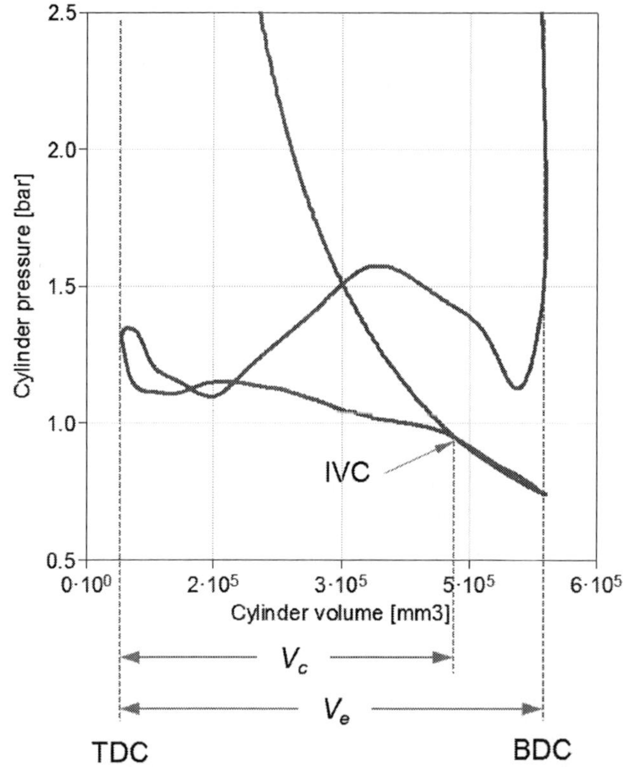

Figure 11 – Example data from Magma engine at 2000 rev/min, 8 bar BMEP (equivalent) showing the difference between effective compression volume (V_c), based on IVC point, and expansion volume (V_e)

5.4.2 Gas Exchange Efficiency at Light Load

In addition to thermodynamic efficiency, part-load fuel consumption is heavily influenced by the efficiency of the gas exchange process. A well-established alternative method to control airflow without the flow losses associated with an inlet throttle is to use EIVC. The result of a shorter, lower lift inlet cam, operating at advanced timing for early closing, is demonstrated in Fig. 12. Here Magma test data is compared to the baseline build (which has a conventional inlet cam profile) at a light-load operating point. During the inlet stroke the cylinder pressure is higher for the Magma engine (with the same mass flow), resulting in reduced pumping work and higher gas exchange efficiency.

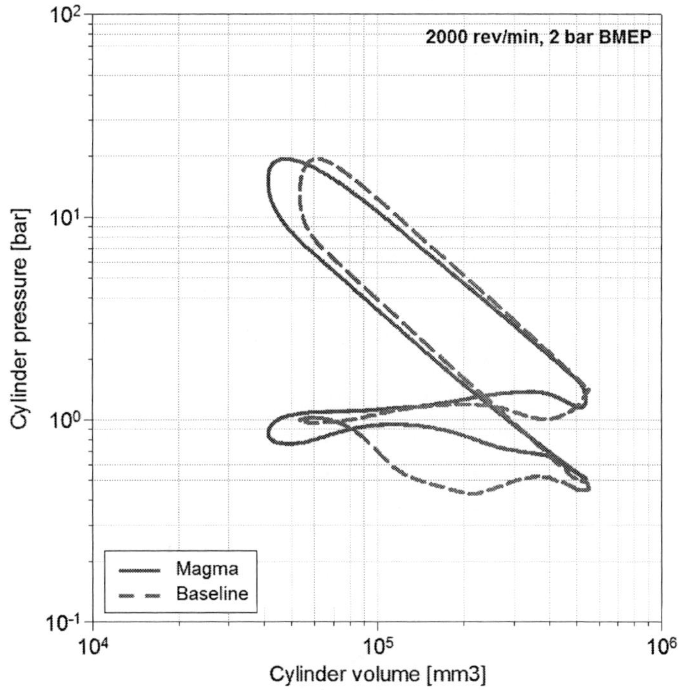

Figure 12 – Cylinder pressure vs cylinder volume at 2000 rev/min, 2 bar BMEP equivalent

At 8 bar BMEP (equivalent) the inlet manifold pressure is close to atmospheric or slightly positive, as shown in Fig. 13, so the benefit available from de-throttling is small.

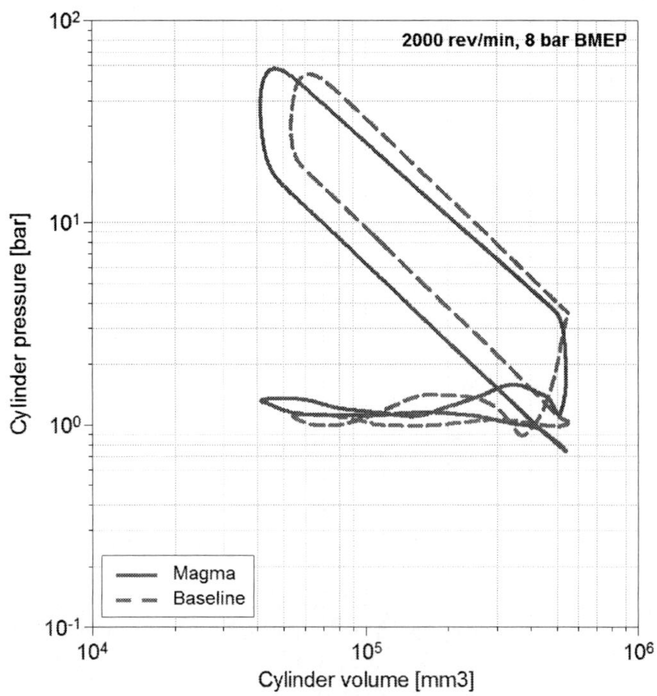

Figure 13 – Cylinder pressure vs. cylinder volume at 2000 rev/min, 8 bar BMEP equivalent

5.4.3 *Drive-Cycle Analysis*

In order to put the fuel consumption benefits in context, analysis was carried out on a drive-cycle basis. Vehicle simulation generated the speed/load traces shown in Figs. 14 and 15 for the WLTC and FTP-75 cycles.

Weighting factor sets were then generated so that these key points could represent both cycles when the single-cylinder results were applied as either a 1.5 litre three-cylinder variant (L3) or a 2.0 litre four-cylinder variant (L4). For clarity, the key points are at very similar BMEP levels (the FMEP between the L3 and L4 being close), but of course the torque at those BMEP levels for the two engines is quite different. Figures 14 and 15 show the speed torque traces for the WLTC and FTP75 on a speed / torque basis, with the L3 and L4 key-points overlaid.

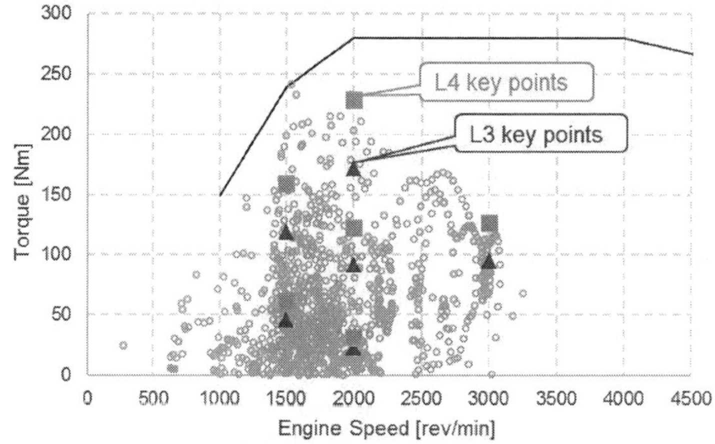

Figure 14 – Speed vs BMEP trace for WLTC test cycle derived from vehicle simulation tool, showing key points for L3 and L4 torques

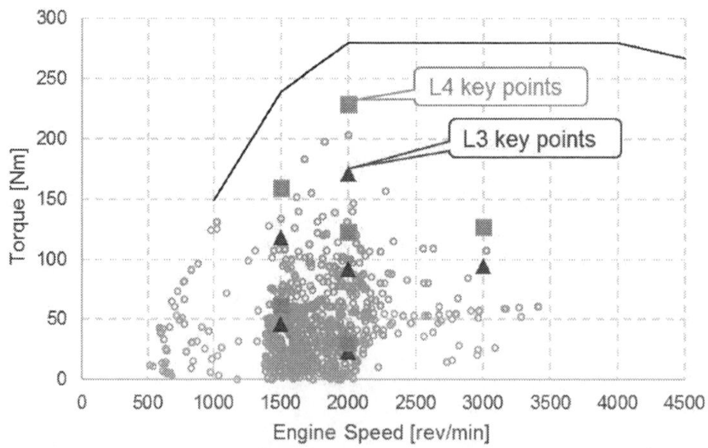

Figure 15 – Speed vs BMEP trace for FTP-75 test cycle derived from vehicle simulation tool, showing key-points for L3 and L4 torques (same BMEPs)

The vehicle model was not varied between the engine types. The model was constructed so as to be suitable for a 1.5 litre L3 with 23.4 bar BMEP peak torque (~280 Nm) and 93 kW/litre (140 kW). The L4 variant is a lower rated engine at 17.6 bar BMEP and 70 kW/litre, without the two-stage boosting system. This variant was considered in order to separate the Miller cycle fuel consumption effect from the full Magma concept which includes engine downsizing.

Table 5 shows the fuel consumption improvements at similar BMEP levels for the L3 Magma and L4 variants when compared with exactly the same points on the baseline engine BSFC map. This is the improvement due to the high compression ratio and Miller cycle effects only.

The benefits when the BSFCs are compared at the same torque levels (those of the L3 engine BMEPs) are more striking (Table 6). Here the benefit includes both the compression ratio, Miller cycle, and downsizing effects.

Table 5 – BSFC results at each part-load key-point compared to the baseline engine

Engine speed [rev/min]	**Nominal BMEP** [bar]	**L3 improvement vs. baseline** [%]	**L4 improvement vs. baseline** [%]
1500	4	9.0	9.4
1500	10	3.6	3.8
2000	2	8.1	8.2
2000	8	6.3	6.5
2000	14	5.0	5.0
3000	8	1.8	1.8

Table 6 – BSFC results at each part-load key-point compared to the baseline engine

Engine speed [rev/min]	**Torque** [Nm]	**L3 improvement vs. baseline** [%]
1500	46.0	17.7
1500	118.7	5.3
2000	23.4	21.8
2000	91.9	10.6
2000	171.3	6.3
3000	95.0	7.1

Comparing the Magma drive-cycle fuel consumption to the baseline engine gives the improvements shown in Table 7.

Table 7 – Reduction in cycle fuel consumption using Magma test data

Drive cycle	**Magma L3 improvement** [%]	**Miller L4 improvement** [%]
WLTC	12.5	6.3
FTP-75	16.4	7.5

These figures demonstrate the high potential of the Magma concept in terms of fuel consumption improvement. An important part of the benefit arises from the fact that a 2.0 litre L4 engine can be replaced with a higher rated 1.5 litre L3 Magma engine. An additional benefit of the Magma concept is that it is a combination of proven individual components, arranged in an innovative system.

6 MULTI-CYLINDER ENGINE SPECIFICATION

Following completion of the single-cylinder engine testing a multi-cylinder engine (MCE) specification has been developed. A 1-D gas-dynamics model of the SCE was validated using the test data described above, with a particular focus on knock modelling. A MCE model was then generated in order to refine the boosting system, charge cooling and valvetrain specifications. The primary specifications are shown in Table 8.

Table 8 – Multi-cylinder Magma engine specification

Number of cylinders	3
Swept volume [L]	1.5
Peak torque [Nm]	280
Specific torque [Nm/L]	187
Rated power [kW]	140
Specific power [kW/L]	93
Fuel injection system	Direct
Compression ratio	13:1
Boosting system	Supercharger (LP) and turbocharger (HP)
Charge cooling	ACAC (intercooler) and WCAC (aftercooler)
Valvetrain	Dual VVT with either intake CPS or CVVL

This work again concluded that a supercharger and turbocharger in series would provide the best means of delivering the necessary boost pressure and air flow-rate without excessive exhaust back-pressure. The long-term objective of full-map stoichiometry along with the high Magma boost pressures mean that the charge cooling requirements of this engine are considerable. An air charge-air cooler (ACAC) following the supercharger along with a water charge-air cooler (WCAC) following the turbocharger compressor provides a good combination of compressor outlet temperatures and transient response, while limiting the impact on the vehicle.

7 SUMMARY/CONCLUSIONS

- The Magma concept has been assessed using a single-cylinder research engine over a wide range of conditions
- Early intake-valve closing (EIVC) has again been shown to be effective both in mitigating knock and improving fuel consumption
- Increased air motion is required to preserve acceptable combustion parameters with an EIVC strategy
- A Net IMEP equivalent to 23.6 bar BMEP on a multi-cylinder engine has been achieved with a geometric compression ratio of 13:1
- Part-load key-point fuel consumption benefits between 1.8 and 9.4% have been achieved at equivalent BMEP
- Drive-cycle analysis predicts fuel consumption benefits of 12.5% and 16.4% for WLTC and FTP-75 respectively when the baseline four-cylinder engine is replaced by a downsized three-cylinder Magma specification
- The fuel consumption benefit of the increased compression ratio and Miller cycle application in an engine of the same size as the baseline is 6.3% and 7.5% for WLTC and FTP-75 respectively
- The specification for a 1.5 litre three-cylinder Magma engine has been proposed

8 ACKNOWLEDGMENTS

The authors would like to thank the directors of Ricardo plc for permission to publish this paper. We would also like to acknowledge the significant contributions to the work made by Nicolas Fortino, Srinivasa Sugash Raja, Eric Klos, Michael Campbell, Nigel Fox, Matthew Heaney and Tiago Carvalho.

9 REFERENCES

1. Osborne, R. J., Pendlebury, K. J., Stokes, J., Dalby, J. and Rouaud, C.: "The Magma Engine Concept," JSAE Annual Congress 2015.
2. Pendlebury, K. J., Osborne, R. J., Downes, T. A. and O' Brien, S., "Development of the Magma Combustion System," JSAE Annual Congress 2016.
3. Miller, R. 'High Expansion, Spark Ignited, Gas Burning, Internal Combustion Engines', USA Patent No. 2773490 (1956).
4. Schutting, E., Dumböck, O., Kraxner, T. and Eichlseder, H., "Thermodynamic Consideration of the Miller Cycle on the Basis of Simulation and Measurements", ATZ Live 2016.
5. Zhao, J. "Research and application of over-expansion cycle (Atkinson and Miller) engines - A review", Applied Energy, Volume 185, Part 1, 1 January 2017.
6. Watanabe, I., Kawai, T., Yonezawa, K. and Ogawa, T.: "The New Toyota 2.0-Liter Inline 4-Cylinder ESTEC D-4ST Engine - Turbocharged Direct Injection Gasoline Engine", 23rd Aachen Colloquium Automobile (2014).
7. Hikichi, K., Watanabe, I., Shinagawa, K., and Kudo, M.: "The New Toyota Inline 4 Cylinder 1.2L ESTEC D-4T Engine", 36. Internationales Wiener Motorensymposium (2015).
8. Aiyoshizawa, E., Hori, K.: "The New Nissan High Efficient 4-Cylinder 1.6L GDI turbocharged engine with low pressure EGR - Evolution for Lower Fuel Consumption combined with High Output Performance", 35. Internationales Wiener Motorensymposium (2014).
9. Wurms, R., Budack, R., Grigo, M., Mendl, G., Heiduk, T., Knirsch, S., "The New Audi 2.0l Engine with Innovative Rightsizing - A Further Milestone in the TFSI Technology", 36. Internationales Wiener Motorensymposium (2015).
10. "Optimization of the Combustion Process as Demonstrated on the New Audi 2.0L TFSI," 24th Aachen Colloquium Automobile and Engine Technology 2015.
11. Eichler, F., Demmelbauer-Ebner, W., Theobald, J., Stiebels, B., "The New EA211 TSI evo from Volkswagen", Internationales Wiener Motorensymposium 2016.
12. Demmelbauer-Ebner, W., Middendorf, H., Birkigt, A., Ganzer, M., Hagelstein, D., Persigehl, K., "EA211 TSI evo –The New 4-Cylinder Gasoline Engines from Volkswagen", 25th Aachen Colloquium Automobile and Engine Technology 2016.
13. Kishi, K., Satou, T.: "The New Nissan High Efficient 1.2L 3cyl GDI Supercharged Engine Enables 95 g/km CO2 Emissions and High Driving Performance", 33. Internationales Wiener Motorensymposium 2012.
14. Pendlebury, K. J., Osborne, R. J., Stokes, J., Dalby, J. "The Gasoline Engine at 2020," 23rd Aachen Colloquium Automobile and Engine Technology 2014.
15. Sandoval, D. and Heywood, J., "An Improved Friction Model for Spark-Ignition Engines," SAE Technical Paper 2003-01-0725, 2003, doi:10.4271/2003-01-0725.
16. Patton, K., Nitschke, R., and Heywood, J., "Development and Evaluation of a Friction Model for Spark-Ignition Engines," SAE Technical Paper 890836, 1989, doi:10.4271/890836.
17. Johansson, B., "Path to High Efficiency Gasoline Engine", DEER 2010.

Analysis of the impact of exhaust valve profile on the scavenging and combustion process in a 2-stroke Boosted Uniflow Scavenged Gasoline (BUSDIG) engine

X Wang, J Ma, H Zhao
Centre for Advanced Powertrain and Fuels, Brunel University London, UK

ABSTRACT

In order to take advantage of the 2-stroke cycle operation but avoid the drawbacks of conventional ported 2-stroke engine designs, a novel 2-stroke Boosted Uniflow Scavenged Direct Injection Gasoline (BUSDIG) engine was proposed and designed to achieve aggressive engine down-sizing and down-speeding. The uniflow gas scavenging minimises the air short-circuiting through multiple intake ports and exhaust valves in the cylinder head. The intake ports in the BUSDIG engine are integrated to the cylinder liner and designed to create desirable large-scale flow motion. The exhaust valves in the cylinder head are equipped with a variable valve timing actuator and a cam profile switching device. At part-load operation, a low lift exhaust cam profile can be adopted to effectively retain a part of hot residual gas from previous cycle in the cylinder and achieve the hybrid spark ignition (SI) - controlled auto-ignition (CAI) combustion. In order to understand the impact of the low exhaust valve lift profile and the opening timing on the scavenging process and combustion performances, the three-dimensional (3D) computational fluid dynamics (CFD) simulations are performed with a validated SI-CAI hybrid combustion model. It is found that the exhaust valve opening (EVO) timing of the low lift design can be used to effectively control the mass of the trapped residual gas and hence the subsequent scavenging process and intake charge mass. The CAI combustion can be achieved with an appropriate residual gas fraction (RGF) by adjusting the EVO timing. The current work demonstrates the potential of using a low exhaust valve lift design to achieve efficient clean combustion in the 2-stroke BUSDIG engine. The results would be useful for the development of efficient and high power density engines for a range of applications.

Keywords: computational fluid dynamics, 2-stroke engine, uniflow, variable valve actuation, controlled auto-ignition, hybrid combustion.

ABBREVIATIONS

3D	three dimensional
AIA	axis inclination angle
aTDC	after top dead center
BUSDIG	boosted uniflow scavenged direct injection gasoline
CAI	controlled auto-ignition
CE	charging efficiency
CFD	computational fluid dynamics

CTR	cross tumble ratio
DI	direct injection
DR	delivery ratio
EL	exhaust valve lift
EVC	exhaust valve closing
EVO	exhaust valve opening
IMEP	indicated mean effective pressure
RGF	residual gas fraction
SD	scavenge port opening duration
SE	scavenging efficiency
SI	spark ignition
SOA	swirl orientation angle
SPC	scavenge port closing
SPO	scavenge port opening
SR	swirl ratio
TDC	top dead center
TE	trapping efficiency
TKE	turbulence kinetic energy
TR	tumble ratio
VVA	variable valve actuation

1. INTRODUCTION

The engine downsizing and down-speeding technologies can be used to achieve higher efficiency and reduce the CO_2 emissions from internal combustion engines. However, the application of engine down-sizing and down-speeding by increasing boost is hampered in the 4-stroke gasoline engine due to increased peak cylinder pressure and knocking or low speed preignition (LSPI) combustion. As the 2-stroke engine doubles the firing frequency, it has a lower indicated mean effective pressure (IMEP) and lower peak in-cylinder pressure than a 4-stroke counterpart at the same output. Therefore, the 2-stroke operation can have great potentials to achieve both downsizing and down-speeding. As there is a large overlap between the intake and exhaust processes in the 2-stroke operation, a large amount of hot residual gases can be retained in the cylinder to achieve controlled auto-ignition (CAI) or homogeneous charge compression ignition (HCCI) combustion (1-3). The variable valve actuation (VVA) system can be applied to control the amount of residual gases, as well as the CAI combustion, by adjusting the intake and exhaust valve timings, and the 2-stroke CAI combustion operation can be achieved over a wide range of engine speed and load conditions, including idle operations that could not be achieved with 4-stroke operations (4, 5).

The scavenging process in a 2-stroke engine directly controls the in-cylinder flow motion, charge mixture formation and the subsequent combustion performance. During the 2-stroke scavenging process, the fresh intake charge scavenges the burnt gases out of the cylinder and fills the cylinder for the next cycle. The conventional 2-stroke engines suffer from the fuel/air short-circuiting phenomenon, in which some of the intake fresh mixture would flow directly into the exhaust port during the scavenging process (6), due to the large overlapping period of intake and exhaust process. Compared to the cross and loop scavenge engines, the uniflow scavenged 2-stroke engine shows superior scavenging performance and can operate with little air short-circuiting (7-9). The intake scavenge ports in a uniflow scavenged engine are integrated to the cylinder liner and controlled by the movement of piston top while exhaust valves are placed in the cylinder head. Therefore, the bore distortion caused by uneven thermal loading in the conventional ported 2-stroke engine could be avoided in the uniflow scavenged engine.

In order to overcome the fuel short-circuiting in the conventional 2-stroke engine, the direct fuel injection can be applied after the scavenging process during the intake and exhaust overlap period (6). However, it was found that the injection timing significantly affects the combustion phase and duration in the 2-stroke CAI operation (10) and the 2-stroke CAI operation modes produce more CO and HC emissions due to the shortened mixing time from the fuel injection to ignition (11, 12). Although the direct injection can avoid the fuel short-circuiting, the air short-circuiting cannot be completely eliminated at high boost pressures because of the overlap period of intake and exhaust processes. The short-circuited air can skew the reading of the lambda sensor in the exhaust, resulting in an enriched in-cylinder air/fuel mixture and hence higher CO and uHC emissions when the exhaust lambda value is set to one (1, 5). Although the stochiometric combustion can be achieved by modulating the fuelling rate or boost pressure, the air short-circuiting would lead to lean mixture in the exhaust and reduce the conversion efficiency of a three-way catalyst (6).

As the exhaust valves are placed in the cylinder head in the uniflow scavenged 2-stroke engine, the variable valve actuation (VVA) technology can be applied to adjust the exhaust valve timings and hence the scavenging process under different boost pressures at various engine speeds. In order to reduce the high CO and HC emissions caused by the insufficient mixing with direct injection (1, 5) and avoid the fuel/air short-circuiting, a low exhaust valve lift was applied in this study to avoid the air/fuel short-circuiting and achieve the CAI combustion with conventional port fuel injection by retaining a part of hot residual gas in a 2-stroke boosted uniflow scavenged direct injection gasoline (BUSDIG) engine (13, 14). In order to clarify the impact of the exhaust valve profiles on the scavenging process and the resultant in-cylinder residual gas fraction and flow motions, the three-dimensional (3D) computational fluid dynamics (CFD) engine simulations with different exhaust valve opening timings (EVO) were performed and analysed in detail at the motored conditions. Then, the combustion processes with different EVO timings were investigated by applying a previously developed spark ignition (SI) - controlled auto-ignition (CAI) hybrid combustion model.

2. SPECIFICATION OF THE BUSDIG ENGINE

The design of the cylinder head, piston shape and scavenge ports of the BUSDIG engine is schematically shown in Figure 1. A pent roof cylinder head was incorporated in the current design to accommodate two exhaust valves, a centrally mounted spark plug and high pressure direct injection (DI) injector. A shallow bowl was included in the piston top to guide the fuel jets from the injector and also to avoid the interference with the spark plug. Twelve evenly distributed scavenge ports were integrated to the cylinder block to introduce the fresh charge and scavenge out the residual burnt gas. Each scavenge port occupies a 20° segment on the cylinder circumference and the interval between two adjacent scavenge ports is fixed at 10°.

In terms of the design of the scavenge port, there are mainly four key design parameters, e.g. the axis inclination angle (AIA), swirl orientation angle (SOA), scavenge port opening (SPO) timing and scavenge port height (SPH). As shown in Figure 1, the AIA and SOA were fixed at 90° and 20° respectively in this study, based on the previous study (14). The SPH was kept at 14 mm and the SPO at 130 ^{0}CA. In this study, a low exhaust valve lift (EL) of 3 mm with opening duration of 42 °CA was applied in order to minimize the overlap between exhaust and intake opening. The exhaust valve opening timing (EVO) were gradually delayed from 84 °CA to 124 °CA after top dead center (aTDC) to study its impact on the scavenging and combustion process. Figure 2 shows the exhaust valve lift profiles and normalized scavenge port opening area profiles. The other engine specifications are shown in Table 1.

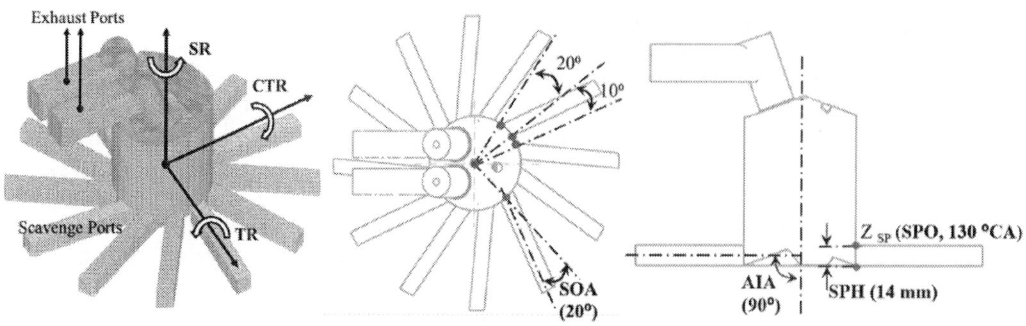

Figure 1. Schematic diagram of the engine design.

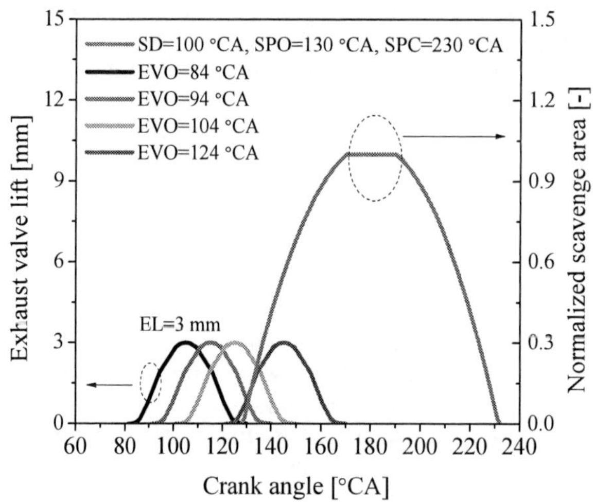

Figure 2. Schematic diagram of the exhaust valve lift profiles and normalized scavenge port opening area profiles.

Table 1. Engine specifications.

Bore	80 mm
Stroke	100 mm
Connecting rod	180 mm
Displacement	0.5 L
Compression ratio	14:1
Cylinder head	Pent roof with 2 exhaust valves, an injector and a spark plug
Piston	Shallow bowl

3. NUMERICAL MODELS

In this study, the simulations were performed with the CFD software STAR-CD (15). The Reynolds-Averaged Navier Stokes (RANS) approach was applied with RNG k-ε turbulence model. The heat transfer was implemented through the general form of the enthalpy conservation equation for the fluid mixture. The Angelberger wall function was used for the simulation of the wall heat transfer.

In order to predict the combustion process in the BUSDIG engine, a set of models for the premixed flame propagation and auto-ignition combustion were employed in this study. The three-zones extended coherent flame model (ECFM3Z), which can consider premixed flame propagation, diffusion flame propagation and auto-ignition combustion, was adopted as the framework of the combustion model. The flame surface density equation was used to describe the flame propagation process and predict the reaction rate of the flame propagation. The average flame surface density is defined as the local area of flame per unit of volume (m^{-1}), which is used to describe the intensity of flame propagation. The tabulated chemistry approach was adopted to predict the auto-ignition of the unburned charge. Chemical kinetic calculations under various thermodynamic and dilution conditions were performed with a reduced gasoline surrogate mechanism (16) to construct the tabulated database of the auto-ignition delay time. The tabulated database is coupled with the CFD simulations by look-up tables. With the tabulated chemistry approach, the auto-ignition tendency was defined to explicitly describe the close degree of fresh mixture from auto-ignition in each cell, and the value of 0 indicates no tendency to auto-ignition and the value of 1 indicates the occurrence of auto-ignition.

During the calculation, the reaction regime of each cell is determined by the average flame surface density and the auto-ignition tendency. The available fuel/air mixture in a cell will be consumed by the flame propagation according to the flame surface density equation when the local average flame surface density of the cell is greater than 0. By contrast, the available fuel/air mixture in a cell will be consumed by auto-ignition combustion according to the tabulated chemistry approach if the auto-ignition tendency of the cell achieves 1. The application of above models enables the prediction of SI combustion, CAI combustion and the SI-CAI hybrid combustion. The detailed modelling and validation of the hybrid combustion basing on the above concept can be found in a previous paper (17).

The ES-ICE software was used to generate the moving mesh for simulations. The arbitrary sliding interface (ASI) was applied between the scavenge ports and the cylinder liner to control the attachment and detachment with the piston movement. The opening and closure of the scavenge ports were determined by the movement of the piston top. ASI was also applied to control the connectivity between exhaust domains and cylinder domain with the movement of exhaust valves. The sensitivity study on the mesh quality showed that an average grid size of 1.6 mm was sufficient to achieve a convergent result (14).

4. SIMULATION CONDITIONS

In order to achieve convergent results, the multi-cycle simulations were applied in this study, and the initial and boundary conditions for the 3-D CFD simulations with motored and fired conditions are shown in Table 2. In the cases with motored conditions, the mixture components for the initial conditions, inlet and outlet boundaries are pure air, i.e. O_2 and N_2. In the cases with fired conditions, the initial mixture components in the cylinder and exhaust ports are pure burned gas, i.e. CO_2, H_2O and N_2, while the initial mixture components in the scavenge ports and the inlet boundary are stoichiometric fuel/air mixtures, i.e. C_8H_{18}, O_2 and N_2. All the simulations were carried out from 60 ^{0}CA after top dead center (aTDC). The crank angle used in this paper is referenced to TDC. The engine speed was fixed at 2000 rpm for all the simulations.

The Pressure-Implicit with Splitting of Operators (PISO) algorithm was used to solve the equations. The equations of momentum, turbulence kinetic energy and turbulence dissipation were discretized with the monotone advection and reconstruction scheme (MARS). The upwind differencing scheme (UD) and central differencing scheme (CD) were applied to discretize the temperature and density equations, respectively. The residual tolerance for the momentum, turbulence kinetic energy and turbulence dissipation was set at 0.001 while the residual tolerance for pressure and temperature was set at 0.0001 to achieve good compromise between convergence and computational time.

Table 2. Simulation conditions.

	Motored conditions	Fired conditions
Initial conditions		
Cylinder temperature [K]	400	Various (1450 ~ 1540)
Cylinder pressure [bar]	5	Various (9.2 ~ 11.5)
Intake temperature [K]	300	350
Exhaust temperature [K]	400	800
Exhaust pressure [bar]	1.05	1.06
Boundary conditions		
Intake temperature [K]	300	350
Intake pressure [bar]	1.6	1.6
Exhaust temperature [K]	400	800
Exhaust pressure [bar]	1.05	1.06
Cylinder head temperature [K]	340	440
Piston top temperature [K]	360	522
Cylinder liner temperature [K]	320	384

5. RESULTS AND DISCUSSIONS

5.1 Effect of EVO timing on the scavenging process (motored conditions)

Figure 3 shows the effect of the EVO timing on the in-cylinder charge mass. The delayed EVO timing leads to shorter blowdown duration and the exhaust process is significantly weakened as shown in Figure 3. The retained in-cylinder mass would hinder the intake scavenge process and the delayed EVO timing from 84 ^{o}CA to 104 ^{o}CA leads to lower peak intake mass flow rate, as shown in Figure 4. However, the subsequent backflow process offsets the intake process, and the total in-cylinder charge mass at the end of scavenging is similar for EVOs from 84 ^{o}CA to 104 ^{o}CA.

The further delay of the EVO timing to 124 ^{o}CA leads to a distinct scavenging process due to significantly shorter blowdown duration (only 6 ^{o}CA) but longer overlap between the intake and exhaust profiles, as shown in Figure 2. The short blowdown duration leads to only slight decrease of the in-cylinder charge after that the intake fresh charge from the scavenge ports gradually scavenges out the residual gas and fill the cylinder. The overall scavenging process is delayed due to the more trapped residual gas in the cylinder at the early stage of scavenging. As the scavenging proceeds, the increase of scavenging area and exhaust lift leads to stronger scavenging process due to the higher pressure difference between the intake and exhaust, which in turn results in even higher peak intake scavenge flow rate than the earlier EVOs. However, the overall retarded scavenging process and the subsequent backflow leads to lowest in-cylinder charge mass at the end of scavenging, as shown in Figure 3.

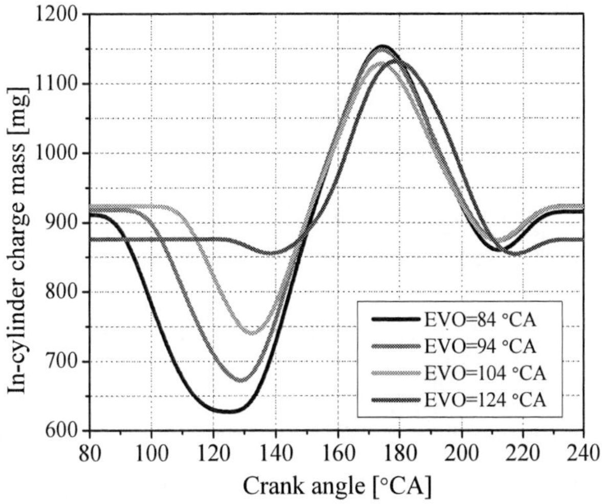

Figure 3. Impact of the EVO timing on the in-cylinder charge mass.

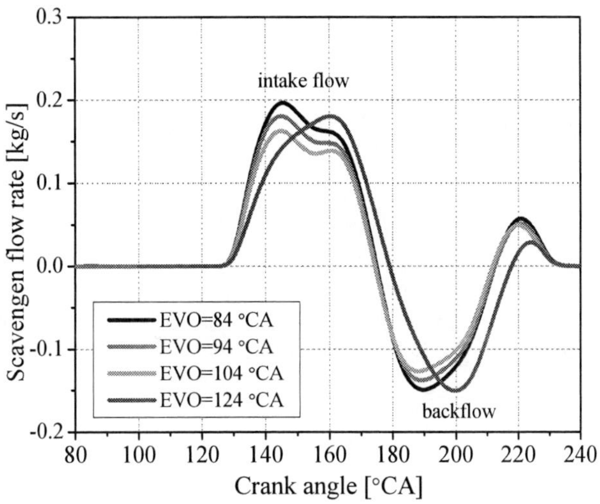

Figure 4. Impact of EVO timing on the intake scavenge flow rate.

In order to understand the impact of the EVO timing on the scavenge process, the in-cylinder residual gas fraction distributions just before the EVC (1st row), after EVC (2nd row) and after IVC (3rd row) for each EVO are shown in Figure 5. As the EVO delays, the incoming fresh charge from the scavenge ports located at the bottom of the cylinder gradually approaches to the exhaust valve just before the EVC. The air short-circuiting is avoided even for the latest EVO timing at 124 °CA, as there is no air distribution observed in the exhaust ports just before the EVO timing, as shown by the RGF distributions in the first row of Figure 5. However, it is believed that the further delay of EVO timing beyond 124 °CA would lead to air short-circuiting as there is some air mixture just beneath the exhaust valves before EVC with the EVO timing of 124 °CA. A large amount of fresh charge starts to fill the cylinder after the closing of exhaust valves and pushes the residual gas to the top of cylinder until the end of the scavenging at IVC.

In order to quantify the scavenging performance, four scavenge parameters, i.e. delivery ratio (DR), trapping efficiency (TE), scavenging efficiency (SE) and charging efficiency (CE), are defined as following:

$$\mathrm{DR} = \frac{\text{delivered fresh charge mass}}{\text{reference mass}} \tag{1}$$

$$\mathrm{TE} = \frac{\text{mass of delivered fresh charge retained in the cylinder}}{\text{total mass of delivered fresh charge}} = \frac{\mathrm{CE}}{\mathrm{DR}} \tag{2}$$

$$\mathrm{SE} = \frac{\text{mass of delivered fresh charge retained in the cylinder}}{\text{total mass of trapped cylinder charge}} \tag{3}$$

$$\mathrm{CE} = \frac{\text{mass of delivered fresh charge retained in the cylinder}}{\text{reference mass}} \tag{4}$$

The reference mass in above equations is calculated by the displaced volume multiplied by the ambient air density.

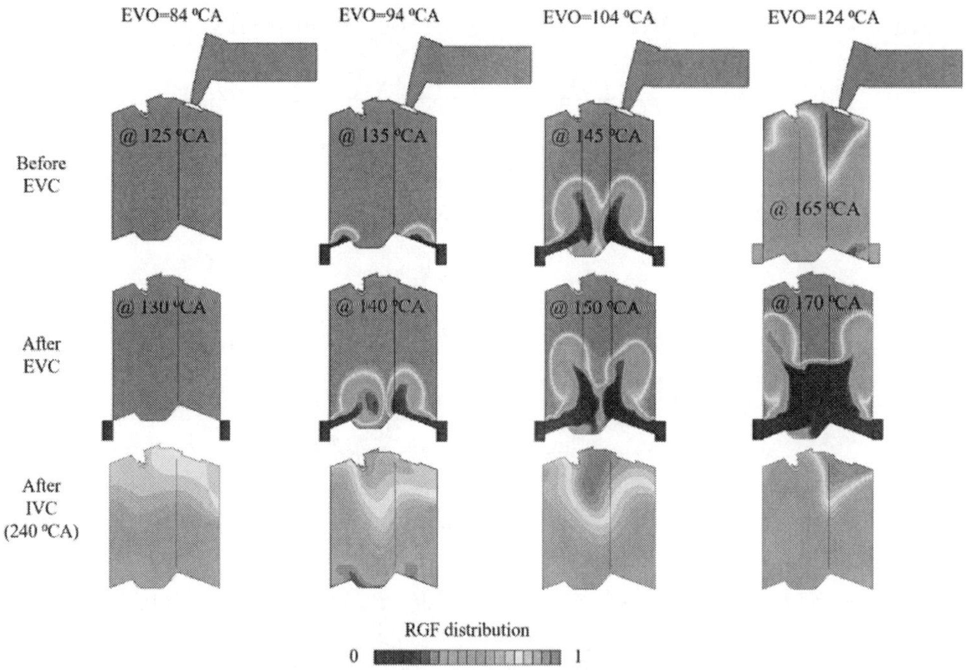

Figure 5. Impact of EVO timing on the evolution of residual gas fraction (RGF) distributions during scavenging.

Figure 6 shows the effect of the EVO timing on delivery ratio (DR), charge efficiency (CE) and scavenge efficiency (SE). As there is no short-circuiting for all EVOs, as indicated in Figure 5, the TE is 1 and DR equals to CE for all cases. As the EVO timing delays from 84 °CA to 104 °CA, both DR/CE and SE show monotonous decreasing trends. Specifically, the DR/CE decrease from 0.66 at 84 °CA to 0.53 at 104 °CA, and SE from 0.43 to 0.35. The reason can be attributed to the shorter blowdown duration with a delayed EVO timing, which increases the scavenging resistance as more residual gas is trapped in the cylinder during the scavenging process. This then leads to both lower DR/CE and SE.

As the EVO delays to 124 °CA, the larger overlap between intake and exhaust process leads to enhanced scavenging process of the in-cylinder residual gas, as shown in Figure 4, thanks to the larger pressure differences between intake and exhaust. This in turn leads to slightly higher DR/CE and SE of EVO 124 °CA than that of EVO 104 °CA.

The scavenging process also affects the in-cylinder large scale flow motions. The in-cylinder swirl ratio (SR) is defined by the following equation (18):

$$SR(\theta) = \frac{\sum_i^m v_i(\theta) r_i(\theta) V_i(\theta) \rho_i(\theta)}{\frac{2\pi n}{60} \sum_i^m r_i(\theta)^2 V_i(\theta) \rho_i(\theta)} \quad (5)$$

where n is engine speed [rpm], θ the crank angle, i the cell number, $V_i(\theta)$ the cell volume, $\rho_i(\theta)$ the cell density, $v_i(\theta)$ and $r_i(\theta)$ are the tangential velocity and radius respectively in the cylindrical coordinate with z axis as the swirl axis.

Similarly, the tumble ratio (TR) and cross tumble ratio (CTR) are calculated by replacing the Z axis in Equation (1) by the X axis and Y axis, respectively, as illustrated in Figure 1.

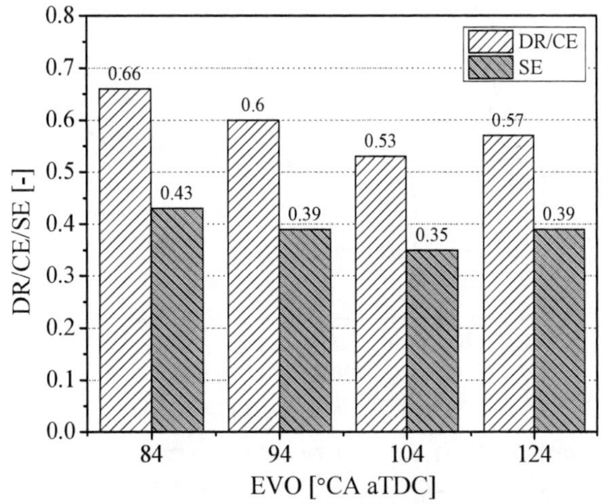

Figure 6. Impact of EVO timing on the delivery ratio (DR), charge efficiency (CE) and scavenge efficiency (SE).

Figure 7 shows the evolutions of the SR, TR and CTR with different EVO timings. The earliest EVO timing of 84 °CA shows stronger swirl flow motions throughout the cycle. The delay of EVO from 94 °CA to 124 °CA leads to lower swirl ratio during and after the scavenging, but produces similar SR level at the top dead center (TDC). In terms of the in-cylinder tumble and cross tumble flow motions, generally they are very weak throughout the cycle, especially around the TDC position due to the substantial decay, for all the EVOs. However, it is still noted the later EVO timing could produce higher peak tumble or cross tumble ratio somewhere during the scavenging, and EVO 104 °CA obtains highest cross tumble ratio around TDC and EVO 124 °CA obtains highest tumble ratio around TDC.

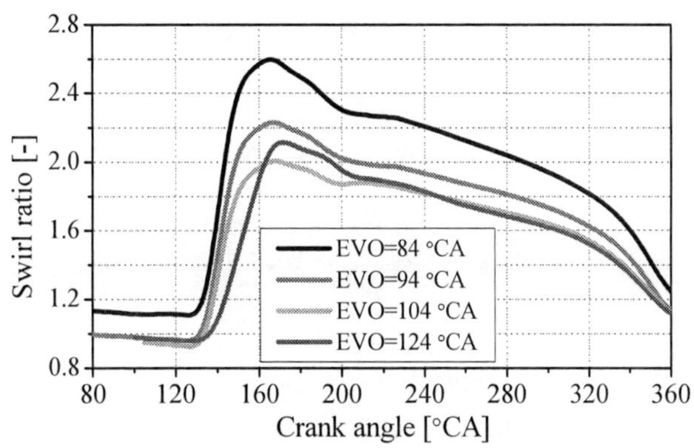

(a) swirl ratio

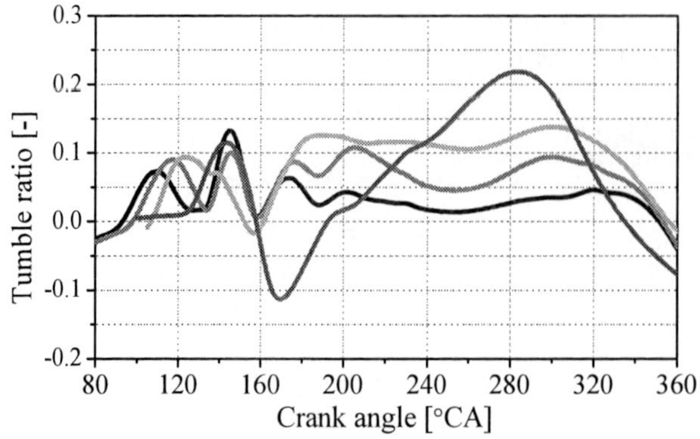

(b) tumble ratio

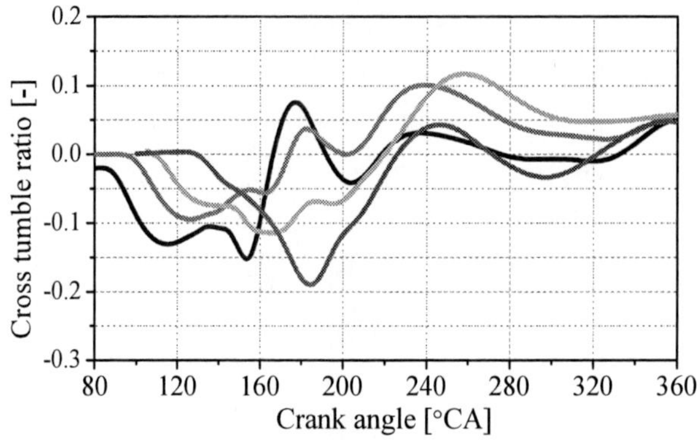

(c) cross tumble ratio

Figure 7. Impact of EVO timing on the in-cylinder (a) swirl ratio, (b) tumble ratio and (c) cross tumble ratio.

Figure 8 and Figure 9 compare the in-cylinder distributions of RGF and turbulence kinetic energy (TKE) respectively for each EVO timing. It is confirmed from Figure 8 that the RGF distribution pattern with more residuals at the top of cylinder persists till the end of compression stroke. The high RGF around the spark plug could affect the initialization and stability of the spark ignitions. Figure 9 indicates that TKE levels for both earliest EVO of 84 °CA and latest EVO of 124 °CA are relatively higher than the intermediate EVOs due to their stronger scavenging flow and swirl/tumble flow motions.

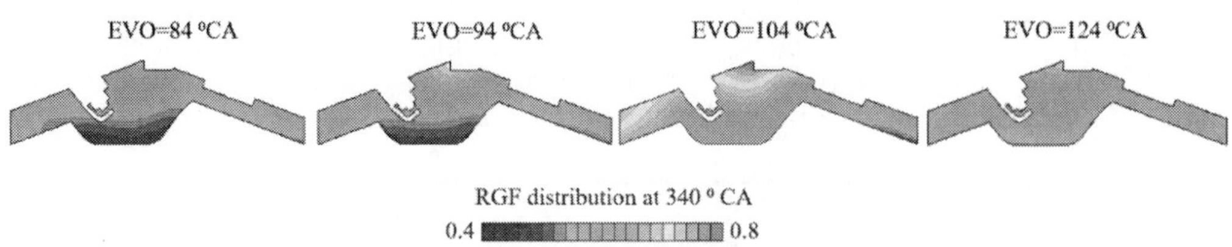

Figure 8. In-cylinder distributions of RGF at 340 °CA for each EVO timing.

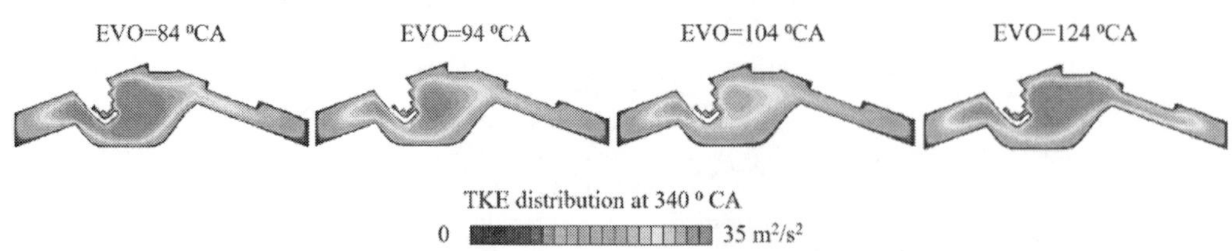

Figure 9. In-cylinder distributions of turbulence kinetic energy (TKE) at 340 °CA for each EVO timing.

5.2 Effect of EVO timings on the combustion process (fired conditions)

In this section, the combustion models are applied to investigate the impact of EVO timings on the combustion process in the BUSDIG engine. Figure 10 shows the pressure traces of multi-cycle simulations with EVO of 84 °CA and 104 °CA. It is noted that the fired 2-stroke operation simulation is extremely sensitive to the initial conditions and multi-cycle simulations are necessary to obtain the convergent simulation result. In this study, the convergent result can be obtained at the ninth cycle for both EVO of 84 and 104 °CA. As the EVO timing delays from 84 °CA to 104 °CA, the in-cylinder fresh charge is reduced, as indicated by the previous analysis of the scavenging performances, this leads to significantly lower peak in-cylinder pressure from 85 bar with EVO of 84 °CA to 68 bar with EVO of 104 °CA. However, it is noted the early EVO timing of 84 °CA leads to early combustion process and negative work during the compression stroke, producing slightly lower indicated mean effective pressure (IMEP) of 4.1 bar than that of EVO 104 °CA with 4.2 bar IMEP.

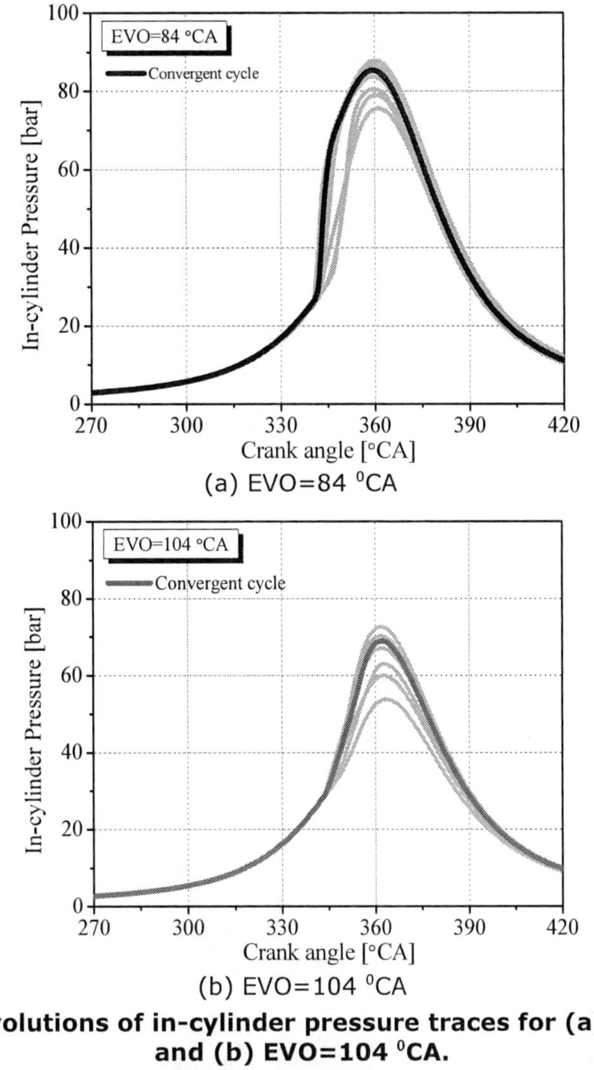

(a) EVO=84 °CA

(b) EVO=104 °CA

Figure 10. Evolutions of in-cylinder pressure traces for (a) EVO=84 °CA and (b) EVO=104 °CA.

Figure 11 compares the heat release rate (HRR) profiles with EVO of 84 °CA and 104 °CA. In both cases, there is no flame propagation process characterized by a slow heat release rate after the spark timing (340 °CA), which means the CAI combustion is achieved for both EVO timings of 84 and 104 °CA. It is noted the auto-ignition occurs right at the spark timing and reaches to its peak value within 5

^{0}CA and the overall combustion duration is around 10 ^{0}CA. The too-early combustion phasing leads to significant negative work and reduces the IMEP of EVO 84 ^{0}CA. In comparison, the combustion process with EVO 104 ^{0}CA is much more moderate and commences at around 4 ^{0}CA after spark timing. The subsequent auto-ignition combustion lasts over 20 ^{0}CA, but the main combustion process is completed before TDC, which also deteriorates the IMEP.

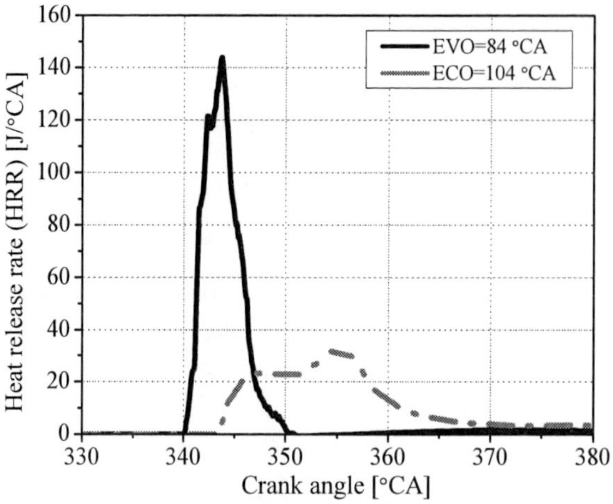

Figure 11. Comparison of the heat release rate (HRR) profiles with EVO of 84 ^{0}CA and 104 ^{0}CA.

In order to understand the different combustion behaviors with EVO 84 and 104 ^{0}CA, Figure 12 compares the average values of RGF, in-cylinder temperature and in-cylinder pressure at 340 ^{0}CA aTDC for each EVO timing. It is noted that the RGF value for late EVO of 104 ^{0}CA is slightly higher, which in turn leads to higher value of in-cylinder temperature. As the overall combustion is much weaker with EVO of 104 ^{0}CA, the average pressure at 340 ^{0}CA is slightly lower. The above analysis of the in-cylinder compositional and thermodynamic conditions indicates that the auto-ignition combustion process is much more sensitive to the dilution effect of in-cylinder residual gas when EVO delays from 84 to 104 ^{0}CA, and the overall more diluted mixture slows down the heat release rate with EVO 104 ^{0}CA although the average temperature is higher.

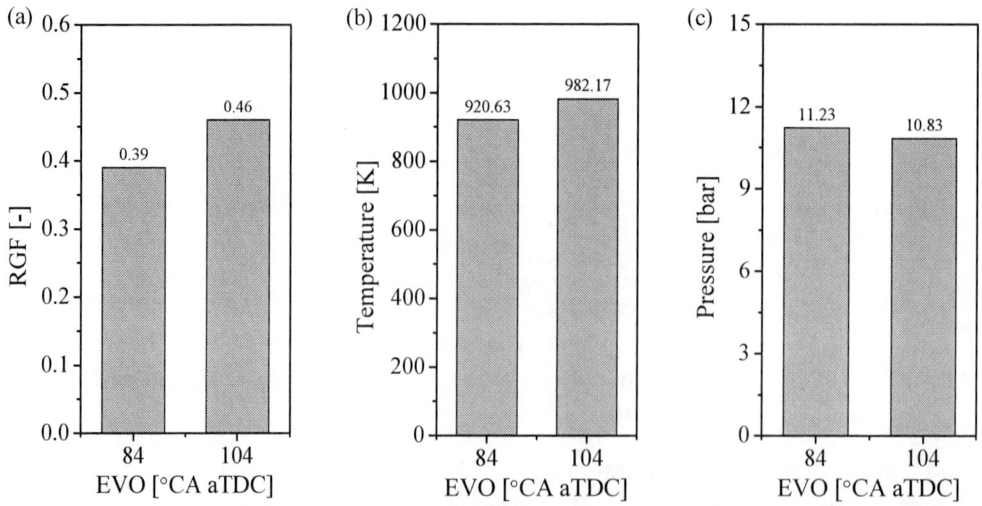

Figure 12. Comparison of the average values of (a) RGF, (b) in-cylinder temperature and (c) in-cylinder pressure at 340 ^{0}CA aTDC with EVO of 84 ^{0}CA and 104 ^{0}CA.

Figure 13 compares the in-cylinder distributions of RGF, temperature and auto-ignition tendency at 340 ^{0}CA for each EVO timing. The high temperature region mainly concentrates in the central region due to less heat transfer loss for each EVO, but the late EVO of 104 ^{0}CA has significantly higher in-cylinder temperature. In comparison, the high RGF region mainly concentrates at the top of cylinder and decreases along the cylinder. It is noted the RGF is extremely higher for late EVO of 104 ^{0}CA that the lowest RGF region at the piston top has slightly higher RGF value than that of the highest RGF region with EVO of 84 ^{0}CA. The over-diluted condition leads to significantly lower auto-ignition tendency, as defined by the close degree of fresh mixture from auto-ignition, in the case with EVO 104 ^{0}CA and the highest auto-ignition tendency appears at the right corner beneath the exhaust valve where the RGF is lowest. In comparison, in the case with EVO of 84 ^{0}CA, the overall auto-ignition tendency is higher and the highest auto-ignition tendency appears at the piston bowl, which also facilitates the propagation of the combustion.

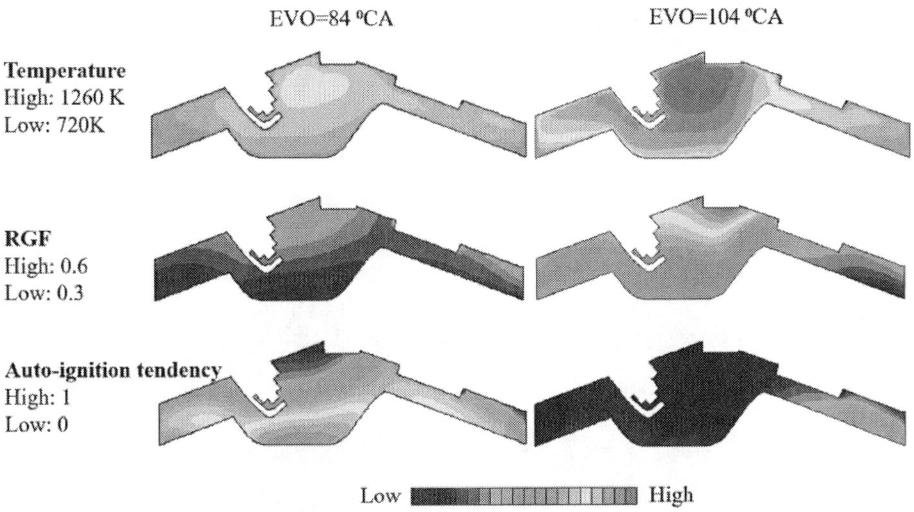

Figure 13. Comparison of the in-cylinder distributions of (a) RGF, (b) temperature and (c) auto-ignition tendency at 340 ^{0}CA aTDC with EVO of 84 ^{0}CA and 104 ^{0}CA.

In this study, the simulation of the combustion process with the later EVO timing of 124 ^{0}CA was also performed. However, it is found the simulation with such a late EVO timing could not achieve the convergence after 9-cycle consecutive simulations. Two distinct combustion modes, i.e. strong combustion and weak combustion cycle, are observed from the pressure traces of the simulation, as shown in Figure 14, and the combustion process jumps between the two modes alternately. Figure 15 shows the intrinsic mechanism of the variations between the strong and weak combustion cycles when applying a late EVO timing of 124 ^{0}CA. A strong combustion cycle produces higher in-cylinder temperature and a large amount of hot residual gas would be retained in the cylinder due to the shorter blowdown duration with a delayed EVO timing. The intake fresh charge for the next cycle is then reduced as a large amount of residual gas occupies the cylinder volume and block the intake charge, which then leads to a weak combustion cycle due to less available fuel/air mixture for combustion. The reduced total in-cylinder charge mass benefits the scavenging process in the next cycle and more intake fresh charge could be retained in the cylinder at the end of scavenging. Then a strong combustion process of more fresh charge can be achieved by either CAI combustion or SI-CAI hybrid combustion thanks to the trapped hot residual gas. The multi-cycle simulations of the combustion process with the late EVO of 124 ^{0}CA indicates that the strong variations of the fresh mixture and hot residual gas between cycles could produce the potential high cycle-to-cycle variations in the future experimental research.

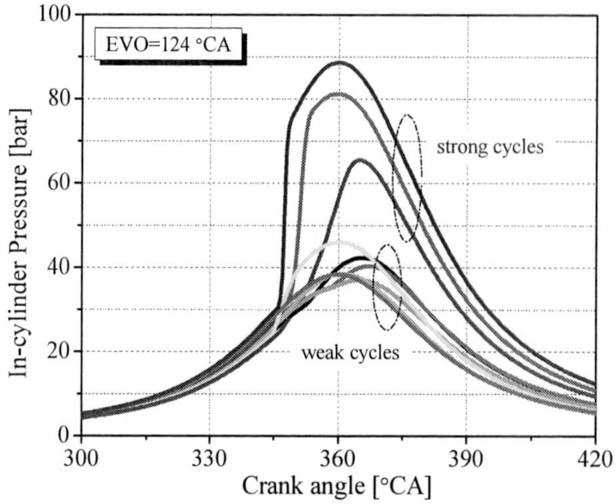

Figure 14. Evolutions of in-cylinder pressure traces with EVO of 124 ^{0}CA.

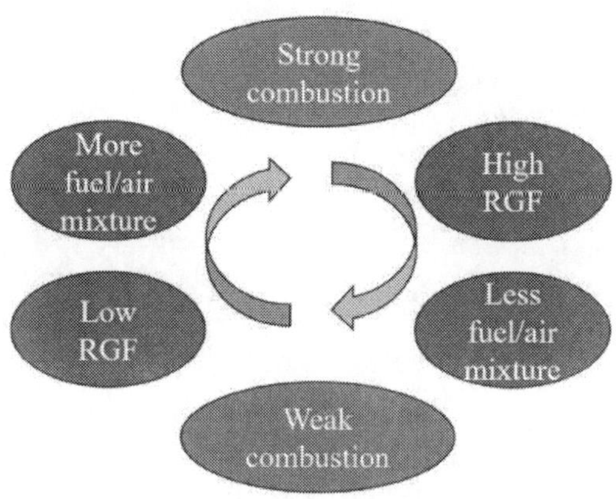

Figure 15. Mechanism of the variation of combustion process with EVO of 124 ^{0}CA.

6. SUMMARY AND CONCLUSION

In this study, the impacts of the low exhaust valve lift profile and the opening timing on the scavenging process and combustion performances in a 2-Stroke Boosted Uniflow Scavenged Gasoline (BUSDIG) engine are investigated by three-dimensional (3D) computational fluid dynamics (CFD) simulations with a validated SI-CAI hybrid combustion model. The main conclusions are summarized as follows:

(1) A low exhaust valve lift profile with a lift of 3 mm and EVO timing ranging from 84 to 124 ^{0}CA was successfully applied in a 2-stroke uniflow scavenged engine to avoid the air/fuel short-circuiting during the scavenging process. The low exhaust valve lift enables the efficient controlled auto-ignition (CAI) combustion to be achieved with premixed fuel/air mixture.

(2) The EVO timing can effectively control the scavenging performances. As there is no short-circuiting for all the studied EVOs, the trapping efficiency (TE) is 1 and the performance of three-way catalyst will not be affected by the presence of excess

oxygen. The delivery ratio (DR) and the charge efficiency (CE) gradually decrease from 0.66 at EVO 84 °CA to 0.53 at EVO 104 °CA, and scavenge efficiency (SE) from 0.43 to 0.35. The further delay of EVO to 124 °CA leads to slight increases of DR/CE and SE due to stronger scavenging process with a larger overlap between intake and exhaust process.

(3) The swirl flow motion is strongest for the earlies EVO timing of 84 °CA. The delayed EVO timing produces higher peak tumble ratio or cross tumble ratio somewhere during the scavenging. EVO 104 °CA leads to the highest cross tumble ratio around TDC and EVO 124 °CA results in the highest tumble ratio around TDC.

(4) The complete controlled auto-ignition (CAI) combustion can be achieved with the low exhaust valve lift with EVO timings of 84 °CA and 104 °CA. The peak in-cylinder pressure decreases from 85 bar at EVO 84 °CA to 68 bar at EVO 104 °CA, and IMEP from 4.1 bar to 4.2 bar. The auto-ignition combustion process becomes much more sensitive to the dilution effect of in-cylinder residual gas when EVO delays from 84 to 104 °CA. The overall more diluted mixture slows down the heat release rate with EVO 104 °CA although the average temperature is higher.

(5) The fired 2-stroke operation simulation is found to be extremely sensitive to the initial conditions and multi-cycle simulations are necessary to obtain the convergent simulation result. The combustion process with the latest EVO timing of 124 °CA alternates between two distinct combustion modes of fast and slow burning, because of large variations of the fresh mixture and hot residual gas between cycles with a later EVO timing.

FUNDING

The authors gratefully acknowledge the financial support by the Engineering and Physical Sciences Research Council (EPSRC). The data of this paper can be accessed from the Brunel University London data archive, figshare at https://doi.org/10.17633/rd.brunel.5371525.v1.

REFERENCES

(1) Zhang, Y., Nora, M.D. and Zhao, H., Comparison of Performance, Efficiency and Emissions between Gasoline and E85 in a Two-Stroke Poppet Valve Engine with Lean Boost CAI Operation. 2015, SAE Technical Paper 2015-01-0827.
(2) Zhang, Y. and Zhao, H., Investigation of combustion, performance and emission characteristics of 2-stroke and 4-stroke spark ignition and CAI/HCCI operations in a DI gasoline. Applied Energy, 2014. 130: p. 244-255.
(3) Zhang, Y. and Zhao, H., Optimisation of boosting strategy for controlled auto-ignition combustion in a four-valve camless gasoline direct injection engine running in two-stroke cycle. International Journal of Engine Research, 2014. 15(7): p. 850-861.
(4) Zhang, Y., et al., CAI combustion of gasoline and its mixture with ethanol in a 2-stroke poppet valve DI gasoline engine. Fuel, 2013. 109: p. 661-668.
(5) Zhang, Y., et al., 2-Stroke CAI combustion operation in a GDI Engine with Poppet Valves. 2012, SAE Technical Paper 2012-01-1118.
(6) Zhang, Y. and Zhao, H., Measurement of short-circuiting and its effect on the controlled autoignition or homogeneous charge compression ignition combustion in a two-stroke poppet valve engine. Proceedings of the Institution of Mechanical Engineers, Part D: Journal of Automobile Engineering, 2012. 226(8): p. 1110-1118.

(7) Abthoff, J., et al., The 2-Stroke DI-Diesel Engine with Common Rail Injection for Passenger Car Application. 1998, SAE Technical Paper 981032.
(8) Laget, O., et al., Preliminary Design of a Two-Stroke Uniflow Diesel Engine for Passenger Car. SAE Int. J. Engines, 2013. 6(1).
(9) Mattarelli, E., Rinaldini, C.A. and Baldini, P., Modeling and Experimental Investigation of a 2-Stroke GDI Engine for Range Extender Applications. 2014, SAE Technical Paper 2014-01-1672.
(10) Zhang, Y., et al., Effects of injection timing on CAI operation in a 2/4-stroke switchable GDI engine. SAE International Journal of Engines, 2011. 5(2011-01-1773): p. 67-75.
(11) Zhang, Y., et al., Experiment and analysis of a direct injection gasoline engine operating with 2-stroke and 4-stroke cycles of spark ignition and controlled auto-ignition combustion. 2011, SAE Technical Paper 2011-01-1774.
(12) Benajes, J., et al., Impact of injection settings operating with the gasoline Partially Premixed Combustion concept in a 2-stroke HSDI compression ignition engine. Applied Energy, 2017. 193: p. 515-530.
(13) Ma, J. and Zhao, H., The Modeling and Design of a Boosted Uniflow Scavenged Direct Injection Gasoline (BUSDIG) Engine. 2015, SAE Technical Paper 2015-01-1970.
(14) Wang, X., Ma, J. and Zhao, H., Evaluations of Scavenge Port Designs for a Boosted Uniflow Scavenged Direct Injection Gasoline (BUSDIG) Engine by 3D CFD Simulations. 2016, SAE Technical Paper 2016-01-1049.
(15) Cd-adapco, Methodology, STAR-CD VERSION 4.14, 2010. 2010.
(16) Mehl, M., et al., An approach for formulating surrogates for gasoline with application towards a reduced surrogate mechanism for CFD engine modeling. Energy & Fuels, 2011. 25(11): p. 5215-5223.
(17) Wang, X., et al., Numerical simulation and validation of SI-CAI hybrid combustion in a CAI/HCCI gasoline engine. Combustion Theory and Modelling, 2013. 17(1): p. 142-166.
(18) Wang, X., Xie, H. and Zhao, H., Computational study of the influence of in-cylinder flow on spark ignition–controlled auto-ignition hybrid combustion in a gasoline engine. International Journal of Engine Research, 2015. 16(5): p. 795-809.

Design and development of a jet ignition engine for stable ultra-lean operation

M P Bunce, H R Blaxill
MAHLE Powertrain LLC, USA
D C Gurney
MAHLE Powertrain Ltd, UK

ABSTRACT

Driven by legislation, social consciousness, and market demand, the automotive sector has been seeking technologies to significantly increase the efficiency of the internal combustion engine in order to conserve fuel resources. Lean combustion in spark ignition (SI) gasoline engines is a well understood method for increasing brake thermal efficiency (BTE). Near-lean operation ($1<\lambda<1.3$), while increasing thermal efficiency, also produces increases in engine-out emissions of nitrogen oxides (NO_x). Combustion stability degrades with enleanment, which results in a lean limit for SI engines. Consequently, homogeneous SI combustion is generally limited to near-lean operation. Ultra-lean operation ($\lambda \geq 2$) has demonstrated the ability to simultaneously increase thermal efficiency and significantly reduce NO_x emissions. Jet Ignition, a pre-chamber-based combustion system, is one such enabling technology.

Several previous studies from the authors [1,2] have proven that a Jet Ignition concept can achieve stable ultra-lean operation in a light duty engine, producing a minimum brake specific fuel consumption (BSFC) value below 200 g/kWh with simultaneous engine-out NO_x emissions less than 100 ppm. Ultra-lean combustion generally and Jet Ignition specifically place unique demands on the engine and supporting systems. Lower exhaust enthalpy resulting from colder exhaust temperatures associated with ultra-lean operation necessitates changes to the boost system. This study seeks to quantify the changes needed to a base engine and boost system to optimize the resulting Jet Ignition engine. A baseline multi-cylinder Jet Ignition engine is tested and performance and emissions data is presented. This data is input to 1D simulations used to evaluate the efficacy of a variety of commercially available boost systems for this ultra-lean application. The boost system evaluation is performed on a dedicated 3-cylinder Jet Ignition engine that will be optimized specifically for Jet Ignition operation. This engine will be assembled and tested, and will be used to confirm the simulation results presented here in a subsequent study.

1 ABBREVIATIONS AND DEFINITIONS

BMEP	Brake mean effective pressure
BSFC	Brake specific fuel consumption
BTE	Brake thermal efficiency
CAD Model	Computer-aided design model
CO	Carbon monoxide

CO_2	Carbon dioxide
COV	coefficient of variance of IMEPg
DI	direct injection
ECU	Engine control unit
g/kWh	Grams per kilowatt-hour
HC	Hydrocarbons
MJI	MAHLE Jet Ignition®
NO_x	Nitrogen oxides
O2	Oxygen
PFI	Port fuel injection
ppm	Parts per million
SI	Spark ignition
VTG	Variable turbocharger geometry
λ	lambda

2 INTRODUCTION

2.1 Background

The perpetual desire to conserve fuel is driving strong demand for increased efficiency in SI engines. A method being increasingly explored to accomplish this goal is gasoline lean combustion [3-5]. Homogeneous lean combustion ($1 < \lambda < \sim 1.6$) in modern SI engines has been proven to increase thermal efficiency. It has also, however, been shown to increase NO_x emissions in the near-lean region, necessitating highly efficient NO_x aftertreatment in order to adhere to criteria emissions regulations [6,7]. Homogeneous ultra-lean combustion with $\lambda > 1.6$ has demonstrated the ability to both increase thermal efficiency and significantly reduce NO_x emissions due to the colder cylinder temperatures innate to combustion with high levels of dilution [7,8]. This has the potential to not only increase overall engine efficiency but also to enable the reduction in scope and/or efficiency requirements of the NO_x aftertreatment solution.

The major limitation in developing lean and ultra-lean combustion systems is the less favorable ignition quality of the mixture. This has necessitated the development of higher energy ignition sources [9,10]. A pre-chamber combustor application is one such technology, having been researched extensively [11-15]. SI engines that utilize pre-chambers generally retain the spark plug but relocate it to the pre-chamber and repurpose it as the ignition source for the air-fuel charge present in the pre-chamber. Products from this combustion event are then transferred to the main chamber through an orifice(s) or valve, thermo-chemically igniting the main chamber air-fuel charge. This effectively establishes a secondary combustion event as the main chamber ignition source, providing greater ignition energy compared to that of a standard single-point spark plug.

Pre-chamber combustion concepts have demonstrated the potential for stable main chamber combustion at higher levels of dilution than are allowable in typical SI engines [16]. They have also demonstrated the ability to counter the loss of ignitability of the lean charge to an extent. This capability results in an extension of the lean limit of the engine. As the air-fuel charge is further enleaned, the portion of this charge entering the pre-chamber becomes regressively ignitable with a standard spark plug. Auxiliary fueling in the pre-chamber can compensate for the enleanment of this incoming charge, extending the lean limit further and enabling the engine to operate in an ultra-lean combustion mode. Direct injection (DI) fuel injector technology has removed one of the major technological barriers and renewed interest in this concept [15], as have modern machining techniques that enable smaller pre-chamber volumes than were previously allowable.

2.2 Jet Ignition

MAHLE Jet Ignition® (MJI) is an auxiliary fueled pre-chamber concept that has been under development for several years [17-19]. The concept incorporates elements studied in previous pre-chamber research including: small pre-chamber volume (< 5% of main combustion chamber clearance volume) for minimizing crevice volume and heat loss, small orifice diameter to promote a high degree of flame quenching, and auxiliary fueling in the pre-chamber to allow separate fueling strategies for pre-chamber and main chamber.

A prototype low-flow DI fuel injector provides a separate fueling event in the pre-chamber. This allows precise, effectively de-coupled control over the mixture in both chambers. This low-flow DI injector also enables the use of a common liquid gasoline for both pre-chamber and main chamber injection. Historically, auxiliary fuel injected in the pre-chamber is gaseous due to metering repeatability and impingement issues with liquid gasoline. Both of these issues have been mitigated through pre-chamber injector development (not described in this study). Up to 3% of total system fuel is injected via auxiliary pre-chamber injection. The remainder is delivered to the main combustion chamber conventionally with port fuel injection (PFI) or DI fueling utilizing off-the-shelf fuel injectors.

A high degree of flame quenching is accomplished by limiting the diameters of the orifices in the nozzle. This quenching aspect of jet ignition is what differentiates it from a torch ignition system [18,20]. Pre-chamber combustion creates a pressure spike in the pre-chamber, forcing contents into the main chamber via the orifices in the nozzle. These resulting jets initiate main chamber combustion through chemical, thermal, and turbulent effects. The chemical effect is a product of the radical species present in the jets. These species are highly reactive and readily ignite the air-fuel mixture present in the main chamber. Containing partially or fully burned combustion products, these jets are at an elevated temperature when they enter the main chamber, above the auto-ignition temperature of gasoline, thereby providing a potential thermal trigger for main chamber combustion. Finally, the jets emerge at a velocity that is presumed to be proportional to the pressure resulting from pre-chamber combustion. The velocity allows the jets to penetrate into the main chamber, entraining the air-fuel mixture in the main chamber as they proceed. This turbulent effect ensures interaction between the turbulent radical jets and the charge producing suitable local mixture conditions for ignition and subsequent flame propagation.

The multi-orifice nozzle produces multiple, distributed ignition sites throughout the main chamber, resulting in short burn durations. A previously published study [2] examines the extent to which the ignition sites can be distributed and the importance of this distribution to thermal efficiency.

The jet ignition system is designed to be a low-cost, practical ultra-lean combustion enabling technology. A rendering of pre-chamber placement in a typical cylinder head is shown in Fig. 1.

Ultra-lean combustion as enabled by the above described Jet Ignition concept creates unique thermal management challenges for modern SI engines. Due to the significant increase in diluent introduced into the cylinder, mean gas temperatures in the main chamber are reduced. Indeed one of the major pathways for the resulting thermal efficiency increase is reduced heat loss both to the exhaust and to the cylinder walls. The reduction in heat loss to the exhaust and thus exhaust enthalpy poses a challenge to a turbocharger application. The turbine located in the exhaust stream relies on the enthalpy of the exhaust gas for operation, especially at high loads. Reduced exhaust enthalpy at ultra-lean conditions could result in unacceptable turbocharger performance, necessitating a shift to costly and complex boost system components.

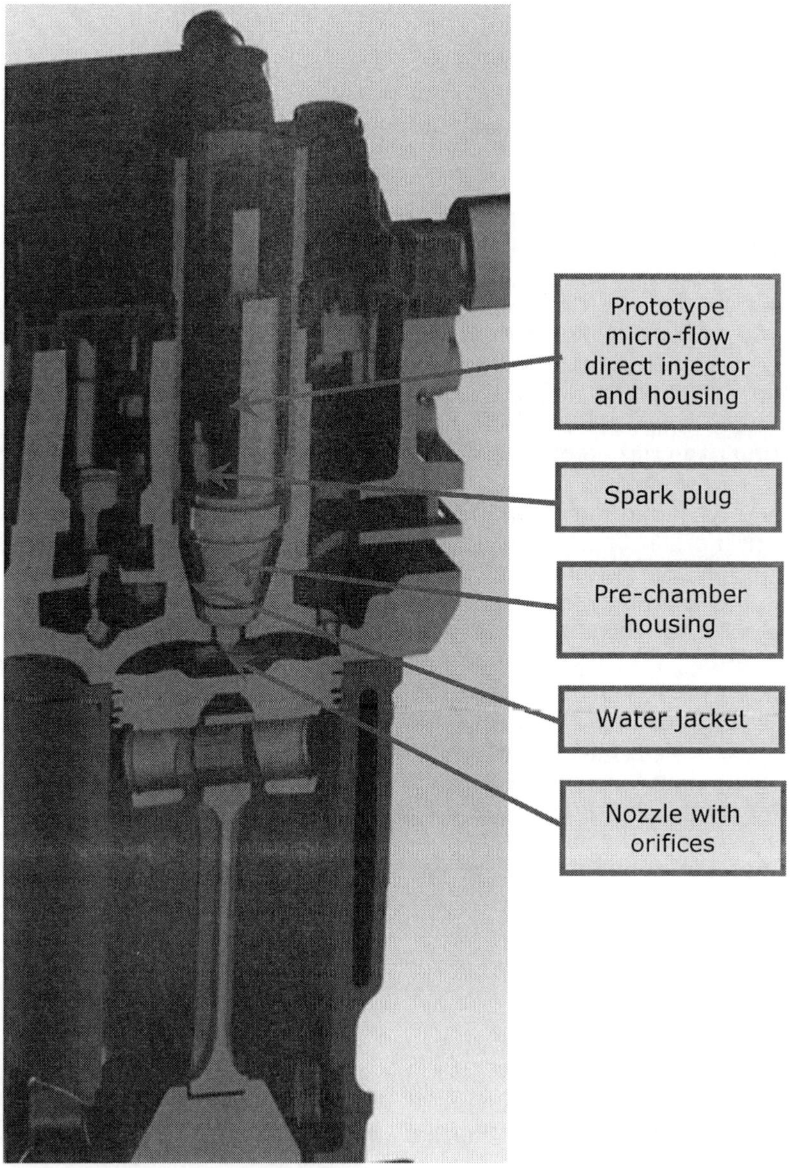

Figure 1. CAD model rendering of a partial cutaway of the pre-chamber assembly in a cylinder head.

2.3 Objective

The objective of this study is to quantify the impact of the Jet Ignition concept on a modern SI engine in order to design a more optimized dedicated Jet Ignition engine. The optimized engine seeks to address the boost system concerns identified in testing of a modern SI engine incorporating Jet Ignition, specifically the challenges associated with reduced exhaust enthalpy resulting from enleanment. Addressing these concerns is intended to move the MJI concept closer to a production-ready state.

3 APPROACH

Data from a multi-cylinder Jet Ignition engine is presented to illustrate the performance and emissions benefits of Jet Ignition-enabled ultra-lean combustion and to quantify the exhaust enthalpy challenge. This engine is primarily based on production engine hardware with some minor modifications in order to incorporate the pre-chamber assemblies. This engine has not been optimized for the jet ignition concept, and simply represents an existing production engine with the cylinder head modified only to the extent that the pre-chamber assembly can fit within it. Because of the challenges associated with lower exhaust enthalpy, the production turbocharger is removed from this engine and is replaced with a boost rig and back pressure valve to emulate nominal turbocharger operation.

Data from this engine is then used as inputs to 1D GT Power analysis, 3D Computational Fluid Dynamics (CFD) conjugate heat transfer analysis, and 3D CFD combustion analysis. The purposes of these analyses are to 1) translate the Jet Ignition concept to an alternative dedicated engine platform, and 2) to optimize the platform design to meet the requirements of the Jet Ignition concept.

This study presents preliminary data contributing to the design of the dedicated Jet Ignition engine platform. The effort is ongoing and intended future studies will demonstrate the remaining steps needed to produce the optimized platform.

4 EXPERIMENT

4.1 Baseline Engine

The baseline multi-cylinder engine evaluated in this study is a 4-cylinder based primarily on Ford production engine architecture. The engine incorporates an identical pre-chamber assembly into each cylinder. The cylinder head is modified to physically accommodate the pre-chamber assembly package. The base DI main chamber configuration is converted to PFI exclusively for this study. As detailed in [1] the compression ratio has been increased from 10:1 to 14:1 due to the knock mitigation properties of Jet Ignition and the adjusted full load target. Engine specifications are listed in Table 1.

Table 1. Baseline engine specifications.

Bore	87.5 mm
Stroke	100 mm
Compression ratio	10:1 (base), 14:1 (tested)
Swept volume	0.601 L per cylinder
Number of cylinders	4
Injection	PFI main chamber (converted from DI), DI pre-chamber
Fuel	Indolene (liquid, both chambers)

Figure 2 shows this engine in the test cell. Figure 3 shows a pre-chamber assembly installed in Cylinder 1.

Figure 2. Baseline multi-cylinder engine in the test cell.

Figure 3. Pre-chamber assembly installed in baseline multi-cylinder engine.

The engine is controlled according to speed and brake mean effective pressure (BMEP). At each operating condition, the engine control unit (ECU) varies throttle position and main chamber fuel quantity to achieve the commanded BMEP at the commanded overall λ. Commanded λ is increased in a stepwise manner. This effectively enleans the main chamber only, as the pre-chamber fueling quantity is adjusted manually, increasing with main chamber enleanment in order to maintain main chamber coefficient of variance (COV) values less than 3%, consistent with the COV limitation of most production light duty gasoline engines. This proceeds until the main chamber fails to ignite consistently (COV > 3%) and the engine reaches its lean limit. Lambda values of 2 or higher are typically achieved.

As the engine is enleaned gradually to λ=2, the naturally aspirated engine is no longer capable of achieving the same high-load BMEP that can be achieved at λ=1, and therefore BMEP reduces as the engine is enleaned. The addition of a boost rig and back pressure valve emulates turbocharged operation. The boost rig is used to maintain BMEP as the engine is enleaned at wide open throttle. Delta pressure across the engine at various speed / load conditions is matched to values from benchmarked production turbocharged light duty engine applications. In lieu of the detailed evaluation of the ability of production turbochargers to provide boost under ultra-lean conditions that follows from this stage of the experiment, this method is believed to provide a reasonably accurate approximation of a lean boost application. Parasitic losses due to additional oil and coolant circuit pump demand for the turbocharger are not included in the data described herein but are believed to constitute a minimal impact to BSFC.

The engine is operated in a "lean boost" configuration, meaning that the target full load curve is similar to that of a comparable displacement naturally aspirated engine (in this case a production 2.5L naturally aspirated engine is benchmarked). The addition of a boost system enables the engine to operate ultra-lean at the target full load curve. The target lambda value range at full load conditions if $1.7 \leq \lambda \leq 2.0$.

In order to provide a comparison with a standard SI engine, a spark plug insert replaces the pre-chamber assembly in the cylinder head. In this configuration, the engine operates as a typical SI engine, with enleanment limited to values typical of PFI SI engines. Note that this engine operates at the elevated compression ratio of the jet ignition configuration (14:1), well above that of the donor SI base engine (10:1). The increased engine efficiency due to this increased compression ratio is offset at mid and high loads by significant knock limitations which restrict spark advance.

Data presented is at the point of minimum BSFC in the λ sweep for each speed and BMEP condition. This applies for all data acquired at 2 bar BMEP and higher. Data acquired below 2 bar BMEP in this study adheres to a minimum COV requirement (with an upper bound of 3%) rather than a minimum BSFC requirement; hence lean limit extension for BSFC reduction in this region is not an expressed goal of this study.

For both engines, DI fuel pressure for the auxiliary pre-chamber fuel is provided externally in the test cell. Injection pressure in the pre-chamber is 100 bar. The friction penalty associated with operating such a pump on engine is minor due to the relatively low injection pressure requirement compared to typical DI systems and the low fuel flow requirement for the pre-chamber injector. Using the above injection pressure and the maximum pre-chamber fuel flow value observed in the dataset, and assuming conservative overall fuel pump efficiency of 60%, the resulting impact on the minimum BSFC value presented here is less than 0.1 g/kWh. This penalty is therefore disregarded for the purposes of this study and is not taken into account in the results presented.

Commanded λ is controlled via a wide-band oxygen (O_2) sensor located in the exhaust manifold. This sensor reading is verified with a calculated λ from measured exhaust emissions. Exhaust emissions are measured using a Horiba MEXA-7100 emissions bench that contains carbon dioxide (CO_2), carbon monoxide (CO), hydrocarbons (HC), O_2, and NO_x analyzers.

4.2 Dedicated Engine

The MAHLE DI3 Downsizing demonstrator engine is selected as the basis of the dedicated Jet Ignition engine due to the authors' familiarity with the engine, access to the design and underlying analyses, and manufacturer agnostic nature of the platform. The engine is a 3-cylinder engine that nominally achieves >30 bar peak BMEP. While nominally a 1.2 liter displacement, a 1.5 liter variant of the DI3 with an elongated stroke has been developed, and this version is used as the basis for the dedicated engine described here. The engine is repurposed here for a boosted ultra-lean application, with a target peak BMEP of approximately 18 bar. Development of the DI3 engine is well documented [21,22]. Table 3 lists the specifications of the nominal DI3 engine (1.2L). The DI3 is depicted in Fig. 4.

Table 2. DI3 engine specifications.

DI3 Specifications		
Configuration	-	In-line 3 Cylinder
Capacity	cm^3	1200
Bore	mm	83.0
Stroke	mm	73.9
Compression Ratio	-	9.3 : 1
Peak Power	kW	120 (5000-6000 min^{-1})
Peak Torque	Nm	286 (1600-4000 min^{-1})
Valve timing	-	Dual VVT (hydraulic)
ECU	-	MAHLE flexible ECU

Figure 4. DI3 engine.

4.3 Analysis

A 1D GT Power analysis has been performed to determine the ability of a variety of production turbochargers to achieve the target full load curve of the dedicated engine. The full load performance curve of a high efficiency production engine of similar displacement (VW 1.5 TSI Evo) has been used as a reference, rather than as a hard target.

The airpath geometry was mostly carried over from the existing DI3 engine. The intake port was redesigned to ensure its suitability for MJI combustion and the flow coefficients adjusted accordingly. Valve lift profiles were taken from the existing DI3 with cam timing optimised for each speed/load point. Turbocharger models were based on steady state maps from MAHLE's turbocharger map database.

The compression ratio was set at 14:1 to match the previous testing and the combustion model was based on measured cylinder pressure data from the baseline MJI engine. This was processed to give maps of 50% burn point and 10-90 burn duration varying with speed, load and λ. The model was run at $\lambda=2$ across the whole engine map for all the results shown here. The interaction of the turbine match on combustion through increased back pressure was investigated and it was found that - within the range considered to be a reasonable turbo match - the variation in residuals with back pressure was low. This is attributed to the 3 cylinder engine configuration which avoids interaction between the blowdown pulses and reduces the sensitivity to back pressure. In this way, the model incorporates the efficiency and exhaust enthalpy data from the baseline engine, adjusting for the 3-cylinder configuration.

5 RESULTS

5.1 Baseline Engine

The BSFC map of the baseline multi-cylinder Jet Ignition engine is presented in Fig. 7. The minimum BSFC location is at 8 bar BMEP from approximately 1900 rpm to 2300 rpm. BSFC values achieved at multiple operating points within this area are at or below 200 g/kWh. There are similarly low BSFC values in an area bounded by 4 bar and 8 bar BMEP and 800 rpm and 1500 rpm. A hypothesis for the apparent de-coupling of these low BSFC regions is the general lack of data at points in between these regions, meaning that the 200 g/kWh BSFC region could be larger than that indicated in Fig. 5. The majority of the map above 4 bar BMEP contains BSFC values at or below 215 g/kWh.

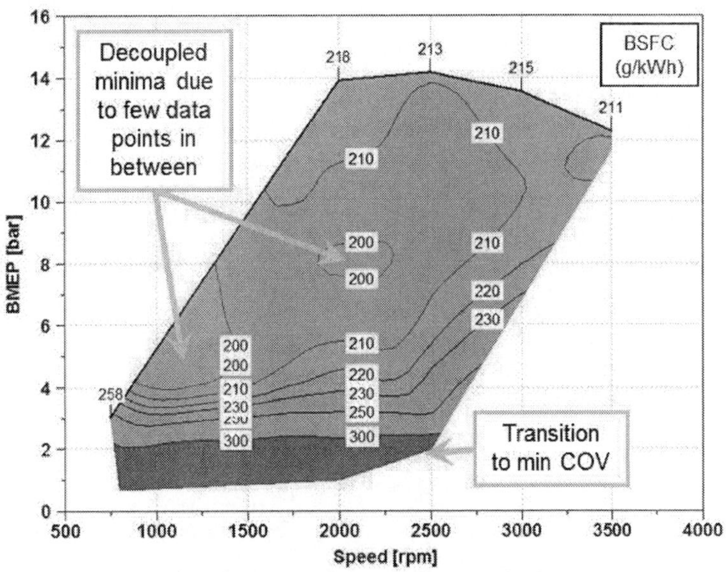

Figure 5. Baseline multi-cylinder engine BSFC map.

The low BSFC values achieved are products of the ability of jet ignition to effectively extend the lean limit of the engine into the ultra-lean region. Figure 6 displays the λ map utilized in this study in order to achieve minimum BSFC results while restricting engine average COV to < 3%. The decreasing enleanment at higher BMEP values is due primarily to a limitation of the boosting system used in the experiment.

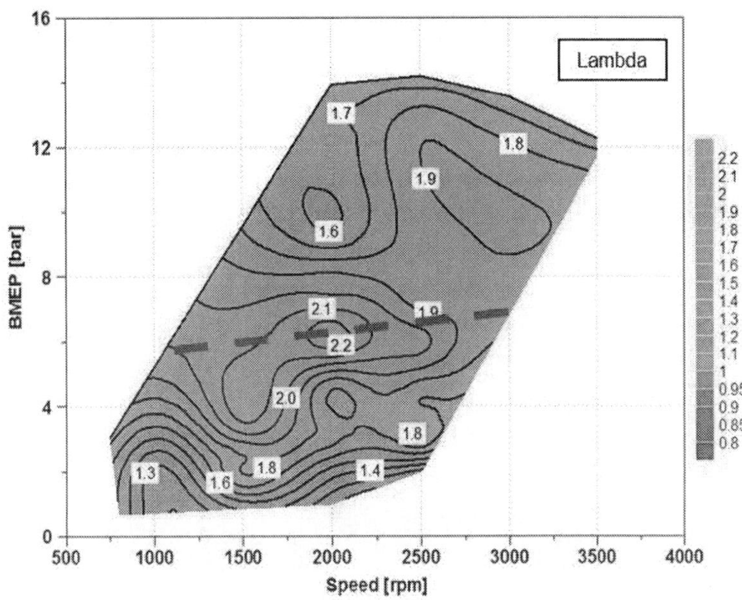

Figure 6. Baseline multi-cylinder engine λ map.

A secondary cause for the decreasing enleanment at higher BMEP values may also be the application of increasing back pressure as boost is added to the engine. The dashed line in Fig. 6 denotes the wide-open throttle line of the baseline engine. Notably, this line occurs at unusually low BMEP values relative to the displacement of the engine. This is due to the significant enleanment of the engine requiring excess air, effectively illustrating the extent of the dethrottling that is possible with an ultra-lean engine. This effect is also illustrated in Fig. 7. Achieving BMEP values above the wide-open throttle line, therefore, requires boosting. As is described in the previous section, back pressure via a back pressure valve is applied when the engine is boosted in order to maintain a constant delta pressure across the engine. The addition of back pressure increases the residual exhaust gas remaining in the cylinder. Quader et al. have demonstrated that the presence of residuals is detrimental to further enleanment of SI engines [3].

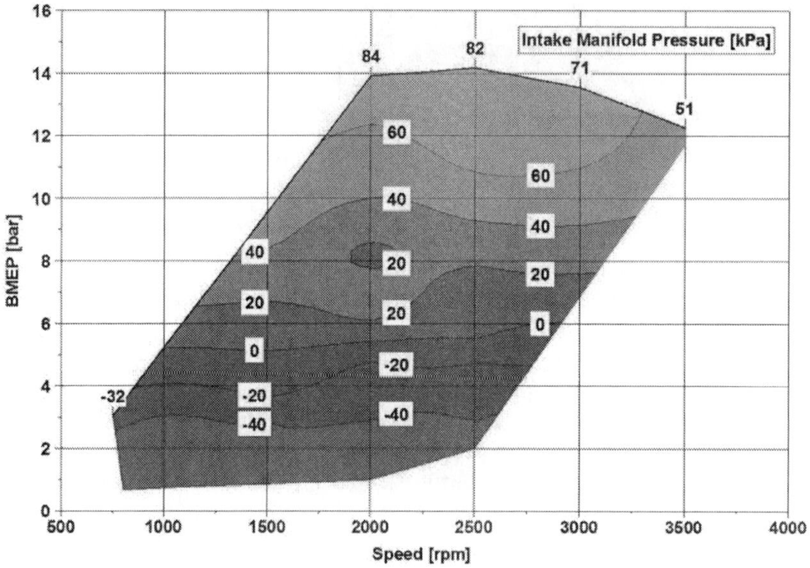

Figure 7. Baseline multi-cylinder engine intake manifold pressure map.

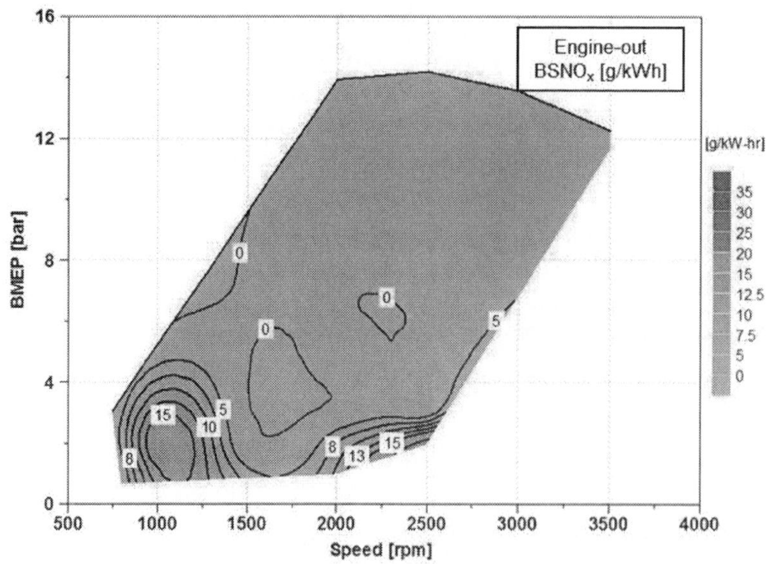

Figure 8. Baseline multi-cylinder engine BSNO$_x$ map.

Significant NO_x reductions are evident with ultra-lean operation. Figure 8 shows engine-out brake specific NO_x ($BSNO_x$) data throughout the engine map. The majority of the map produces < 1 g/kWh NO_x with slight elevations at all BMEP at speeds < 1300 rpm due to the requirement to reduce λ at those conditions. Brake specific HC (BSHC) emissions are displayed in Fig. 9. The BSHC levels are generally comparable to stoichiometric levels. This is believed to be a product of the relatively high combustion efficiency resulting from the high ignition energy jet ignition concept producing a relatively short burn duration (Fig. 10).

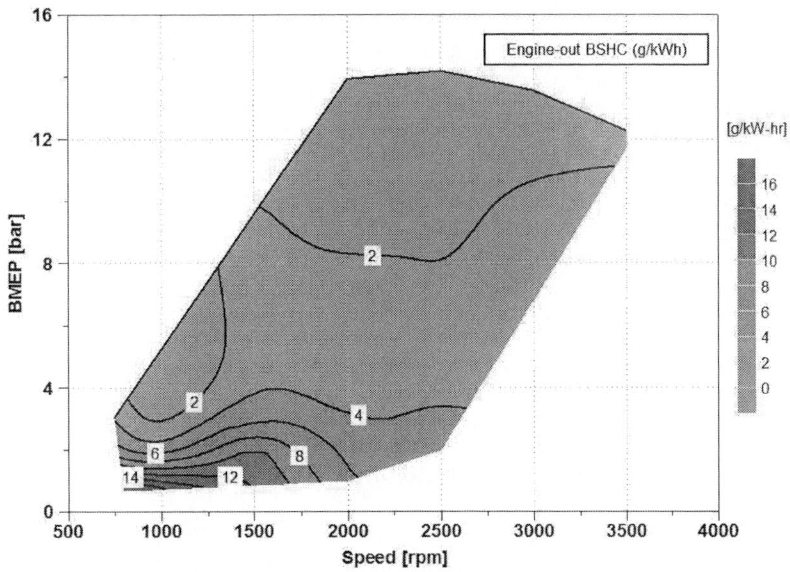

Figure 9. Baseline multi-cylinder engine BSHC map.

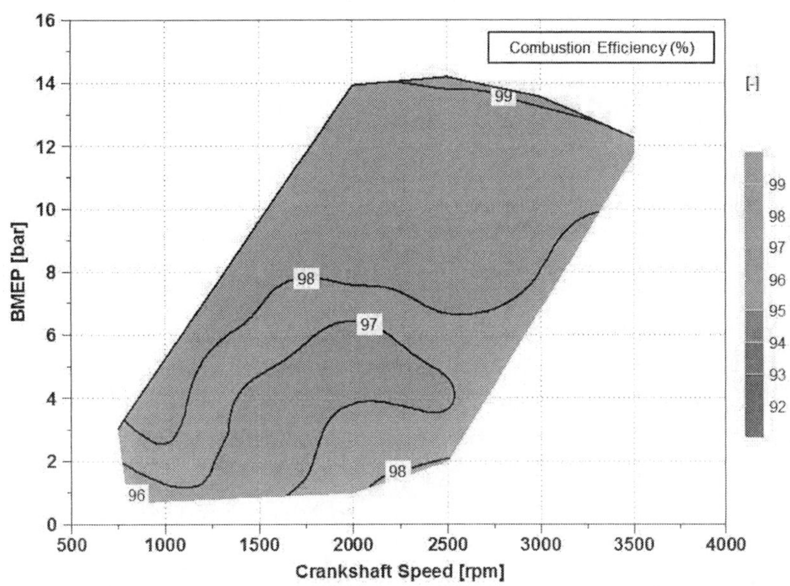

Figure 10. Baseline multi-cylinder engine combustion efficiency map.

The NO_x map represents significant reductions from levels at stoichiometric conditions. Because NO_x is mostly dependent on mean gas temperature and oxygen availability, normalized NO_x reductions follow a nearly identical trend vs. λ at most steady state conditions, as is illustrated in Fig. 11.

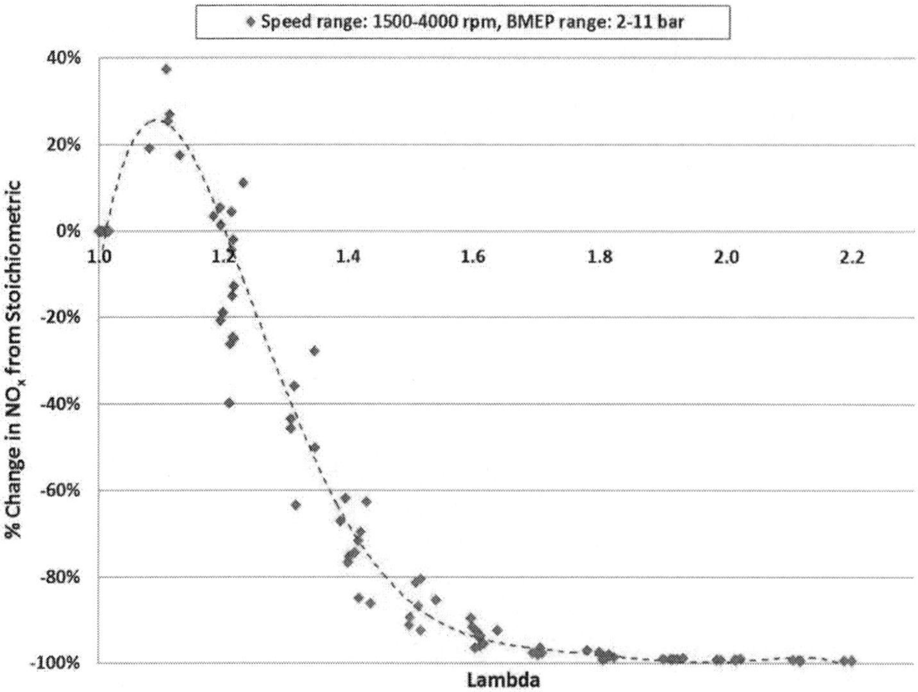

Figure 11. Percentage reduction in NO_x emissions from stoichiometric conditions.

As was previously stated, the excess diluent present in ultra-lean combustion contributes to reductions in exhaust gas temperature. As shown in Fig. 12, there is a linear reduction in exhaust manifold temperature as the engine is enleaned. A >30% reduction in exhaust temperature is observed at λ=2 vs. the level at stoich for multiple steady state points. Figure 13 shows the resulting map of exhaust manifold temperature and Fig. 14 displays this in terms of exhaust enthalpy.

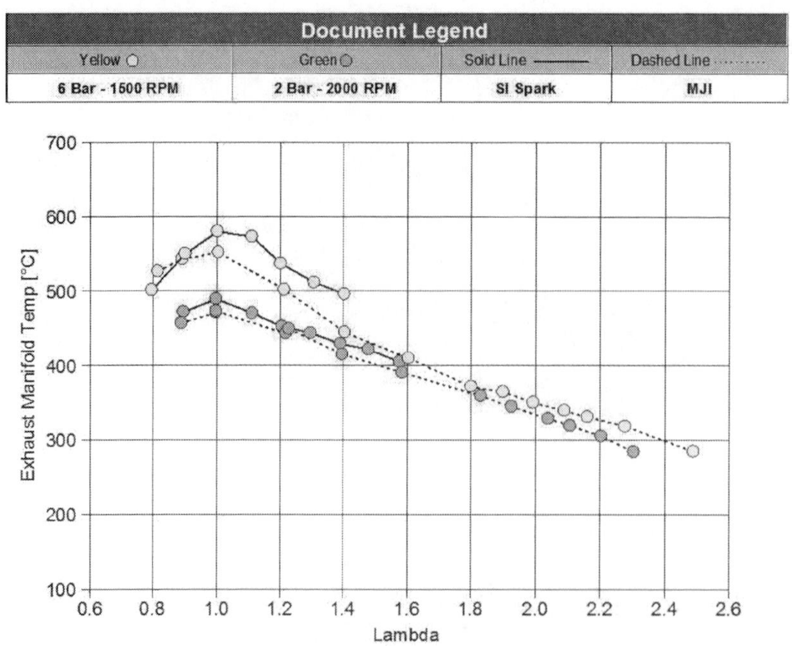

Figure 12. Exhaust manifold temperature vs. λ.

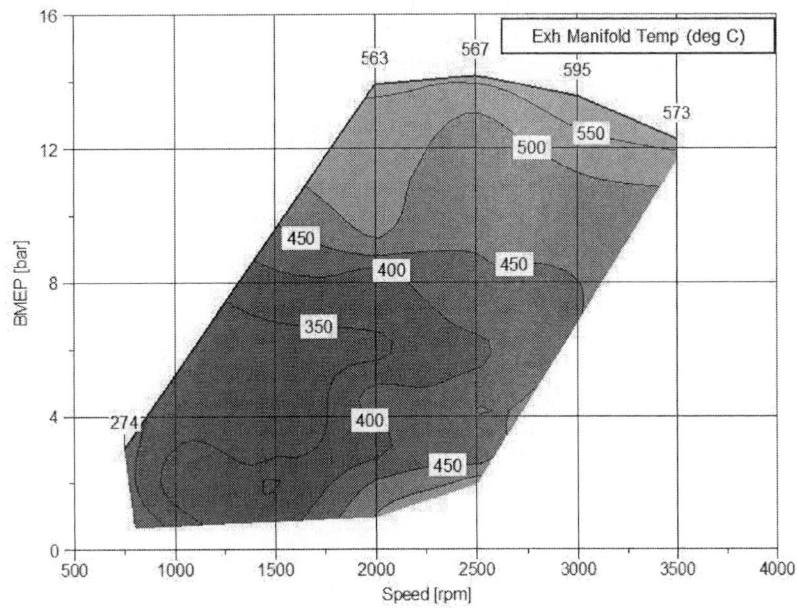

Figure 13. Baseline multi-cylinder exhaust manifold temperature map.

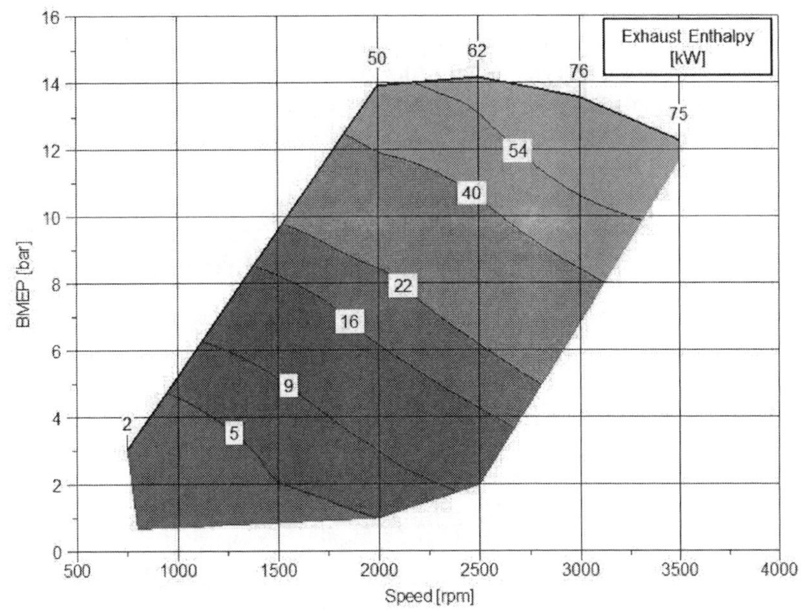

Figure 14. Baseline multi-cylinder exhaust enthalpy map.

5.2 Boost System Analysis

Data acquired on the baseline multi-cylinder engine was input to the GT Power model to evaluate various boost systems as applied to the dedicated 3-cylinder Jet Ignition engine. Figure 15 shows the reference full load curve for this engine. While the peak target BMEP of 17.5 bar is less than the DI3 30 bar BMEP capability, it is believed to be sufficient for a subcompact or compact vehicle, or as part of a hybrid vehicle.

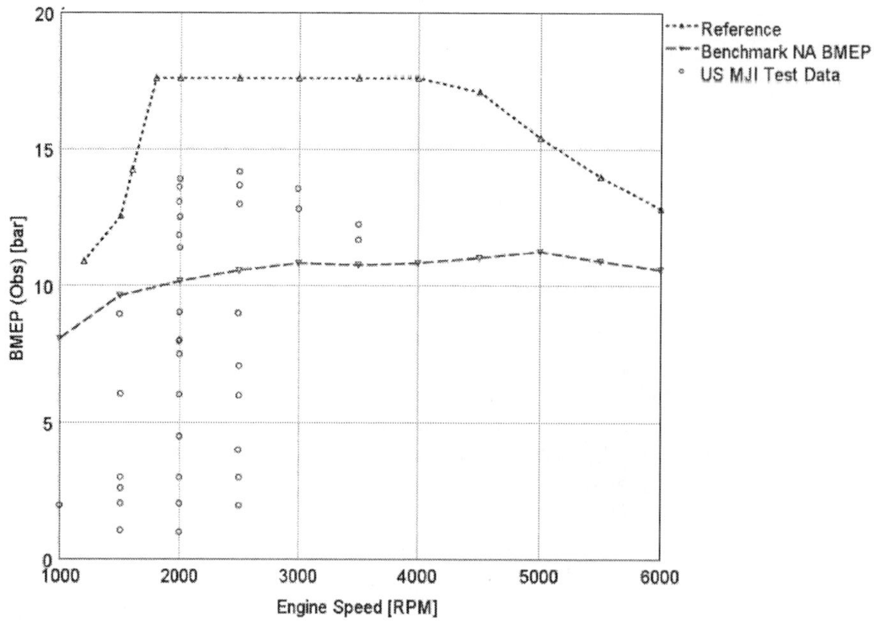

Figure 15. Reference full load curve for dedicated Jet Ignition engine.

Table 3 lists the boost systems evaluated in this study. The systems varied in terms of functionality and configuration, including fixed geometry turbochargers, variable geometry turbochargers (VTG), and multi-stage boost systems. These systems were evaluated using multiple criteria that included ability to meet full load curve requirements, predicted resulting peak BTE, and system complexity as a surrogate for cost.

Table 3. Boost systems evaluated.

Manufacturer	Application	Configuration	Peak Power (kW)	Peak BMEP (bar)	Peak BMEP speed (rpm)	Predicted Peak BTE @ 3000rpm
BMTS	2.0L diesel	Turbo VTG	76	17.2	2500	42.2%
BMTS	MAHLE DI3 CNG VTG	Turbo VTG	66.5	19.6	2500	41.7%
BMTS	MAHLE DI3 Demo Vehicle	Turbo fixed geometry	71.7	17.3	2500	41.3%
MHI	Dynamic downsizing	Turbo fixed geometry	96	17.6	3500	42.7%
-	1.0L gasoline	Turbo fixed geometry	59.5	17.5	2000	39.8%
Rotrex / MHI	C15-20 / TD04 11GFT	Turbo and supercharger	96	17.6	1800	42.7%
Aeristech / MHI	DI3 E-boost / TD04 13GFT	Turbo and e-boost	96	17.6	1800	42.7%

Figure 16 shows the predicted performance of the evaluated boost systems, with Figs. 17-19 showing the performance of individual boost systems. The fixed geometry turbochargers all struggle to achieve the full load BMEP requirement at low speed, due primarily to the reduced exhaust enthalpy present. The smallest turbocharger, from a 1.0l gasoline application, has acceptable low speed torque but is too small for the high airflow required in this application. The small turbine required for good low speed torque was also found to give high pumping work, curtailing the peak BTE of the engine. Conversely, the fixed geometry turbocharger from the MAHLE Dynamic Downsizing engine [20,21] is well suited to meeting the full load requirements at higher engine speeds and also results in one of the highest predicted BTE values of the systems evaluated. This is a relatively large turbocharger that produces positive pumping work in the peak efficiency region of the engine map. Unfortunately this turbocharger is the least suited for the low speed full load demands as it struggles to maintain boost pressure at low engine speeds and loads. This results in achievement of peak BMEP at 3500rpm which is 1000rpm higher than most of the other systems. Based on these results, a fixed geometry turbocharger match was quickly discounted and variable geometry or two stage boosting systems considered.

The VTG turbochargers are better able to match the full load requirements at low speed as the VTG vane size can be closed to increase turbine work, maximizing the low speed airflow. The 2.0L diesel VTG in particular appears to match the full load curve reasonably well. Peak power is below the reference curve but the value of 76kW is still considered acceptable. Note that peak power is achieved at $\lambda=2$ and is limited by the maximum airflow achievable with a single compressor. In addition, the compressor and turbine are operating at reasonably high efficiency in the peak efficiency region of operation and give one of the higher BTEs predicted by the model for any given boosting system.

The two-stage boost systems match the full load curve requirements more precisely, with the supercharged/electrically assisted compressor better able to meet the high boost requirement at low speed when exhaust enthalpy is low. This allows the use of a larger turbine and low pressure compressor to meet the high speed performance target.

This work focused on the extremes of the full load performance prediction; peak power and low speed torque. Further work is necessary to properly manage the transition between the two boosting systems. The Aeristech compressor was simulated with the turbochargers in the same configuration as the MAHLE Dynamic Downsizing demonstrator car, with the Aeristech compressor upstream of the turbocharger compressor. The Rotrex compressor was located downstream of the turbocharger compressor. Interestingly, both configurations gave similar predicted performance values, but with very different operating points on the compressor maps. The downstream supercharger proved easier to control during the transition between supercharged and turbocharged operation than the upstream e-booster although this is attributed to the relative location of the two compressors rather than any significant difference between the Rotrex and Aeristech devices.

Both two-stage systems also achieve the highest BTE of all the options evaluated. This is because at the peak efficiency operating point, the supercharger is operating at a low pressure ratio and effectively not performing any work on the system, thereby minimizing parasitic losses. Peak BTE is achieved at the highest load that can be reached with the supercharger/ebooster inactive. BTE then decreases as engine load increases due to the drive power required for the supercharger stage of the boost system. Note that the e-booster drive power was calculated and assumed to be taken directly from the crankshaft, not taken from a battery.

The simulations show how the key requirement for any boosting system is its ability to operate with the low enthalpy exhaust gas at low engine speeds. The single stage turbocharger options investigated were shown to be incapable of providing sufficient boost under these conditions and were discounted. The two stage boosting systems avoid this limitation to give the best overall performance but at the cost of significant complexity. Instead, a VTG has been selected as the preferred boost system for the dedicated Jet Ignition engine application due to its acceptable adherence to the full load target, resulting high BTE, expected ability to maintain a relatively high BTE at higher engine speeds, and relatively low cost when compared to multi-stage systems.

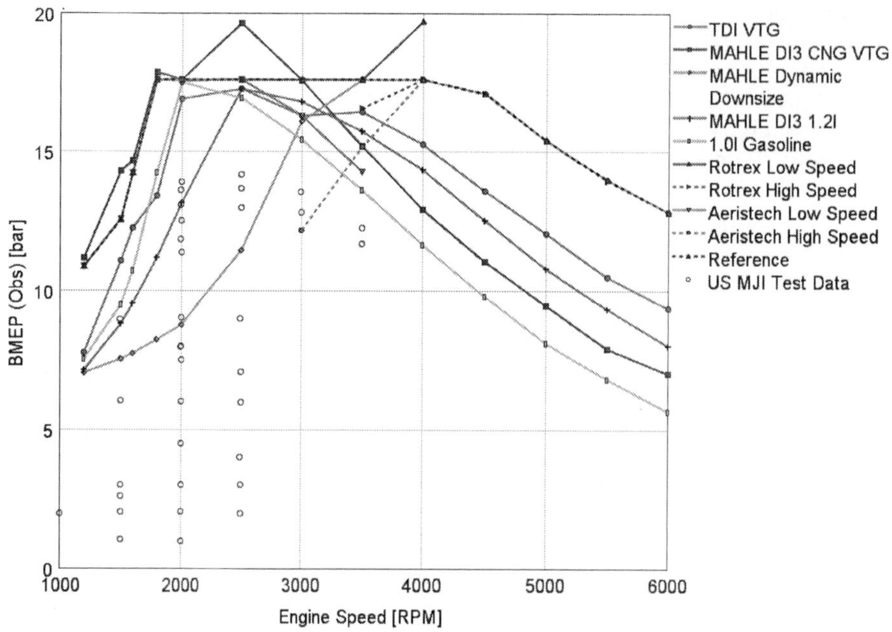

Figure 16. Predicted full load curves for various boost systems applied to the dedicated Jet Ignition engine.

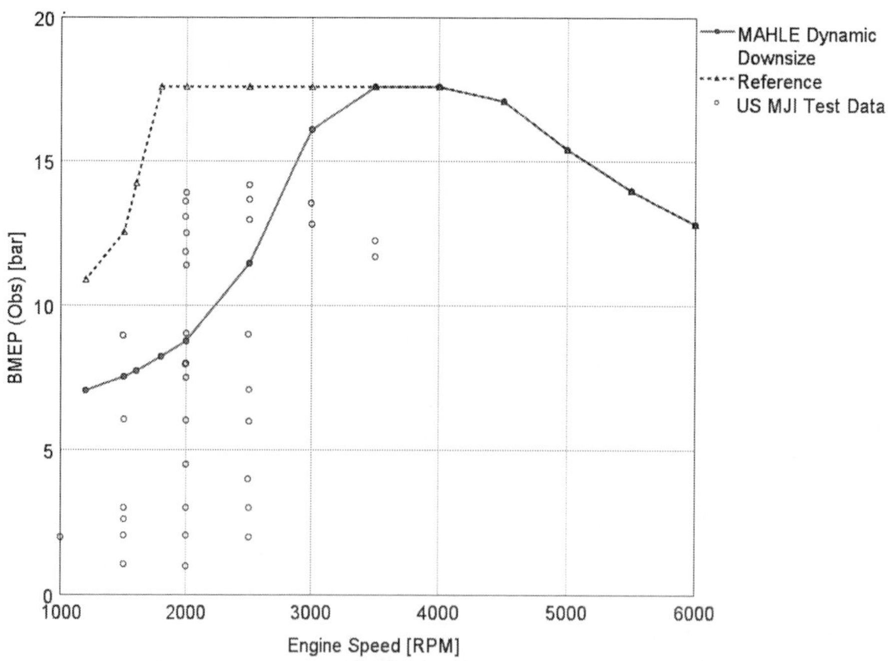

Figure 17. Predicted full load curve for MAHLE Dynamic Downsize Turbocharger (without E-booster) applied to the dedicated Jet Ignition engine.

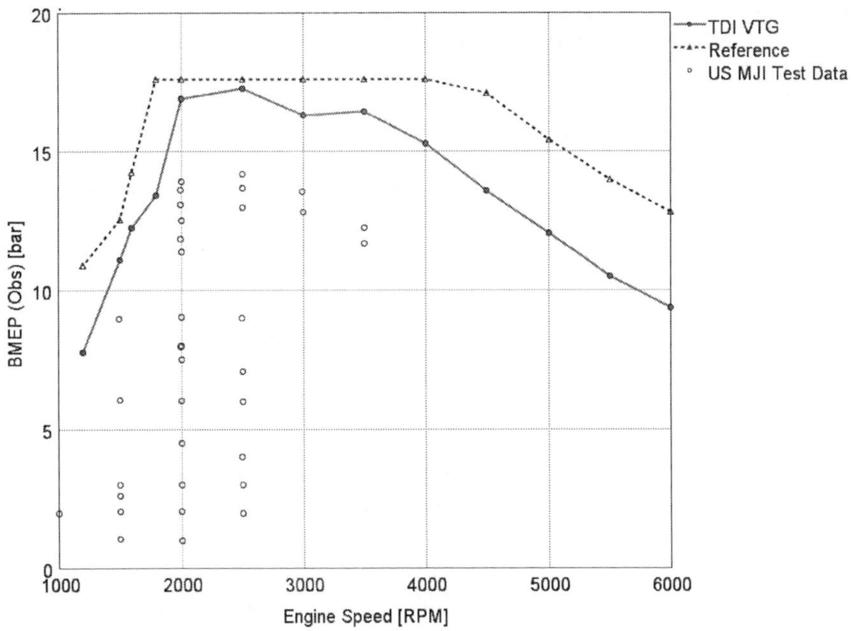

Figure 18. Predicted full load curve for 2.0L TDI VTG applied to the dedicated Jet Ignition engine.

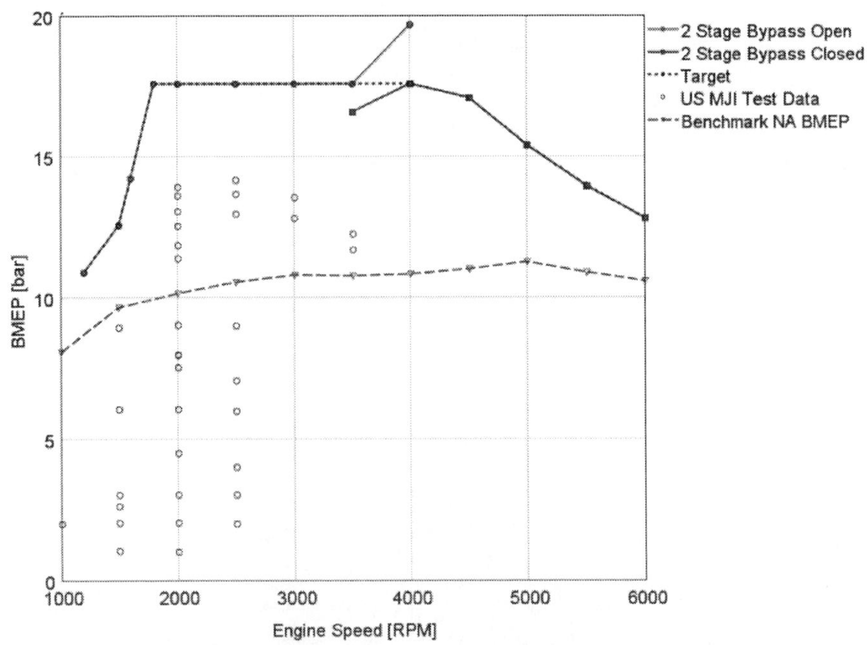

Figure 19. Predicted full load curve for Rotrex C15-20 / Dynamic Downsize turbocharger applied to the dedicated Jet Ignition engine.

The operating point on the compressor maps for the 2.0l VTG and the two stage boosted system with Rotrex supercharger are shown below. The difficulty in achieving the reference performance can be seen from the VTG compressor map. The high flow rate and pressure ratio required to achieve performance at $\lambda=2$ results in the operating point running up the surge line at low speed and hitting the choke line at high speed. The high pressure ratio required at low massflow demonstrates the requirement for the VTG turbine to provide sufficient drive power.

The two stage boosting systems allow the use of a larger low pressure turbocharger, enabling the peak power performance to be met. The low pressure compressor size has been selected to make peak power at peak compressor efficiency. Note how the pressure ratio required of the supercharger is relatively low. The additional airflow provided by the supercharger results in an increase in exhaust gas enthalpy, used by the turbine to drive the low pressure compressor, the operating point of which runs up the surge line at low speed. This 'feedback loop' allows the turbocharger to do the majority of the work, maximizing the efficiency of the system.

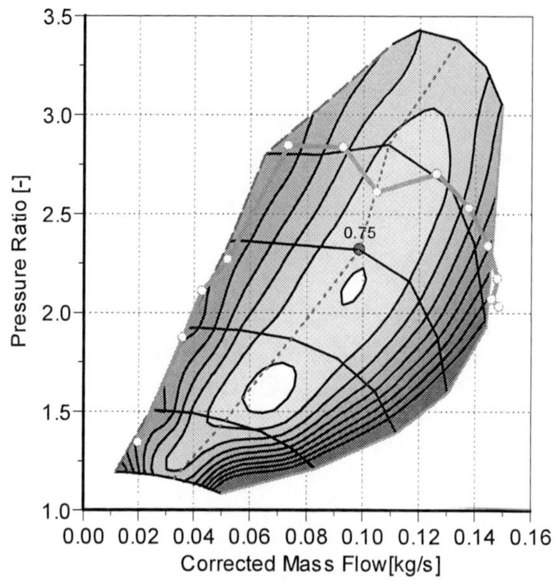

Figure 20. Predicted Operating Point on Compressor Map, TDI VTG.

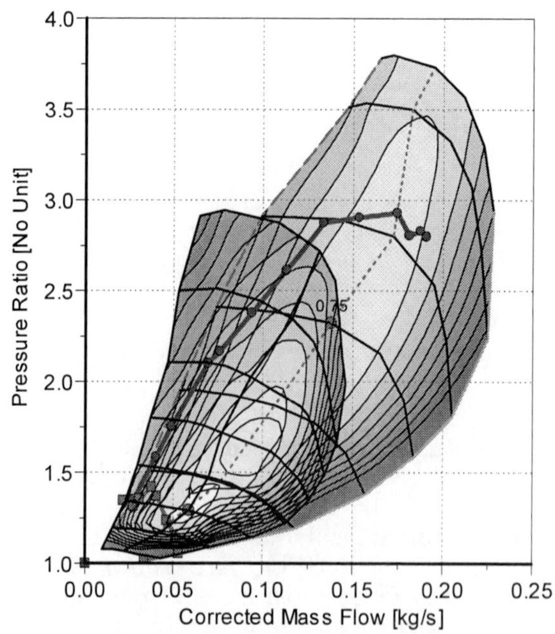

Figure 21. Predicted Operating Point on Compressor Map, Rotrex C15-20 / Dynamic Downsize Turbocharger.

6 CONCLUSIONS

Previous studies [16-19] have demonstrated the potential of jet ignition as an enabling technology for ultra-lean SI combustion to significantly increase thermal efficiency, reduce BSFC, and reduce NO_x emissions. As indicated by these trends, ultra-lean combustion is a substantively different combustion mode than typical stoichiometric operation. As such it introduces system-level concerns that need to be incorporated into the development of a dedicated Jet Ignition engine in order to fully exploit these benefits.

Reduced exhaust enthalpy inherent to ultra-lean combustion poses a challenge for traditional turbochargers. This study quantifies the challenge through baseline engine testing of a production engine platform with design modifications to incorporate the Jet Ignition pre-chamber assembly package. A dedicated Jet Ignition engine platform is selected and data from the baseline engine is input into a 1D GT Power model in order to evaluate the efficacy of multiple commercially available boost systems when applied to the target engine. A VTG boost system is selected based on the ability of the resulting system to achieve the full load targets and maintain a high predicted BTE (>42%) while minimizing system complexity. The selection of a commercially available boost system indicates that the boost requirements of a Jet Ignition engine operating ultra-lean over the complete engine map can be met with existing boost technology and does not necessitate a shift towards systems of significantly increased complexity.

Future work includes full thermo-mechanical analysis on the engine structure, specifically the cylinder head. The engine will be assembled and tested to verify the simulation results and will serve as a platform for further Jet Ignition evaluation.

More generally, the Jet Ignition concept will continue to be developed on various engine platforms, including the engine described in this study. Aside from determining an appropriate boost system for ultra-lean operation, other challenges to be addressed include determining an effective aftertreatment package and optimizing cold start operation.

7 ACKNOWLEDGEMENTS

The authors wish to acknowledge Ian Reynolds and Graham Irlam of MAHLE Powertrain. The authors wish to thank the DOE Vehicle Technologies Office for funding a portion of this study, as well as Steve Wooldridge at Ford Motor Co for his contributions.

8 REFERENCES

1. Bunce, M. and Blaxill, H., "Sub-200 g/kWh BSFC on a Light Duty Gasoline Engine," SAE Technical Paper 2016-01-0709, 2016.
2. Bunce, M., Blaxill, H., Kulatilaka, W., and Jiang, N., "The Effects of Turbulent Jet Characteristics on Engine Performance Using a Pre-Chamber Combustor," SAE Technical Paper 2014-01-1195, 2014.
3. Quader, A. A., "Lean Combustion and the Misfire Limit," SAE Technical Paper 741055, 1974.
4. Husted, H., Piock, W., Ramsay, G., "Fuel Efficiency Improvements from Lean Stratified Combustion with a Solenoid Injector," SAE Technical Paper 2009-01-1485, 2009.
5. Germane, G., Wood, C., Hess, C., "Lean Combustion in Spark-Ignited Internal Combustion Engines – A Review," SAE Technical Paper 831694, 1983.

6. Heywood, J., Internal Combustion Engine Fundamentals, McGraw-Hill, 1988.
7. Yamamoto, H., "Investigation on Relationship Between Thermal Efficiency and NO_x Formation in Ultra-Lean Combustion," SAE Journal Paper JSAE 9938083, 1999.
8. Dober, G. G., Watson, H. C., "Quasi-Dimensional and CFD Modelling of Turbulent and Chemical Flame Enhancement in an Ultra Lean Burn S.I. Engine," Modeling of SI Engines SP-1511, 2000.
9. Ward, M., "High-Energy Spark-Flow Coupling in an IC Engine for Ultra-Lean and High EGR Mixtures," SAE Technical Paper 2001-01-0548, 2001.
10. Qiao, A., Wu, X., "Research on the New Ignition Control System of Lean- and Fast-Burn SI Engines," SAE Technical Paper 2008-01-1721, 2008.
11. Ricardo, H. R., Recent Work on the Internal Combustion Engine, SAE Transactions, Vol 17, May 1922.
12. Gussak, L. A., Karpov, V. P., Tikhonov, Y. Y., "The Application of the Lag-Process in Pre-chamber Engines," SAE Technical Paper 790692, 1979, doi:10.4271/790692.
13. Robinet, C., Higelin, P., Moreau, B., Pajot, O., Andrzejewski, J., "A New Firing Concept for Internal Combustion Engines: "l'APIR"," SAE Technical Paper 1999-01-0621, 1999.
14. Murase, E., Ono, S., Hanada, K., Oppenheim, A., "Pulsed Combustion Jet Ignition in Lean Mixtures," SAE Technical Paper 943048, 1994.
15. Toulson, E., Schock, H., Attard, W., "A Review of Pre-Chamber Initiated Jet Ignition Combustion Systems," SAE Technical Paper 2010-01-2263, 2010, doi:10.4271/2010-01-2263.
16. Attard, W., Toulson, E., Fraser, E., Parsons, P., "A Turbulent Jet Ignition Pre-Chamber Combustion System for Large Fuel Economy Improvements in a Modern Vehicle Powertrain," SAE Technical Paper 2010-01-1457, 2010, doi:10.4271/2010-01-1457.
17. Attard, W., Kohn, J., Parsons, P., "Ignition Energy Development for a Spark Initiated Combustion System Capable of High Load, High Efficiency and Near Zero NO_x Emissions," SAE Journal Paper JSAE 20109088, 2010.
18. Attard, W., Blaxill, H., "A Gasoline Fueled Pre-Chamber Jet Ignition Combustion System at Unthrottled Conditions," SAE Technical Paper 2012-01-0386, 2012, doi:10.4271/2012-01-0386.
19. Attard, W., Blaxill, H., "A Lean Burn Gasoline Fueled Pre-Chamber Jet Ignition Combustion System Achieving High Efficiency and Low NO_x at Part Load," SAE Technical Paper 2012-01-1146, 2012, doi:10.4271/2012-01-1146.
20. Bassett, M., Hall, J., Cains, T., Underwood, M. et al., "Dynamic Downsizing Gasoline Demonstrator," SAE Int. J. Engines 10(3):2017.
21. Bassett, M., Hall, J., Hibberd, B., Borman, S. et al., "Heavily Downsized Gasoline Demonstrator," SAE Int. J. Engines 9(2):729-738, 2016.

VEHICLE AND SIMULATION

The impact of fuel octane sensitivity and RON on the efficiency of a DISI engine

R F Cracknell [1], V Pellicciari [1,3], A Prakash [2], A Aradi [2], C Wang [1,4]
[1] Shell Global Solutions, UK
[2] Shell Global Solutions, USA
[3] University of Bath, UK
[4] University of Birmingham, UK

ABSTRACT

Higher fuel octane allows for spark timing advance and consequently higher torque output and higher engine efficiency. Increased market demand for efficient gasoline vehicles has driven the development of highly downsized, boosted direct injection spark ignition (DISI) engines, which can in particular benefit from higher octane.

Inter-cooling in boosted engines and the introduction of cooled EGR can lead to a much lower temperature for a given pressure in a real engine as compared to the test conditions in the CFR engine used to define the Research Octane Number (RON) and Motor Octane Number (MON) octane rating scales. The relevance of MON to modern engines has been questioned, and there is a growing body of evidence to suggest that for a given RON, it may actually be beneficial to have a high octane sensitivity (RON-MON) or in other words a lower MON.

To investigate this further, tests have been conducted in a single cylinder DISI engine using three different compression ratios, and fuels with two different levels of RON but with three octane sensitivity levels ranging from 5 to 15. Processing the experimental engine data with GT-Drive, the behaviour of a full-vehicle in response to variations in fuel octane quality was predicted. The New European Drive Cycle (NEDC) and Worldwide harmonized Light vehicles Test Cycle (WLTC) were simulated on a Chassis Dynamometer (CD). Increased RON and octane sensitivity were both found to have beneficial effects on engine thermal efficiency and fuel consumption.

1 INTRODUCTION

Driven by increasing market demands for higher fuel efficiency, gasoline engines have drastically increased in complexity over the past two decades to incorporate direct injection (DI), turbocharging, cylinder deactivation, intercooling and EGR. With engine downsizing becoming increasingly popular, fuel octane quality can play a significant role towards improving engine performance and reducing fuel consumption [1, 2].

Under high-load conditions the engine spark must frequently be retarded from maximum brake torque (MBT) timing to prevent knock onset; this prevents engine damage but reduces efficiency [3]. Knock is the auto ignition of the compressed end-gas ahead of advancing flame front initiated by the spark plug. Higher octane fuels allow operation closer to MBT by improving anti-knock behaviour of the fuel/air mixture, and hence can provide gains in thermal efficiency when operating in knock-limited operating conditions [3, 4, 5].

Given its impact on engine efficiency, understanding the octane appetite of modern gasoline engines has become highly relevant to the effective development of automotive fuels [3]. The octane quality is defined using two industry standards: RON and MON, which are measured on a single-cylinder standardized naturally aspirated carburettor engine designed in 1929, known as the Cooperative Fuel Research (CFR) engine [6,7].

The aforementioned advances in engine technology, particularly with the introduction of EGR and inter-cooling in conjunction with boosting, can lead to significantly lower engine temperatures than experienced in the CFR engine for the same pressures [8, 9]. Although it is generally accepted that a higher RON is beneficial to anti-knock behaviour and thus thermal efficiency, the relevance of MON to modern engines is increasingly being challenged [3]. A growing body of data suggests that for a given RON, a lower MON or higher octane sensitivity can benefit engine efficiency and therefore reduce fuel consumption when operating at near-knocking conditions [4].

The Octane Index (OI) is an octane rating characteristic that takes into account the unique operating conditions of an engine, in an attempt to reconcile the differences in temperature and pressure between a real engine and the CFR engine [3].

Equation 1: $\mathrm{OI} = \mathrm{RON} - \mathrm{K} \times \mathrm{S} = \mathrm{RON} - \mathrm{K} \times (\mathrm{RON} - \mathrm{MON})$

Where K is a unique engine operating condition and S=RON-MON is octane sensitivity.

Most investigations into the effects of octane index have found that modern engines exhibit negative K values when operating at near-knocking conditions [4]. A negative K value means that the engine will be more knock resistant using fuels with a higher RON and a higher sensitivity [4]. Previously, studies have been based on fuels where RON and MON are deliberately decorrelated in order to separate the impact of these parameters, and the fuels themselves tend to be bespoke blends rather than fuels from a particular market [10, 11].

To further investigate how increasing octane quality can benefit the end-customer, six E10 fuels with market relevant RON/MON were designed and tested in a single cylinder DISI engine, under a range of speed-load conditions. A GT-drive full-vehicle model was developed using the experimental SCE engine response to the different fuels [12]. WLTC and NEDC testing conditions were simulated and variations in fuel consumption were analysed [13, 14]. Increasing RON and octane sensitivity were both found to have beneficial effects on engine thermal efficiency and fuel consumption.

2 SINGLE CYLINDER ENGINE TESTING

2.1 Fuel matrix

The six fuels tested are reported in Table 1. They all contain approximately 10% ethanol by volume and were blended in Shell Technical Centre Hamburg (STCHA) without any deposit control additives or other performance boosting additives. It is possible to blend fuels of varying sensitivity through controlling the relative amounts of paraffins (having low sensitivity), and high sensitivity components (e.g. aromatics, olefins). Given the study's focus on octane sensitivity, the fuels are split into two subsets of constant RON but varying MON. F2 is representative of regular grade gasoline in the North American market while F5 represents premium grade in the same market. The properties of all fuels were tested using standard methods before being transferred to the single cylinder engine.

Table 1: Fuel Matrix Properties

	RON	MON	Density (Kg/m^3)
F1	92	87	718.0
F2	92	82	742.3
F3	92	78	731.1
F4	98	93	698.6
F5	98	88	730.0
F6	98	83	749.0

2.2 Experimental facilities and single cylinder engine

A 4-stroke single-cylinder AVL DISI research engine was used to carry out all testing, the setup is shown in Figure A1. The Compression Ratio (CR) of the engine can be varied between 9.5:1, 10.5:1 and 11.5:1 by inserting metal sheets of varying sizes between the cylinder liner and crankcase; these are denoted CR1, CR2 and CR3 respectively. The 4-valve pent-roof cylinder head with centrally mounted piezo direct injector and Variable Valve Timing (VVT) system for all four valves make up the engine's combustion system. The spark plug was centrally located within the combustion chamber.

An electric dynamometer was used to maintain the engine operating at a constant speed (±1 RPM) for all power outputs. Real time combustion indication and analysis was achieved using an AVL Indicom system. An IAV FI2RE management system was used to control the engine. A Siemens CATs system controlled the fuel, air, coolant, oil conditioning units and emissions measurement equipment during operation. The CATs system also carried out signal acquisition and recording during the engine's operation.

The in-cylinder pressure was measured using a Kistler pressure transducer, installed in a sleeve on the intake and exhaust bridge. A charge amplifier (ETAS ES630.1) was used to collect pressure measurements. In the range between 30° and 70° CA After Top Dead Centre (ATDC) the data was recorded at a resolution of 0.1° CA, in the remaining range at a resolution of 1° CA. Key measurement locations for operating pressures and temperatures are labelled with P and T respectively in Figure A1.

Boosted air (up to 3 bar) was delivered to the engine using a connected external air handling device. Before entering the engine, the air was dried, filtered and delivered to a 200L conditioning unit where the air pressure and temperature were tightly controlled using a closed-loop approach. An AVL fuel mass flow meter was used to control fuel consumption and individual AVL conditioning systems controlled the fuel, coolant and lubricant during engine operation.

2.3 Experimental procedure

The experimental methodology was separated in two main phases outlined in table A1. In the first phase, F2 and CR3 (11.5) were used to determine the engine's optimization parameters such as optimum injection mapping and valve timings. These remained unchanged throughout all testing thereafter. In the second phase F1-F6 were tested for all compression ratios at a number of speed-load conditions, described in table 2. Tests for F2 CR2 (10.5) were repeated at the start, middle and end of the procedure to verify consistency of results.

Table 2: Speed-Load conditions tested

Engine Speed (RPM)	Engine IMEP (bar)
1000	2, 4, 6.5 ,8, 9.5, 12, 14, 16
1800	2, 4, 6.5 ,8, 9.5, 12, 14, 16, 18, 20, 22
2500	
3500	

Before each experimental set-up, the fuel system was flushed with the new sample and the compression ratio was manually adjusted as previously described. An engine warm up period of about 30 minutes was required to warm up oil and coolant to temperatures of 360K and 355K respectively. Engine leakage was monitored and oil dilution checked using hydrocarbon emissions data. Daily engine operating check points were performed to ensure correct operation of all measurement systems and engine, after which the testing protocol begun.

With each testing condition the spark timing was gradually advanced until MBT was reached, estimated by placing MFB50 at 7.5 ±0.5° ATDC. In some cases the onset of knock prevented reaching MBT advance and the testing was carried out in Knock-Limited Spark Advance (KLSA). At 1800 RPM the tolerated threshold for engine knock was 2 bar cycle-based maximum in-cylinder pressure oscillation. The CATs software was used to take 10 Hz measurements for in-cylinder, exhaust and intake pressures as well as ignition and injection signals, which were recorded onto the AVL Indicom system every 100 cycles.

By interpolating Indicated Specific Fuel Consumption (ISFC) measurements for all operating conditions tested, an ISFC map can be approximated.

2.4 Impact of octane quality on thermal efficiency

During SCE testing, ISFC values under steady-state operations were obtained. The indicated thermal efficiency (ITE) can hence be found using equation 2 below.

Equation 2: $\eta = \frac{1}{ISFC \times LHV}$,

where LHV=fuel lower heating value.

Using the above formula three efficiency metrics were defined: $\eta_{part\text{-}load}$ (ITE at 1800 RPM, 8 bar IMEP), $\eta_{high\text{-}load}$ (ITE 3500 RPM, 12 bar IMEP) and η_{Max} (maximum ITE); these are presented for all fuels at CR2 in Table 3. The first two conditions were selected to represent part-load and high-load engine operation.

Table 3: Indicated Thermal Efficiency for F1-F6 at CR2

Fuel	RON	S	η_{Max}	$\eta_{part\text{-}load}$	$\eta_{high\text{-}load}$
F1	92	5	38	35.1	37.1
F2		10	37.8	35.8	36.8
F3		15	39.1	36.3	38.2
F4	98	5	39	36.9	37.8
F5		10	39.8	36.7	38.9
F6		15	38.9	36.7	39

Using these results, a number of observations can be made in relation to the SCE octane response. Overall thermal efficiency increased with a higher fuel sensitivity and RON, although different trends are observed between the two fuel subsets (F1-F3 & F4-F6). In the first subset (RON92), only small η_{Max} variations were obtained between F1 (38.0%) and F2 (37.8%). For the same 5-point sensitivity increase, a much larger efficiency benefit was obtained between F2 and F3 (η_{Max} increased by 1.3%).

In the second subset (RON98), $\eta_{part\text{-}load}$ remained nearly unchanged (0.2% variation) between F4, F5 and F6 in spite of a 10-point sensitivity increase. On the other hand, a 1.2% increase in $\eta_{high\text{-}load}$ was observed between F4-F6. The K value of an engine varies with its operation on the speed-load map: it is generally accepted that for higher-load operation K decreases [14]. With reference to the fuel Octane Index (OI = RON - K×S), it can therefore be understood why $\eta_{high\text{-}load}$ (high-load operation with lower K value) exhibits a stronger response to sensitivity variations than $\eta_{part\text{-}load}$ (part-load engine operation with higher K).

Increasing RON produced greater benefits in thermal efficiency than observed when increasing Sensitivity. A 6-point increase in RON produced η_{max} gains of +1.0%, +2.0% and +0.2% for fuels of sensitivity 5, 10 and 15 respectively. On the other hand, a 5-point increase in Sensitivity produced η_{max} gains of -0.2%, +1.3% (RON92), +0.8% and -0.9% (RON98).

Throughout the study RON and sensitivity both proved beneficial towards thermal efficiency in a configuration representative of a modern DISI engines. RON was found to be the most relevant octane rating parameter as it produced the greatest impact on engine performance.

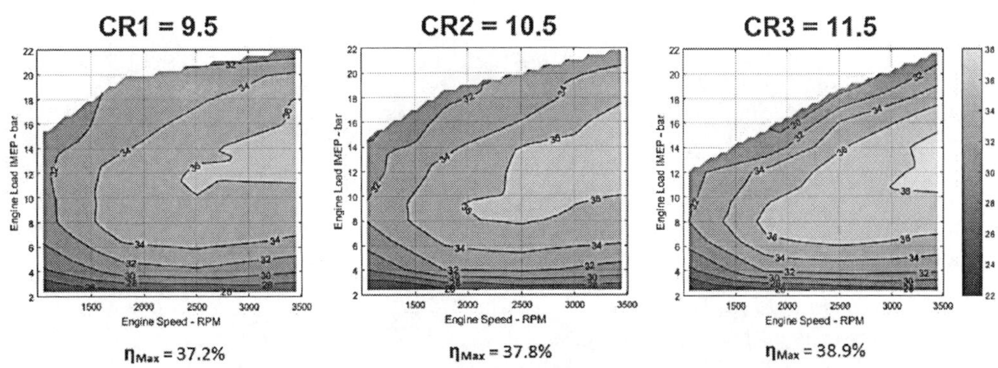

Figure 1: Indicated Thermal Efficiency of F2 for all compression ratios

2.5 Impact of compression ratio on thermal efficiency

Indicated thermal efficiency maps for F2 at all three compression ratios are presented in figure 1; a number of observations can be made from these.

At higher compression ratios the engine becomes more susceptible to knocking, due to higher in-cylinder pressures. The earlier onset of knock reduces the engine's speed-load operating range. In fact, at a speed of 1800RPM, F2 CR1 tolerated a load of 20 bar IMEP, whereas only 14 bar IMEP was possible at CR2 before excessive knocking.

Although the engine operating range is reduced, a higher compression ratio produces significant benefits in maximum thermal efficiency. For F2, increasing CR from 9.5 to 11.5 led to η_{max} gains of 2.7%. The benefits of combining higher octane fuels with higher compression ratios as part of this study are discussed further in reference [3].

3 SIMULATING DRIVE CYCLES USING GT-DRIVE

In the previous section, the study analysed the impact of RON and Sensitivity on engine efficiency under constant speed-load conditions.

In this section, the study analyses the impact of RON and MON on full-vehicle performance using the GT-Drive simulation environment. This approach aims to analyse the impact of fuel octane quality using a parameter that is ultimately more relevant to the end-user: total fuel consumption.

3.1 Model setup

Within GT-Drive a simple vehicle model was established using the tyre, axle, brake, vehicle, differential and transmission standard library components as shown in figure 4. The model was developed with kinematic properties representative of a typical European small-sized family vehicle. The most significant kinematic properties for all subsystems are reported in Table B1.

A map-based engine model predicted engine performance for full-vehicle operation. A Brake Mean Effective Pressure (BMEP) map and a Friction Mean Effective Pressure (FMEP) are required to define engine performance [12]; a Brake Specific Fuel Consumption (BSFC) map is required to compute vehicle fuel consumption.

The FMEP map for full load was adapted from the Audi 3rd generation 1.8L 4-cylinder EA888 engine shown in figure A2. For part loads, FMEP was assumed to be proportional to IMEP. Having obtained the FMEP map, both BSFC and BMEP were calculated. The BSFC maps for F1-F6 at CR2 are reported in figure 2.

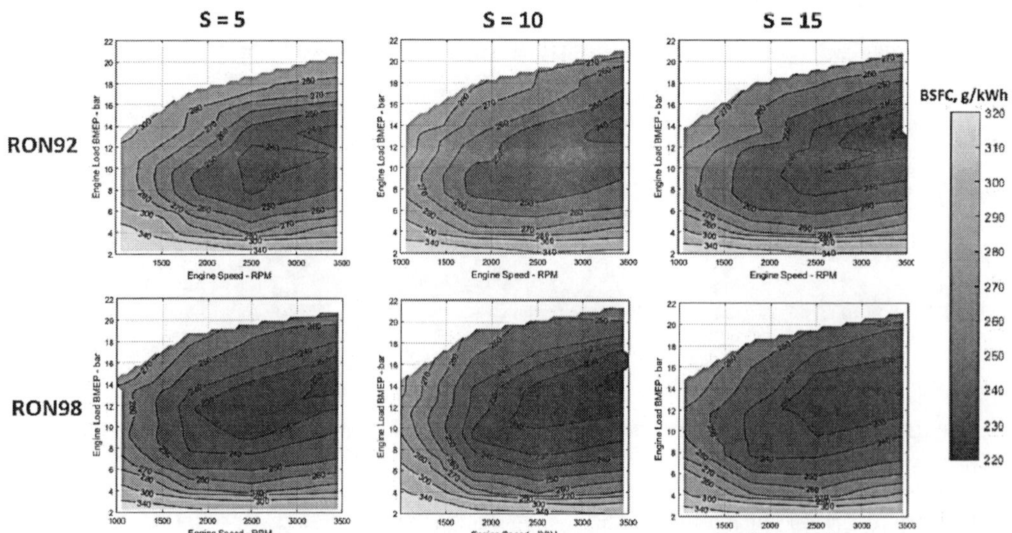

Figure 2: BSFC maps for F1-F6 at CR2

Over a full drive cycle, differences along BSFC maps led to variations in total fuel consumption. Hence, it becomes possible to estimate how fuel octane quality impacts total fuel consumption over a drive cycle without a detailed combustion model. F1-F6 were modelled with equal density and heating value in GT-Drive to obtain a constant volumetric energy density, and hence have no impact on overall fuel consumption. A control system representing a "virtual driver" was designed using the Controller, ICE controller and Driver standard components as shown in Figure 3. The simulations were executed in dynamic forward-facing mode, meaning a set of system inputs (throttle, brake, shift signals, etc.) were used by the PID controller to match a given velocity-time profile (the system output) [15].

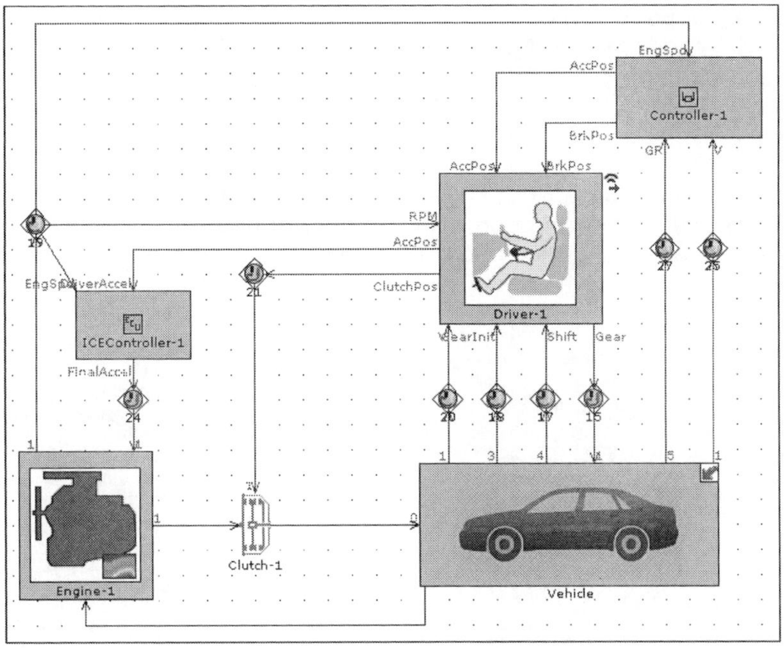

Figure 3: Dynamic, forward model closed-loop controller

Initially, the NEDC velocity-time profile was defined and the full-vehicle drive cycle simulation was executed using the BSFC map for F2CR2. The up- and down-shift speeds were tuned to maintain engine RPM below 3500RPM (maximum speed tested in SCE) and the simulation was repeated for all fuels at CR2. This procedure was replicated for the WLTC class 3 drive cycle. During all simulations fuel consumption was recorded at a sampling frequency of 1Hz.

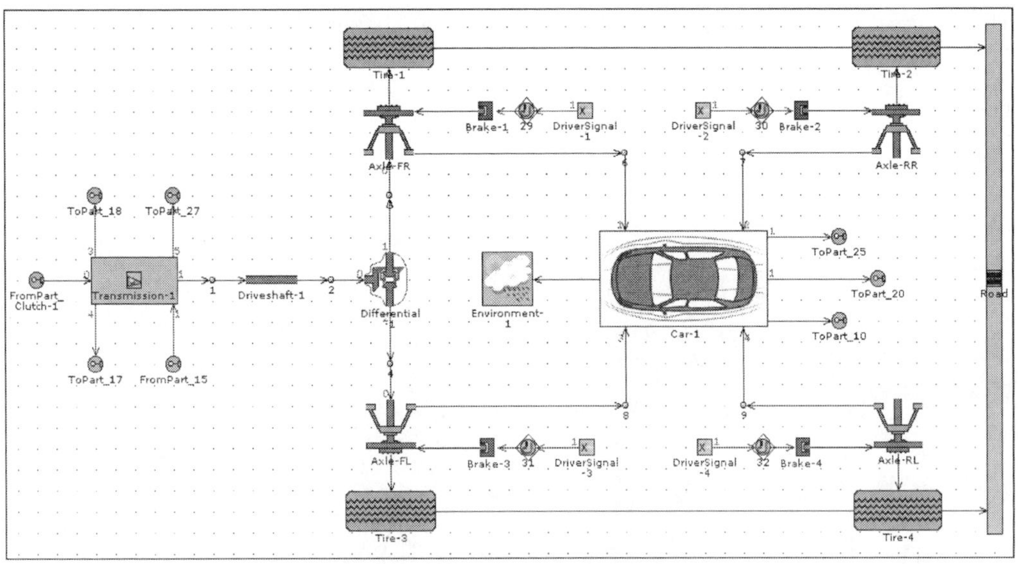

Figure 4: Full-vehicle kinematic behaviour model

3.2 Impact of octane quality on full-vehicle performance

The velocity, gear shift and fuel consumption predicted for NEDC and WLTC simulations are shown in Figure 5 and Table 4. The velocity profile controlled by the "virtual driver" remained within 0.4% of the standard drive cycle requirements.

The simulation supports earlier findings: predicted fuel consumption decreased with increasing RON and sensitivity. At RON92, a 10-point increase in Sensitivity reduced fuel consumption by 3.0% and 1.7% for the NEDC and WLTC respectively. At Sensitivity 5, a 6-point increase in RON reduced fuel consumption by 2.7% and 2.6% for the NEDC and WLTC respectively.

This result is important because it shows that increased fuel sensitivity (lower MON) can lead to lower fuel consumption over a standard drivecycle, even though the K value varies throughout the map and may be positive in places.

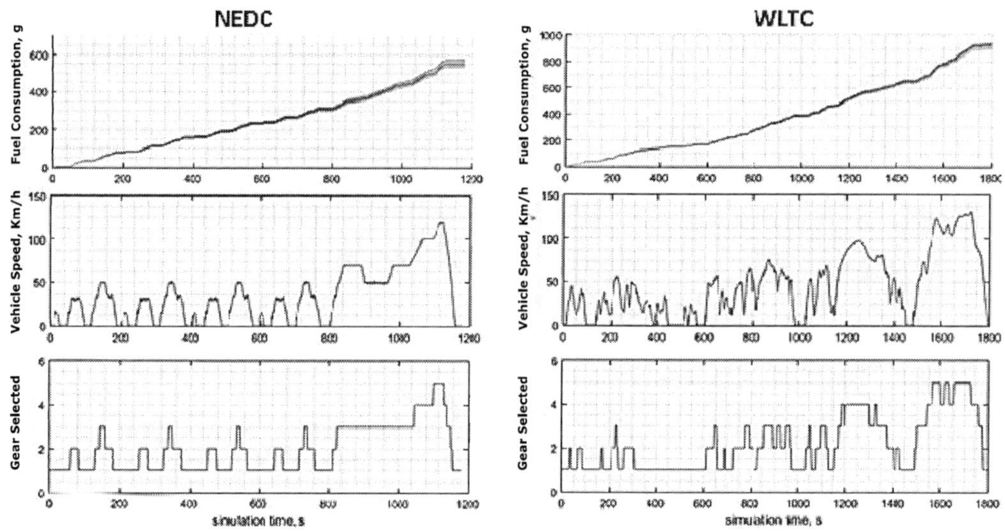

Figure 5: Fuel consumption, vehicle speed and Gear selected during NEDC and WLTC drive cycles for All fuels at CR2

The simulation highlights a significant limitation associated with common drive cycle testing: only a narrow range of speed-load conditions are achieved and there is increased focus on real world driving [18]. Figure 6 shows engine load was limited to 18.3 and 12.9 bar during NEDC and WLTC test cycles, whereas loads up to 22bar were achieved during SCE testing.

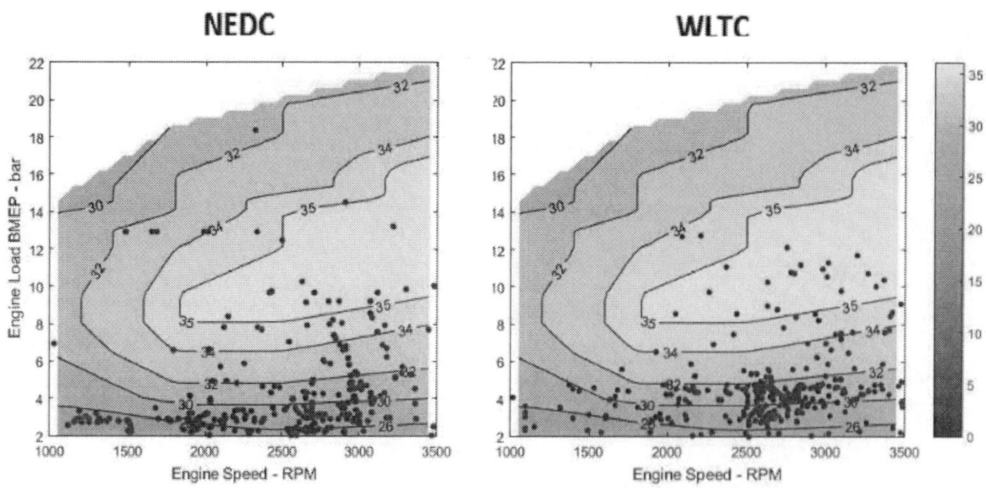

Figure 6: Speed-load points for drive cycle simulations at F2 CR2

Low-load engine operation led to a lower octane response: the simulation predicted a 4.1% (NEDC) and 3.8% (WLTC) reduction in fuel consumption between F1 and F6. By comparison, a notably higher octane response was observed in the SCE for high-load conditions, due to a lower engine K value. A 4.6% increase in relative thermal efficiency for part-load (1800RPM/8bar) and a 5.1% increase for high-load (3500RPM/12bar) operation were observed during SCE testing. Previous work conducted by Shell has shown real world driving is frequently knock limited [18].

Table 4: Normalized fuel consumption for full-vehicle drive cycle simulations

	Normalized Fuel Consumption	
	NEDC	**WLTC**
F1	1.000	1.000
F2	0.995	0.997
F3	0.970	0.983
F4	0.973	0.974
F5	0.963	0.965
F6	0.959	0.962

This study outlines the value of using both SCE and full-vehicle testing. On one hand, steady-state SCE testing can assess a wide range of speed-load conditions and isolate ideal operation for maximum performance, fuel efficiency and octane response and this can be complemented by a drive cycle simulation. On the other hand, full-vehicle testing (whether on a CD or on-road) identifies the speed-load conditions obtained for a particular vehicle setup and driving style and determine what benefits will ultimately be delivered to the on-road driver. The beneficial impact of higher sensitivity on Fuel Economy over the whole drivecycle is noteworthy.

4 CONCLUSION

Six E10 fuels with two different levels of RON and three octane sensitivity levels were used to investigate how fuel octane quality can impact thermal efficiency in modern DISI engines. In the first part of the study, tests were carried out in a single cylinder DISI engine where increasing sensitivity and RON both proved beneficial towards thermal efficiency. A 6-point increase in RON produced gains in peak indicated thermal efficiency of +1.0%, +2.0% and -0.2%, whereas a 5-point increase in Sensitivity produced gains of +0.2% +1.3% for RON 92, and +0.1%, -0.9% for RON 98, respectively.

In the second part, the study presented a methodology to model variations in full-vehicle fuel consumption in response to changing fuel octane quality. GT-Drive was used to simulate the NEDC and WLTC drive cycles. Fuel consumption decreased with increasing RON and increasing sensitivity. A 10-point increase in sensitivity (at RON92) produced a 3.0% and 1.7% reduction in fuel consumption over the NEDC and WLTC cycles respectively, whereas a 6-point increase in RON (at S=5) produced a 2.7% and 2.6% reduction in fuel consumption for the two drive cycles.

5 REFERENCE LIST

1. Xu H, Wang C, Ma X, Sarangi AK, Weall A, Krueger-Venus J. "Fuel injector deposits in direct-injection spark-ignition engines". Progress in Energy and Combustion Science. 2015;50:63-80.
2. Wang C, Daniel R, Ma X. "Comparison of Gasoline (ULG), 2, 5-Dimethylfuran (DMF) and Bio-Ethanol in a DISI Miller Cycle with Late Inlet Valve Closing Time". SAE Technical Paper; 2012-01-1147.
3. Prakash, A., Wang, C., Janssen, A., Aradi, A. et al., "Impact of Fuel Sensitivity (RON-MON) on Engine Efficiency," SAE Technical Paper 2017-01-0799.
4. Remmert, S., Cracknell, R., Head, R., Schuetze, A. et al., "Octane Response in a Downsized, Highly Boosted Direct Injection Spark Ignition Engine," SAE Technical Paper 2014-01-1397.
5. Stone, R. (2012). "Introduction to internal combustion engines". Palgrave Macmillan.
6. ASTM. D2699 - Standard Test Method for Research Octane Number of Spark-Ignition Engine Fuel; American Society for Testing and Materials, 2013.
7. ASTM. D2700 - Standard Test Method for Motor Octane Number of Spark-Ignition Engine Fuel; American Society for Testing and Materials, 2011.
8. Prakash, A.; Cracknell, R.; Natarajan, V.; Doyle, D.; Lobato, P.; Hinojosa, M. "Understanding the Octane Appetite of Modern Vehicles". SAE Technical Paper 2016-02-0834.
9. Kalghatgi, G., "Fuel Anti-Knock Quality - Part I. Engine Studies," SAE Technical Paper 2001-01-3584.
10. Kalghatgi GT. "Fuel anti-knock quality-Part II: Vehicle Studies-how relevant is Motor Octane Number (MON) in modern engines". SAE Technical Paper; 2001-01-3585.
11. Kalghatgi GT, Nakata K, Mogi K. "Octane appetite studies in direct injection spark ignition (DISI) engines". SAE Technical Paper; 2005-01-0244.
12. Bisordi, A., OudeNijeweme, D., Bassett, M., Stansfield, P. et al., "Evaluating Synergies between Fuels and Near Term Powertrain Technologies through Vehicle Drive Cycle and Performance Simulation," SAE Technical Paper 2012-01-0357.
13. Chappell, E., Burke, R., Lu, P., Gee, M. et al., "Analysis of a Diesel Passenger Car Behavior On-Road and over Certification Duty Cycles," SAE Technical Paper 2016-01-2328.
14. Brace, C. J., Burke, R. and Moffa, J., "Increasing accuracy and repeatability of fuel consumption measurement in chassis dynamometer testing". Proceedings of the Institution of Mechanical Engineers, Part D: Journal of Automobile Engineering, 2009. 223 (9): p. 1163-1177.
15. Horrein, L., Bouscayrol, A., Delarue, P., Verhille, J. N., & Mayet, C. (2012). "Forward and backward simulations of a power propulsion system". IFAC Proceedings Volumes, 45(21), 441-446.
16. Samuel, S., Morrey, D., Fowkes, M., Taylor, D. et al., "Numerical Investigation of Real-World Gasoline Car Drive-Cycle Fuel Economy and Emissions," SAE Technical Paper 2004-01-0635.
17. Orlebar, CN., Joedicke, A and Studzinski W. "The effects of octane, sensitivity and K on the performance and fuel economy of a direct injection spark ignition vehicle." SAE Technical Paper 2014-01-1216.
18. Yow, S., Nattrass, S., and Jones, W., "The Application of Telematics to Demonstrate Octane Quality Effects in Real World Driving," SAE Technical Paper 2015-01-1953.
19. Eiser, A., Doerr, J., Jung, M., & Adam, S. (2011). The new 1.8 l TFSI engine from Audi Part 1: Base engine and thermomanagement. MTZ worldwide eMagazine, 72(6), 32-39.

Appendix A – Single Cylinder Testing

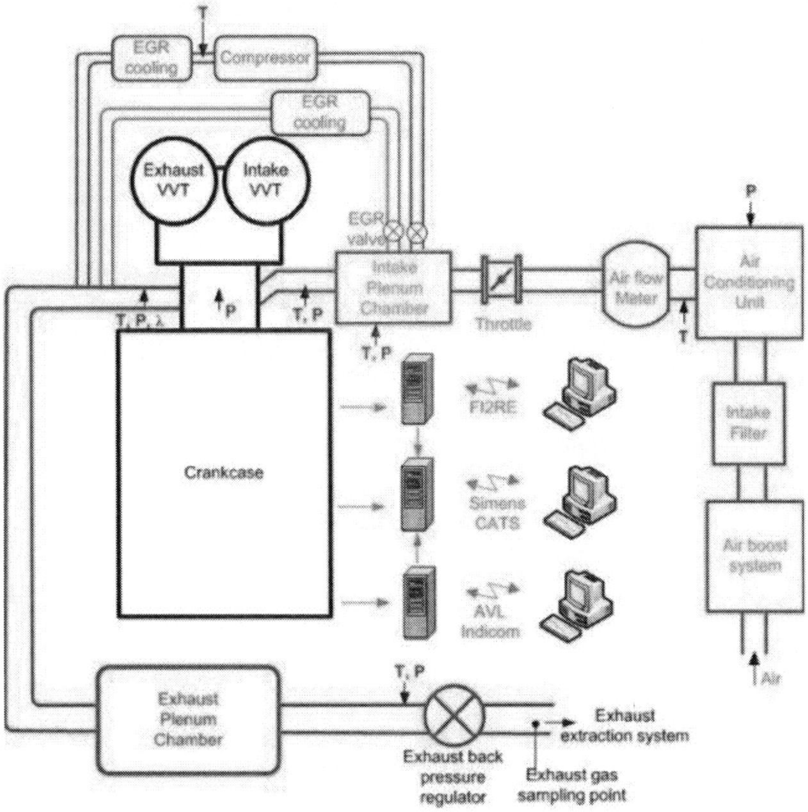

Figure A1: SCE set up, relevant pressure and temperature positions labelled P and T respectively

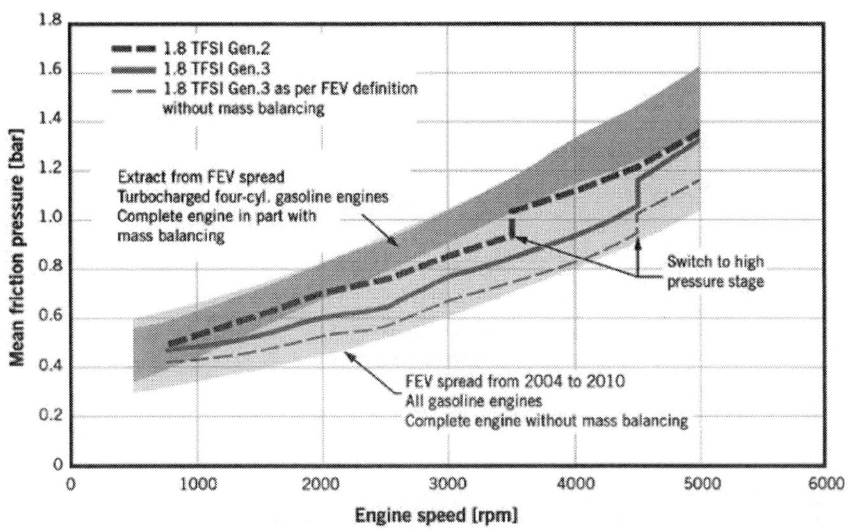

Figure A2: Full-load FMEP map of Audi 1.8L 4-cylinder EA888 engine. 3rd generation is used [19]

Table A1: detailed testing procedure for all fuels

Procedure	Fuel	Compression Ratio
Establish Engine optimization parameters, ex. optimum injection mapping	F2	CR 3
Testing phase	F2	CR1, CR2, CR3
	F6	
	F1	
Ensure experiment repeatability and results consistency	F2	CR2
Testing phase	F5	CR1, CR2, CR3
	F4	
	F3	
Ensure experiment repeatability and results consistency	F2	CR2

Appendix B – GT-Drive Simulation

Table B1: most relevant Gt-Drive full vehicle model

Vehicle Properties	
Mass (incl. cargo & passengers)	1300 kg
Aerodynamics	
Drag coefficient	0.32
Frontal area	0.8 m^2
Centre of pressure (vertical distance from axle)	0
Transmission	
Gear Ratios (1-5)	3.538, 2.125, 1.36, 1.029, 0.72
In-gear efficiency	~0.97
Input moment of Inertia	0.05 kg m^2
Output moment of Inertia	0.05 kg m^2
Gear Ratio Transition Time	0.3 s
Axle	
moment of inertia	1.25 kg m^2
Friction coefficient	0
Differential	
Input Inertia	0.01 kg m^2
Output Inertia	0.01 kg m^2
Final Drive Ratio	3.2
Efficiency	100%
Fuel	
Fuel Density	731.1 Kg/m^3
Fuel Heating Value	42.17 J/kg

Experimental study on combustion characteristics and emission performance of 2-phenylethanol addition in a downsized gasoline engine

J Y Pan, H Q Wei, G Q Shu, H J Xiong, F D Quan, M Z Pan
State Key Laboratory of Engines, Tianjin University, China

ABSTRACT

To ascertain the possibility of 2-phenylethanol (2-PE) as a bio-additive in SI engines, combustion characteristics and emission performance of 2-PE addition were investigated in a downsized gasoline engine. For aspirated conditions with identical intake quantity, there is a decrease for cylinder pressure and heat release rate as 2-PE addition is increased. Ignition delay and combustion duration become postponed, and combustion phasing is retarded. Further, 2-PE addition shows good knocking resistance under various intake pressures, because of its high Octane Number and physicochemical properties. Besides, 2-PE addition results in more unburned hydrocarbon, carbon oxide and particle matter and less nitrogen oxides.

1. INTRODUCTION

With the development of advanced combustion technology, fuel flexibility becomes critically important for internal combustion engines nowadays. For spark-ignited (SI) engines, high-octane quality fuels always play a crucial role in increasing thermal efficiency and reducing CO_2 emissions [1-2]. As a bio-derived component, 2-phenylethanol (2-PE) is considered as a novel octane booster recently. Compared with conventional anti-knock additives (e.g. methyl ter-butyl ether and ethanol), 2-PE not only has abundant and low-cost feedstock known as lignocellulose [3-4], but also possesses relatively larger heating value and higher energy density. Meanwhile, its high sensitivity in Octane Number is also desirable for highly downsizing gasoline engines [5].

Up to now, several studies have been performed to explore 2-PE fuel properties through both fundamental combustion and engine experiments. Using quantum mechanical calculations, Sakai et al. [6] studied the thermal decomposition of 2-PE and found that the barrier height for ring opening reaction is the lowest among all subsequent reactions. Shankar et al. [7] experimentally evaluated combustion characteristics of 2-PE fuel in a modified CFR engine and ascribed its good anti-knock property to both chemical kinetics and charge cooling effect. Meanwhile, they developed an auto-ignition model to isolate chemical kinetic effect and ascertained the essential of hydroxyl group in increasing heat of vaporization. Further, Ratcliff et al. [8] examined the performance characteristics of 2-PE fuel and their impacts on knock resistance and particular emissions in a single-cylinder direct-injection SI engine. The experimental results showed that knock resistance correlates well with its higher Research Octane Number, and a significant increase in particulate matter emissions (factors of 2 to 5) was also observed during engine experiments.

Despite these explorative studies on 2-PE combustion characteristics and emission performance as pure fuels or an addictive, further comprehensive study is still needed before extensive applications of 2-PE fuel as an octane booster into engine combustion. In this work, different blends of 2-PE and gasoline, ranging from 0% to 30% in volume fraction, were adopted, and both naturally aspirated and supercharging conditions are involved. Engine performance including in-cylinder pressure, torque output, knocking characteristics, conventional emissions and particle matter (PM) emissions were investigated and discussed.

2. EXPERIMENTAL APPARATUS AND METHODOLOGY

2.1 Experimental apparatus

Current experiments were carried out on a single-cylinder, four-stroke, direct-injection, SI engine. Engine combustion system consists of a flattop piston and a pent-roof cylinder head with spray-guided design. The engine geometry is 80 mm in bore, 100 mm in stroke, 0.5 L in sweep volume and 10:1 in compression ratio. Further engine specifications can be found in in Table 1. This engine was modified from a Ricardo E6 research engine with robust construction. A six-hole Bosch high drive electric valve (HDEV) 5-type injector was adopted to provide an injection pressure ranging from 6.0 to 14.0 MPa. Injection timing, injection duration and spark timing was controlled by an open ECU manufactured by MoTeC400. The mixture equivalence ratio was determined by a wideband lambda sensor with resolution 0.001, uncertainty ±0.8% and response time within 0.15 s. Engine speed was controlled by a direct current dynamometer, with an accuracy of ±1 r/min. Engine load was measured by a ZEMIC H3-C3-200kg-3B load cell with uncertainty of ±0.5%. A water-cooled piezo-electric pressure transducer (Kistler 6041A) was mounted on engine head to acquire local in-cylinder pressure. Then the pressure data processed by a Kistler 5018 charge amplifier was sent to a SKLE Combustion Analyzer. A pressure sampling with 0.1 CAD resolution was triggered by a photoelectric encoder coupled to crankshaft. Intake air flow was measured using a calibrated laminar gas flow meter. Fuel mass was measured by a fuel consumption meter ranging from 0 to 125 kg/h with an uncertainty of ±0.2%. Coolant and oil temperatures were controlled by a SIEMENCE proportional–integral–differential (PID) controller with an uncertainty of ±3 ℃. Conventional emissions (UHC, CO and NOx) were measured by Horiba Mexa 7200H and particle matter emissions were determined through an Engine Exhaust Particle Sizer (TSI 3090) with a measurement range of 5.6-560 nm. Surrounding environmental temperature and humidity were kept the same with atmospheric conditions. Schematic view of the general engine setup can be found in Fig. 1.

Table 1 SI Engine specifications

Engine type	Single cylinder, 4-stroke
Bore×stroke	80×100mm
Sweep volume	0.5 L
Compression ratio	10:1
Valve mechanism	Dual-overhead camshafts, 2-valve
Combustion system	Spray-guided direct-injection
Injection pressure	10 MPa
Intake valve Open timing	10 bTDC
Exhaust valve Close timing	20 aTDC

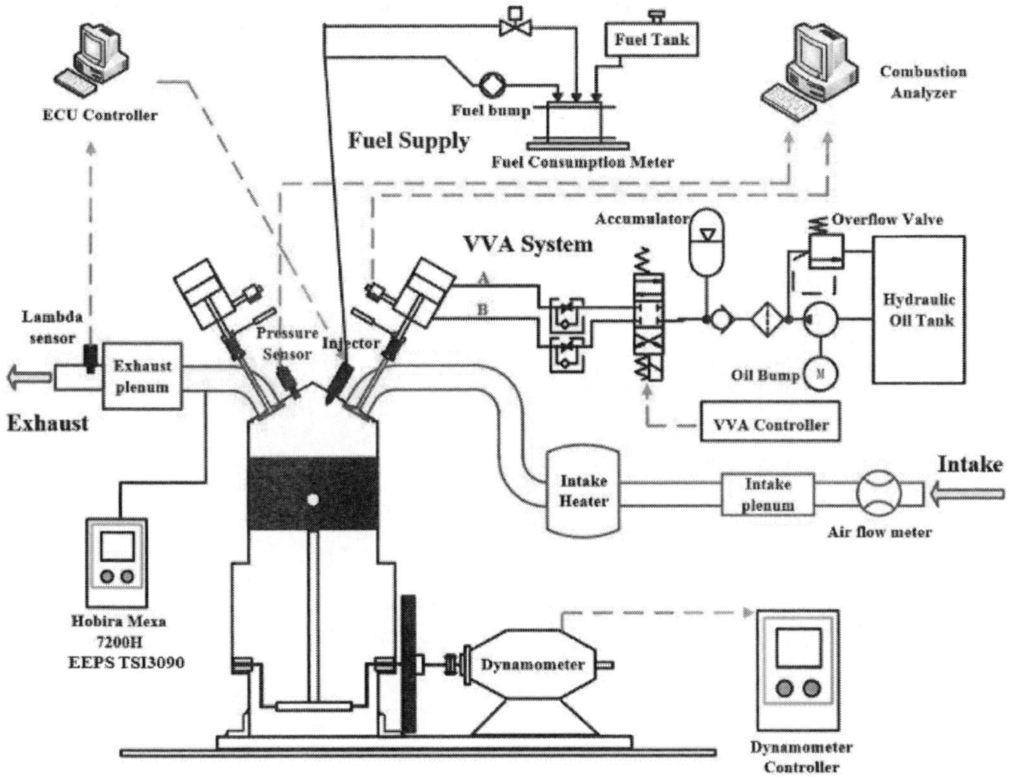

Fig. 1 Schematic view of the SI engine setup

2.2 Methodology

In this experiment, unleaded commercial gasoline supplied by Shell in China was adopted as a reference fuel which has a Research Octane Number of 92. 2-PE fuel with a purity of 99.5% was supplied by Tianjin YuanLi Chemicals Co. Ltd. in China. The comparisons of fuel property between gasoline and 2-PE can be found in Table 2. Four blending cases (i.e. 2-PE0, 2-PE10, 2-PE20 and 2-PE30) were conducted, representing 10%, 20% and 30% by volume of 2-PE addition in gasoline reference fuel, respectively. Engine experiments were performed at a constant engine speed of 1500 r/min with both naturally aspirated and supercharging conditions where intake pressure is P_{in}=1.0, 1.2 and 1.4 atm (denoted by P1.0, P1.2 and P1.4, respectively). Engine knock regime was achieved by advancing spark-ignition timing. Early injection of homogeneous operation was performed by constant start-of-injection at 60 crank angle after the TDC of intake stroke. Equivalence ratio was kept at 1.0 ± 0.01 and fuel injection pressure was 10 MPa for all tests. 200 continuous engine cycles were carried out at each operation point and the average results were utilized to analyze combustion characteristics. In addition, an aluminum alloy was adopted to seal engine cylinder head as an emission sampling valve. This diaphragm cut an electromagnet-actuated tube cutter instantaneously at a pre-set crank angle during a sampling cycle. Then cylinder mixture was discharged from the cylinder into a sampling bag. Simultaneously, the sampled gas mixture was immediately quenched and diluted with high pressure nitrogen at a dilution ratio of 200:1, sampling flow of 10 L/min, and temperature below 52 Celsius degree to prevent any possible additional reactions during sampling processes.

Table 2 Comparisons of fuel properties between gasoline and 2-PE [7, 9]

Fuel	Gasoline	2-PE
Molecular Formula	C4~C12	C8H10O
RON	92	110
Sensitivity	10	20
Density at 300 K [kg/L]	0.76	1.017
Molar mass [kg/mol]	0.111	0.122
LHV [MJ/kg]/ [MJ/L]	43.5/33.06	36.7/37.32
Heat of Vaporization [kJ/kg]	223.2	564.81

3. RESULTS AND DISCUSSIONS

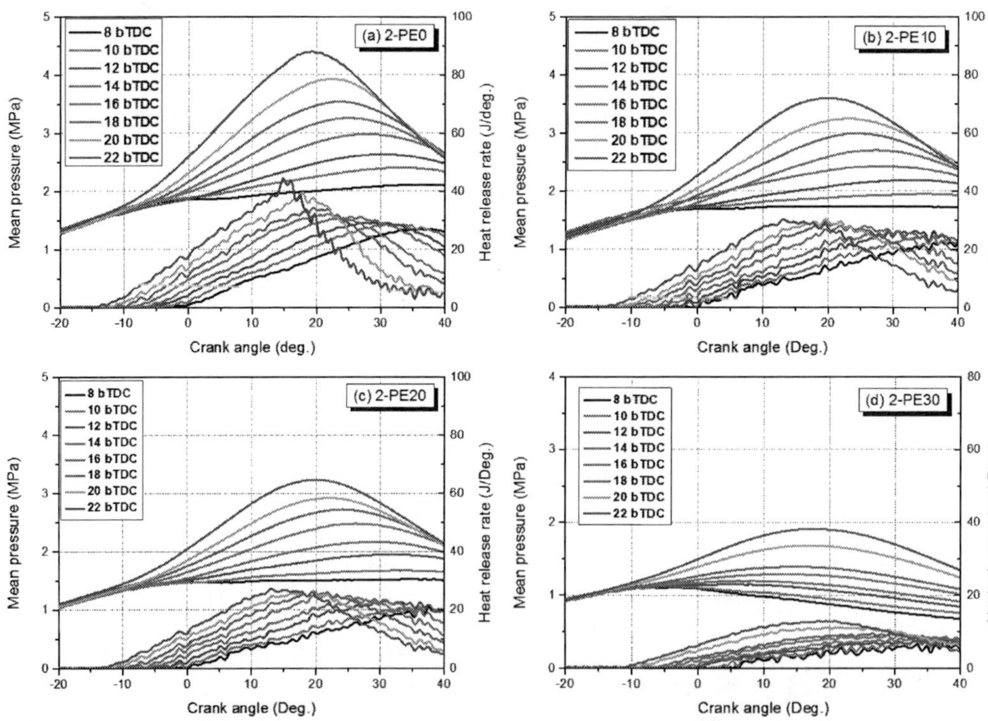

Fig. 2 Mean pressure and heat release rate of 200 cycles at naturally aspirated conditions, with blending level of (a) 0%, (b) 10%, (c) 20% and (d) 30%

To demonstrate the effects of 2-PE addition on engine performance, Fig.2 shows the results of mean in-cylinder pressure and corresponding heat release rate of different 2-PE/gasoline blends at naturally aspirated conditions. It should be noted that current experiments are based on identical intake quantity rather than energy content under different blending levels, and fuel injection quantity is adjusted online to maintain stoichiometric combustion. It is observed from Fig. 2 that under the same spark-ignition timing conditions, combustion phasing (e.g. the location of pressure peak) is postponed and in-cylinder pressure is decreases as the increases of 2-PE/gasoline blends. The deterioration of engine combustion can be alleviated through optimizing spark advance angle, e.g. compared with the pure gasoline (i.e. 2-PE0) scenario with spark-ignition timing t_{sp} =14 bTDC, an approximate 17%

increase in pressure peak (P_{max}) is achieved for 2-PE10 at t_{sp}=22 bTDC. However, this optimization strategy is only effective for low blending level conditions, and when 2-PE addition is increased up to 30%, in-cylinder pressure cannot return back even with sufficiently large spark advance angle. These results indicate that low levels of 2-PE addition in gasoline (no more than 20% here) may be appropriate in order to maintain good engine performance.

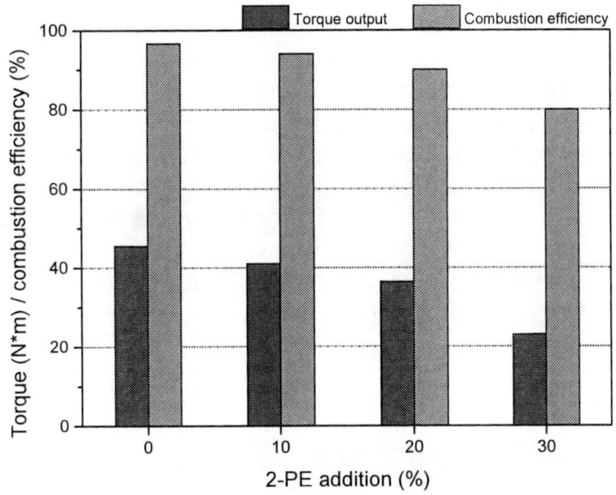

Fig. 3 Torque output and combustion efficiency under optimal spark-ignition timing conditions

Fig. 3 further presents the torque output and combustion efficiency at optimal spark-ignition timing conditions. It shows that as 2-PE addition is increased from 0-30%, engine torque output is declined from 45.46 N·m to 23.07 N·m. Meanwhile, there is an obvious decreasing trend for combustion efficiency, ranging from 96% to 84%. This again demonstrates the deterioration of engine combustion as 2-PE addition into gasoline fuel. Noting that different blends have distinct energy contents in current work, and we can obtained from Table 2 that for given intake quantity, 2-PE has 10% higher energy content than gasoline fuel under stoichiometric conditions. Instead, 2-PE addition leads to obvious decreases in in-cylinder pressure and deteriorating combustion characteristics. Therefore, further analysis on the correlations between physicochemical properties and combustion process will be necessary.

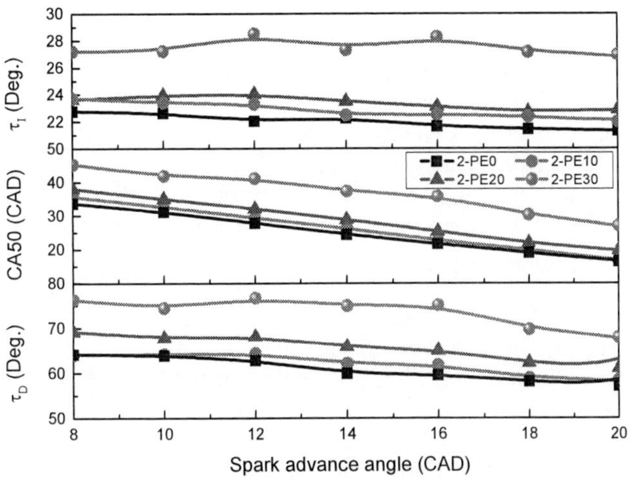

Fig. 4 Ignition delay (τ_I), CA50 location and combustion duration (τ_D) of different blending levels under naturally aspirated conditions

To quantify the variations of combustion characteristics with 2-PE addition, Fig. 4 shows the results of ignition delay time, CA50 location and combustion duration of different 2-PE/gasoline blending levels under naturally aspirated conditions. Here ignition delay time, CA50 location and combustion duration are obtained from heat release history, which are defined as the crank angle interval of 0-10% heat release, the crank angle position of 50% heat release and the crank angle interval of 10-90% heat release, respectively. For the same blending conditions, it shows no obvious variations in ignition delay time at different spark advance angles, while there is an obvious advance of CA50 location and shortage of combustion duration. For the same spark-ignition timing conditions, ignition delay time and combustion duration are postponed and CA50 location is delayed as the increases of 2-PE additions. These variations become more obvious when 2-PE/gasoline blending level reaches over 20%, which correlates well with in-cylinder pressure history. Previous fundamental study [7] has showed that 2-PE fuel possesses relatively lower gas-phase reactivity controlled by an aromatic ring and higher heat of vaporization by hydroxyl group. As a consequence, the engine fueled with 2-PE addition shows slower initial flame kernel formation, turbulence flame propagation and combustion heat release, agreeing well with in-cylinder pressure history shown in Fig. 2.

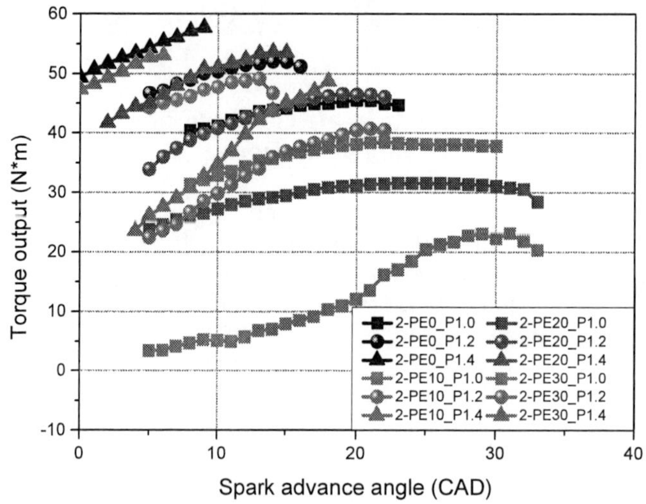

Fig. 5 Engine torque output at different spark-ignition timing and intake pressure conditions

Fig.5 gives the torque output of different spark-ignition timing and intake pressure conditions to elucidate the effects of 2-PE addition on engine performance under various operation conditions. It is observed that for the same blending-level conditions, as intake pressure is raised from P_{in}=1.0 to 1.4 atm, there is a significant increases in engine torque output caused by more energy content in mixture. Meanwhile, the optimal spark-ignition timing shifts to the Top Dead Center, which is generally thought the best operation condition for engine performance. While for the same intake pressure conditions, engine torque output shows some decreases when 2-PE addition is introduced to the gasoline engine. Under low 2-PE/gasoline blending level conditions (e.g. 10%), the disparity in torque output can be compensated through supercharging and advancing spark-ignition timing in actual engines. However, when the 2-PE/gasoline blending level is beyond 30%, engine torque output cannot reach the level of pure gasoline under naturally aspirated conditions, despite high supercharging is adopted. Again the lower gas-phase reactivity of 2-PE fuel results in slow ignition and flame propagation, which further affects engine combustion and torque output. In order to maintain sufficient torque output, low levels of 2-PE addition coupled with supercharging shall be effective approaches.

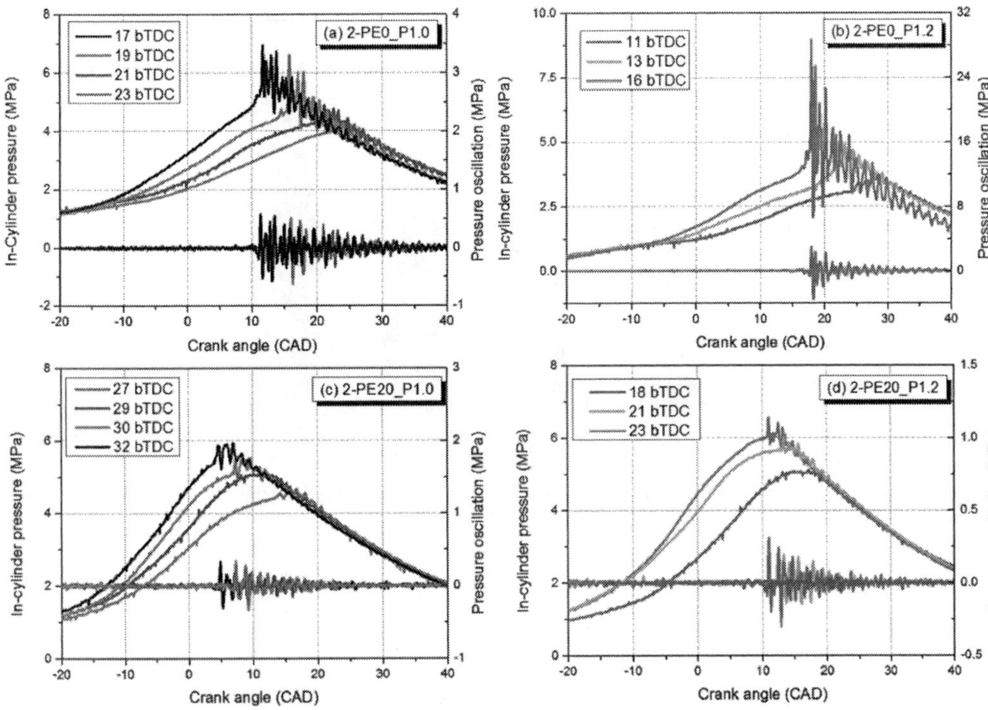

Fig. 6 In-cylinder pressure and pressure oscillation of knocking cases, (a) 2-PE 0%+P_{in} 1.0 atm (b) 2-PE 0%+P_{in} 1.2 atm (c) 2-PE 20%+P_{in}1.0 atm and (d) 2-PE 20%+P_{in}1.2 atm

To further demonstrate above point of view, Fig.6 shows the in-cylinder pressure and pressure oscillation (filtered by 3-25 kHz) under both naturally aspirated and supercharging conditions in which knocking combustion occurs obviously. It is observed that for naturally aspirated conditions, there is an obvious pressure oscillation superimposed on in-cylinder pressure curve when the spark-ignition timing is advanced to t_{sp}=21 bTDC. With the definition by the maximum amplitude of pressure oscillation (MAPO) [10], engine knocking intensity reaches approximately KI=0.5 MPa at t_{sp}=23 bTDC. When 2-PE/gasoline blending level is raised up to 20%, knocking intensity is reduced to KI=0.3 MPa with an extremely early spark-ignition timing (t_{sp}=32 bTDC). When intake pressure of P_{in}=1.2 atm is adopted, a heavy knock intensity of KI=4.0 MPa is observed at the spark-ignition timing of t_{sp}=16 bTDC. Meanwhile, knocking intensity becomes highly sensitive to spark-ignition timing at supercharging conditions. When 20% 2-PE is blended in gasoline reference fuel, knocking intensity is significantly decreased to a lower level at KI=0.25 MPa, even with a very early spark-ignition timing of t_{sp}=23 bTDC. This clearly indicates the good knock resistant of 2-PE octane booster that knocking tendency can be effectively alleviate even with low levels of 2-PE addition in gasoline under early spark-ignition timing and supercharging conditions.

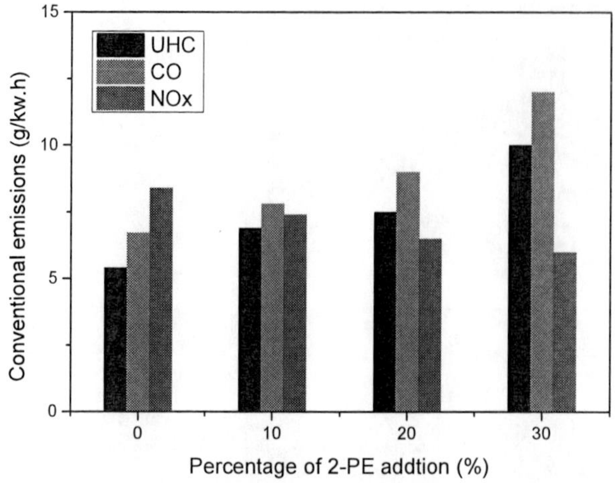

Fig. 7 Conventional emissions of different blending levels under naturally aspirated conditions

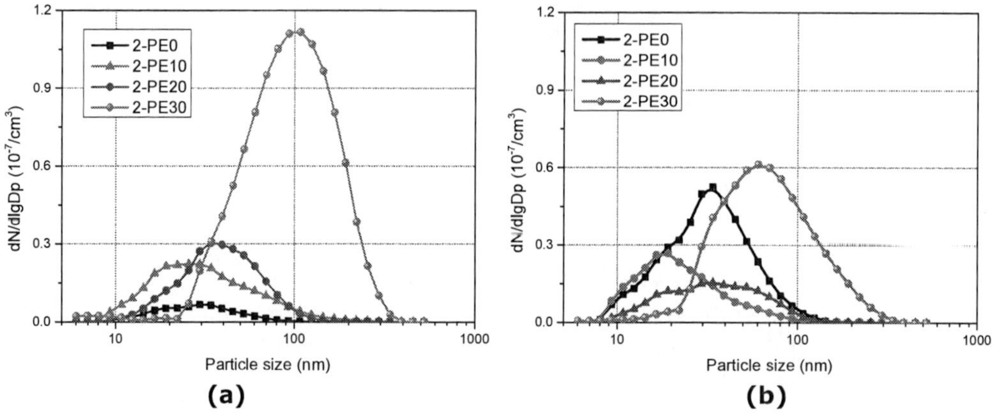

Fig. 8 Size distributions of particle matters under different blending conditions with spark-ignition timing of (a) t_{sp}= 10 and (b) t_{sp}=20 bTDC

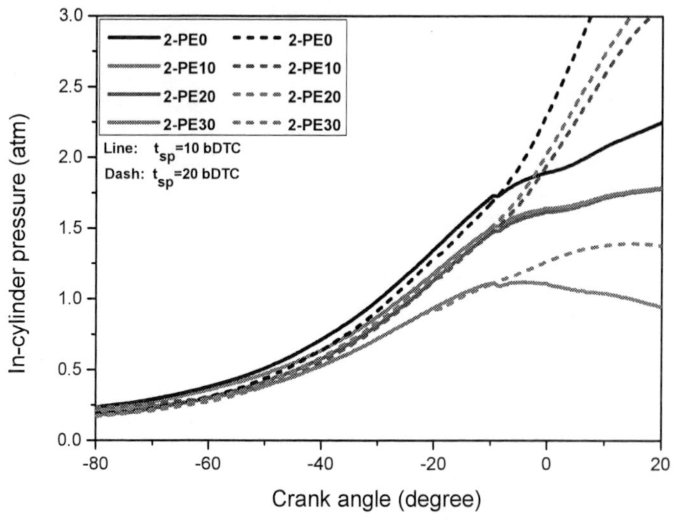

Fig. 9 Thermodynamic state of combustion process under different spark-ignition timing and 2-PE addition conditions

Besides combustion characteristics, emission performance is also very important for the novel bio-derived fuel applied into real engines. Fig. 7 shows the results of conventional emissions, including unburned hydrocarbons (UHC), carbon oxide (CO) and nitrogen oxides (NOx) under naturally aspirated conditions. It is observed that as the increases of 2-PE/gasoline blending level, there is an increasing trend in UHC and CO emission and decreasing trend in NOx emission. This variation tendency becomes more obvious at high blending-level conditions (e.g. 30%). The reason for this is that the deterioration of engine combustion results in incomplete combustion and thus low combustion temperature. Meanwhile, it is noted that compared with pure gasoline scenarios, there is no obvious increases in conventional emissions with low levels of 2-PE addition (e.g. below 20%).

Fig. 8 further shows the size distributions of particle matter under different blending conditions. For the cases of spark-ignition timing at t_{sp}=10 bTDC, it is observed that as blending level is raised, there is an increase in the concentration of particle matter, especially for high blending-level conditions. This is because 2-PE fuel has low volatility and high heat of vaporization, such that in-cylinder thermodynamic state decreases as 2-PE addition is raised up, as shown in Fig. 9. As a consequence, such variation affects air/fuel mixing and gives rise to extremely rich regions, leading to poor combustion and high levels of PM emissions. Compared with gasoline reference fuel, it is 10 times higher in PM emissions when 2-PE addition reaches 30% level. As mentioned earlier, high levels of 2-PE addition will result in the deterioration of engine combustion and even misfire. Meanwhile, it is observed that lower blending level (e.g. 10%) seems to result more particle matters in nucleation mode, while further increases in blending level will lead to more particle matters in accumulation mode. When spark-ignition timing is advanced to t_{sp}=20 bTDC, PM emissions show a decreasing trend as the increases of 2-PE addition, even though the size distributions of particle matter seem similar to that of t_{sp}=10 bTDC case. This is because the spark-ignition timing for pure gasoline cases (i.e. 2-PE0) is near knocking regime where incomplete combustion phenomenon becomes obvious, while when low levels of 2-PE addition (e.g. 10% and 20%) is introduced into gasoline fuel, engine knock is suppressed and combustion process is approved followed by lower PM emissions. Therefore, the high emissions of particular matter caused by 2-PE addition can be effectively suppressed through engine combustion optimization.

4. CONCLUSIONS

In this work, a novel bio-derived component, 2-phenylethanol (2-PE), has experimentally studied in a downsized gasoline engine to explore both combustion characteristics and emission performance. Different 2-PE/gasoline blends and intake pressures have been adopted to investigate the effects of 2-PE addition on engine performance, knock resistance, conventional emissions and particle matter emissions. The conclusions obtained are the followings:

Compared with gasoline reference fuel, 2-PE addition results in obvious decreases in in-cylinder pressure and heat releases rate. Correspondingly, engine torque output and combustion efficiency under optimal spark-ignition timing are declined as 2-PE addition is raised under naturally aspirated conditions. Further analysis on combustion phasing shows that ignition delay and combustion duration are postponed and CA50 location is retarded, especially for high blending-level scenarios. These observations suggest that 2-PE fuel possesses a lower gas-phase reactivity than gasoline fuel. Meanwhile, due to the good knock resistance of 2-PE addition, engine torque output can be improved through optimizing spark-ignition timing and intake pressure.

On conventional emissions, it is found that as the blending level is raised, there is an increase in unburned hydrocarbon and carbon oxide and decrease in nitrogen oxides, which correlates well with the decreasing thermodynamic state when 2-PE addition is introduced. For particle matter emissions, low volatility and high heat of vaporization of 2-PE fuel result in more poor combustion and high levels of particulate. Meanwhile, spark-ignition timing and combustion process also play important role in particle matter emissions. In terms of size distributions, lower 2-PE addition lead to more particle matter in nucleation mode while higher blending level results in more particle matter in accumulation mode.

5. ACKNOWLEDGEMENT

This work was supported by Major Research Plan of National Natural Science Foundation of China (91641203), National Natural Science Foundation of China (Grant No. 51476114, 51706152), China Postdoctoral Science Foundation (2016M590201), Tianjin Natural Science Foundation (17JCZDJC31500), Tianjin Science & Technology Supporting Program (13ZCZDGX04400).

6. REFERENCES

(1) Wang, Z., Liu, H., Reitz, R.D. (2017) *Progress in Energy and Combustion Science* **61**, 78-112. DOI: 10.1016/j.pecs.2017.03.004

(2) Pan, J.Y., Shu, G.Q., Wei, H.Q. (2014) *Combustion Science and Technology* **186**, 192-209. DOI: 10.1080/00102202.2013.857665

(3) McCormick, R.L., Ratcliff, M.A., Christensen, E. (2015) *Energy & Fuels* **29**, 2453-2461. DOI: 10.1021/ef502893g

(4) Atsumi, S., Taizo, H.N., James, C.L. (2008) *Nature* **451**, 86-86. DOI: 10.1038/nature06450

(5) Zhou, L., Boot, M.D., Goey, L.P.H. (2012) *SAE International Journal of Fuels and Lubricants* **5**, 1216-1239. Doi:10.4271/2012-01-1697

(6) Sakai, Y., Ando, H., Oguchi, T. (2013) *Chemical Physics Letters* **556**: 29-34. DOI: 10.1016/j.cplett.2012.11.050

(7) Shankar, V.S.B., Al-Abbad, M., El-Rachidi, M. (2017) *Proceedings of the Combustion Institute* **36**, 3515-3522. DOI: 10.1016/j.proci.2016.05.041

(8) Ratcliff, M., Burton, J., Sindler, P., Christensen, E. (2016) *SAE Int. J. Fuels Lubr.* **9**, 59-70. DOI: 10.4271/2016-01-0705

(9) Elfasakhany, A. (2016) JOURNAL OF CLEANER PRODUCTION **139**, 1057-1067. DOI: 10.1016/j.jclepro.2016.09.016

(10) Shu, G.Q., Pan J.Y., Wei, H.Q. (2013) Applied Thermal Engineering **51**, 1297-1306. DOI: 10.1016/j.applthermaleng.2012.11.039

Development of multi-fidelity powertrain simulation for future legislation

A S van Niekerk, P J Kay, B Drew, N Larsen
University of the West of England, UK

ABSTRACT

Automotive manufacturers are facing increased pressure from more stringent emissions legislation and new legislative driving cycles. This creates a need for increased simulation capabilities. An engine simulation was developed in Ricardo WAVE, and was coupled with lower fidelity Ricardo IGNITE to enable the investigation of emissions and engine performance over different driving cycles. The simulation was validated against experimental engine emissions and performance data, obtained from the test cell, with good agreement. A provisional simulation using a 100 % bio-fuel yielded promising results that agreed with similar studies; thus enabling the impact of future renewables on emissions to be evaluated.

1 INTRODUCTION

Automotive manufacturers are facing more stringent emission legislation with the New European Driving Cycle (NEDC) being phased out and replaced with the World-Harmonised Light Vehicle Test Procedure (WLTP) and the real world driving emissions (RDE) driving cycle criteria (1). Current driving cycles such as the NEDC are optimised to ensure high reproducibility and allows the removal of external influences that can affect the test outcome. The NEDC is seen as unrepresentative of the everyday driving environment. A more comprehensive driving cycle has been developed to ensure that it will be more representative of everyday driving. This development resulted in the WLTP and RDE legislative driving cycles (1). The RDE adds the road as a testing environment that enables it to be more realistic, yet it has the added unknown influences of the road profile, the ambient conditions, traffic conditions and the behaviour of the driver. This imposes extra challenges on car manufacturers in terms of design and calibration to ensure that their cars are compliant and able to achieve the required emissions limits. Car manufacturers need to implement novel technologies and development techniques incorporated with increased simulation capabilities to be able to handle the increased workload, while keeping costs to a minimum.

The concentration of greenhouse gasses (GHG) in earth's atmosphere has been rising steadily and mean global temperatures along with it. Since the Industrial Revolution, an upward trend was noticed in the quantities of GHGs such as carbon dioxide (CO_2) which is released into the atmosphere when burning fossil fuels (2). Consequently, Directive 2009/28/EC (3) lists individual targets for each member state for the share of energy from renewable sources in gross final consumption of energy by 2020. The UK has committed to obtain 15 % of its gross final consumption of energy from renewable sources, which include a 10 % share from transport.

Table 1: Ricardo WAVE simulation information

Label	Description
A	Throttle position (TPS)
B	Engine RPM
C	Target fuel pressure
1.1	Main injection total fuel injected
1.2	Main injection start of injection
1.3	Main injection duration
2.1	Pilot injection total fuel injected
2.2	Pilot injection start of injection
2.3	Pilot injection duration

In addition, the concentration of emissions such as Carbon Monoxide and Nitrogen Oxides from road transportation, which are harmful to human health, are being regulated with the EURO VI standard (4).

It is currently unclear what the impact of using new renewable fuels to meet the UK climate targets will be on emissions from vehicles over the new test drive cycles. To evaluate this a comprehensive simulation is required to evaluate the impact. This paper presents the development of the simulation, including the validation against experimental data.

2 SIMULATION DEVELOPMENT

A simulation platform that can handle increased simulation capabilities is necessary. The simulation environment needs to be flexible enough to accommodate multiple driving scenarios and be able to accurately calculate emissions and other engine performance characteristics. Ricardo Software has developed a suite of software tools that can be used to assist in automotive development and to optimise different configurations. Ricardo IGNITE is a system based simulation package that can be used to focus on complete vehicle system simulation and optimisation (5). It can simulate full duration drive cycle simulations and predict fuel consumption and emissions enabling the exploration of the impact various technologies, such as exhaust gas recirculation, have on overall engine emissions and fuel consumption over different drive cycles. IGNITE also has the capacity to conduct co-simulations with Ricardo WAVE. WAVE provides a fully integrated treatment of time-dependent fluid dynamics and thermodynamics in the form of one-dimensional calculations that enables simulations to be carried out based on intake, combustion and exhaust systems (6). The following sections discuss how the different simulation are configured to successfully simulate a specific engine used in the engine test cell facilities.

2.1 Ricardo WAVE

Figure 1 shows a screen shot of the simulation in Ricardo WAVE. Table 1 lists the description of each label as seen in Figure 1. Table 2 lists all the engine data used in the simulation. Access to the engine operating maps on the ECU is possible and it was used in WAVE to configure the engine simulation. Map values are a function of throttle position (TPS) and engine speed (RPM). A mapless compressor was used to simulate boost as per the targets set out in the ECU data. Multi-pulse injectors were used to deliver the fuel to the engine. A pilot injection as well as a main injection were simulated. The exhaust gas recirculation (EGR) system was replicated using an orifice with variable diameter to control the percentage of exhaust gas being recirculated.

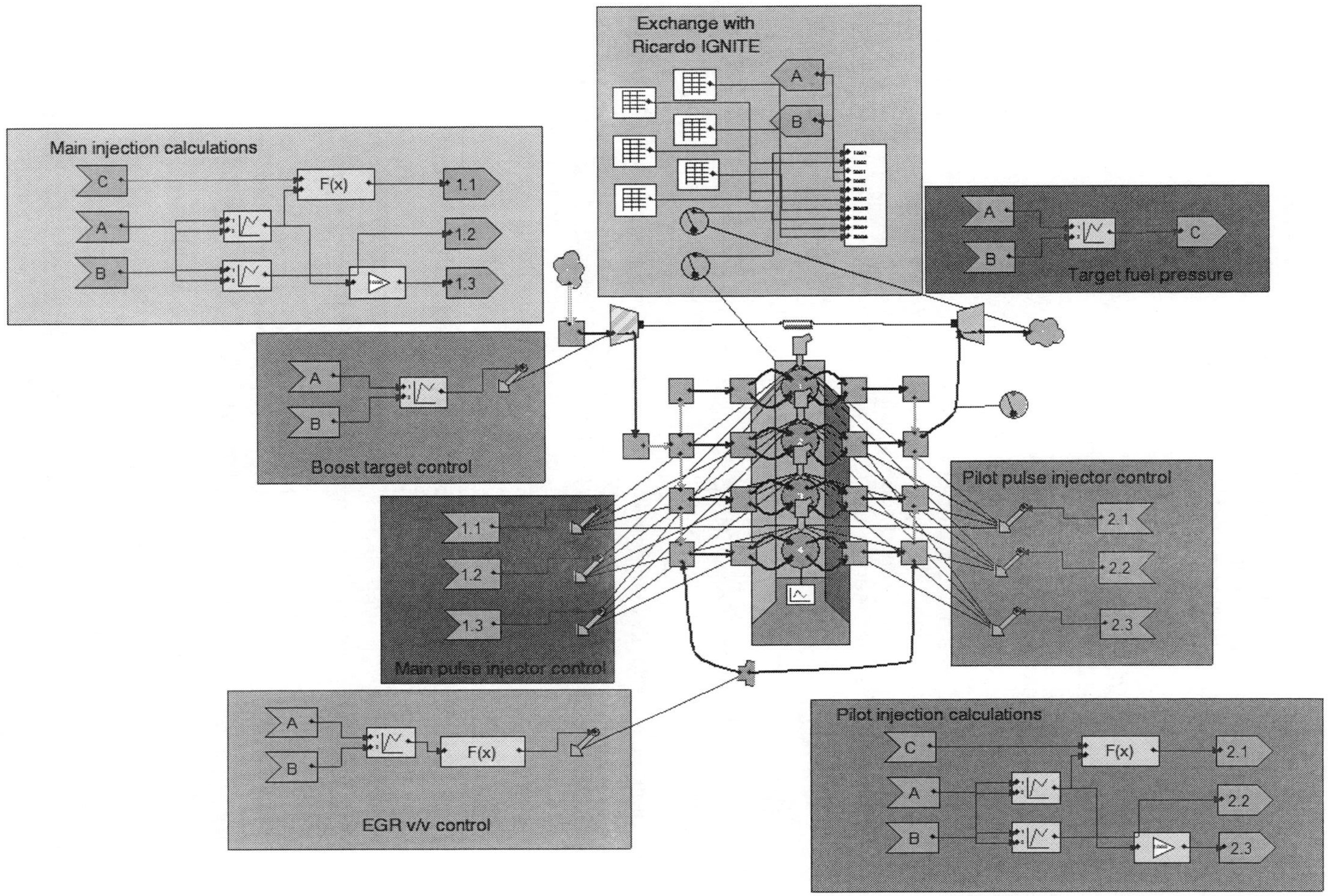

Figure 1: Screen shot of the engine simulation in Ricardo WAVE.

Table 2: Ricardo WAVE engine data

Description	Data
Bore (mm) (7)	89.9
Stroke (mm) (7)	94.6
Connecting rod length (mm)	106
Piston pin offset (mm) (7)	0
Clearance height (mm)	5.7
Compression ratio (CR) (7)	17.5
Number of cylinders (7)	4
Firing order	1-3-4-2

Ricardo WAVE's emission sub-models were activated to enable calculations for CO emissions, HC emissions and NO_x emissions. The diesel jet sub-model was also activated to ensure increased accuracy in fuel spray, heat release and soot emissions calculations.

2.2 Ricardo IGNITE

Figure 2 shows a screen shot of the IGNITE simulation. The use of the lower fidelity Ricardo IGNITE enables the simulation to undergo a legislative driving cycle or any other custom-programmed route. Vehicle characteristics such as frontal area, wheel diameter, engine inertia as well as vehicle weight are also required for the simulation. IGNITE calculates the required engine speed and load in order to follow the defined driving cycle. The input used in the development of the simulation is listed in Table 3. The engine inertia and the rolling resistance coefficient were calculated using the procedure as set out by Kunniger (9). The NEDC driving cycle is pre-programmed in IGNITE and can be selected when choosing the required driving cycle. Other information such as the gear shift strategy, is available in the Council Directive 91/441/EEC (10). The gearbox ratios for the car as well as the final gear ratio is available in the Ford workshop manuals (7). A custom driving cycle option was used to enable the simulation to simulate the WLTP driving cycle. The necessary data required by IGNITE for the WLTP is available on the United Nation Economic Commission of Europe (UNECE) website (1). A Ricardo WAVE engine object was also added to enable co-simulation between the two simulation packages.

Table 3: Ricardo IGNITE input data

Description	Data
Vehicle test mass (kg)	3500
Vehicle frontal area (m^2) (8)	5.6
Wheel radius (m)	0.33
Engine inertia (kgm^2)	0.343
Rolling resistance coefficient	0.01125
Drag coefficient (8)	0.4

3 SIMULATION VALIDATION

Validation of the two simulations were conducted using data obtained from experiments run on the engine dynamometer. Steady state data was obtained using available instruments as well as measurements from the ECU software interface.

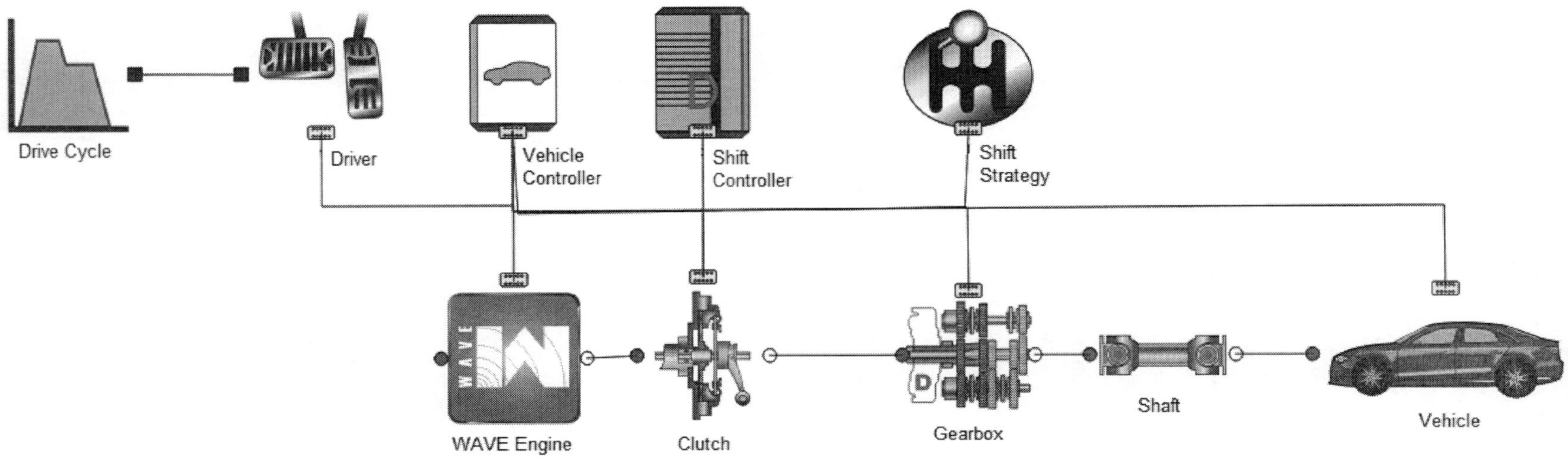

Figure 2: Screen shot of the simulation in Ricardo IGNITE.

3.1 Ricardo IGNITE

The engine speed and load profiles calculated by Ricardo IGNITE were validated using first principle fundamentals in order to establish confidence in the software's results.

3.1.1 Engine speed profile

The engine speed profile was calculated using the following equation:

$$RPM = \frac{30VG_tG_f}{\pi R} \tag{1}$$

The speed of the vehicle, V in meters per second, is available from the driving cycle profile, R is the radius of the wheels in meters and G_t and G_f are the current gear ratio selected and the final drive ratio of the drivetrain, respectively. Figure 3a and figure 3c show the results of the comparison for the NEDC urban cycle and figure 3b and figure 3d show the results of the comparison for the NEDC extra urban cycle. Both graphs show good agreement between the IGNITE generated profiles and the calculated profiles. In figure 3c as well as in figure 3d there are slight overshoots evident just after each acceleration period and deceleration period for the IGNITE calculations. This is caused by the fact that IGNITE is using a PID controller to calculate acceleration and braking for the vehicle to ensure that it accurately follows the programmed driving cycle (5). The overshoot is within the tolerance of ±1km/h set out in the Council Directive 91/441/EEC (10).

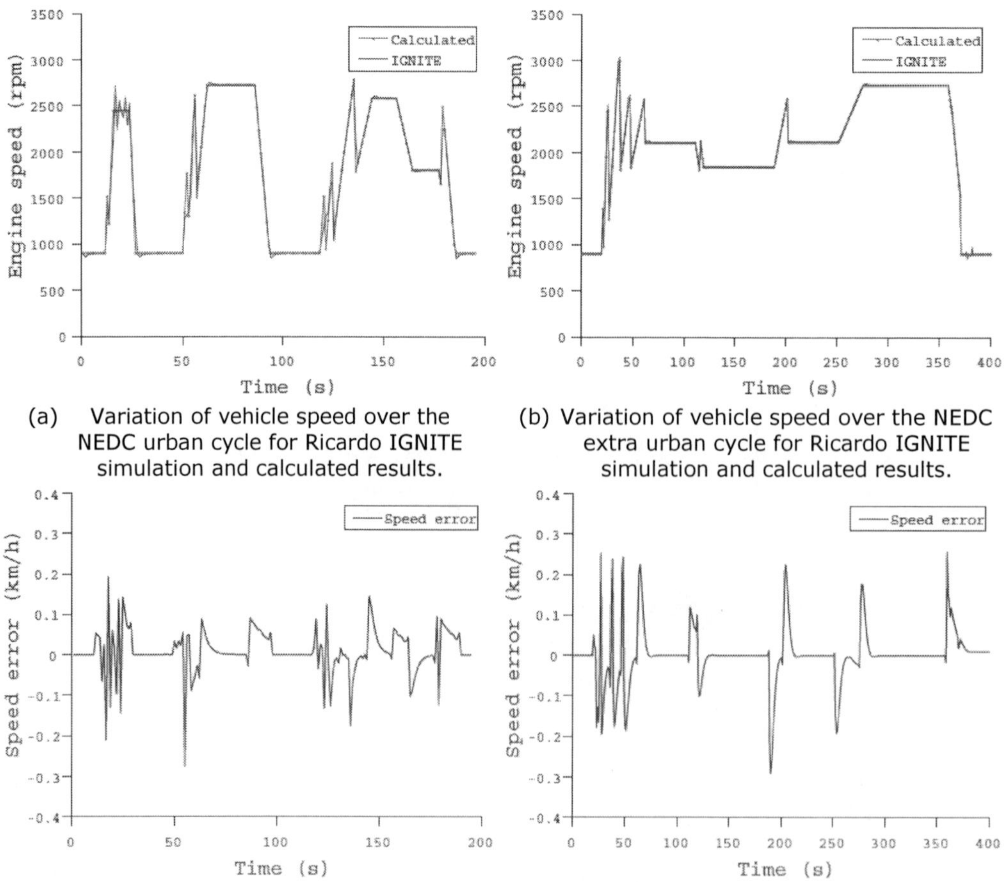

(a) Variation of vehicle speed over the NEDC urban cycle for Ricardo IGNITE simulation and calculated results.

(b) Variation of vehicle speed over the NEDC extra urban cycle for Ricardo IGNITE simulation and calculated results.

(c) IGNITE speed error compared to NEDC urban cycle.

(d) IGNITE speed error compared to NEDC extra urban cycle.

Figure 3: Comparisons between Ricardo IGNITE simulation and calculated results for the NEDC speed profile.

3.1.2 Engine load profile

The engine load requirements as calculated by IGNITE for the NEDC drive cycle was validated using the procedure as set out by Kunniger (9). Figure 4a shows the results of the comparison between the calculated engine load requirements and the IGNITE results for the NEDC urban cycle and figure 4b compares the results for the NEDC extra urban driving cycle. The biggest differences between the compared data sets is during the acceleration and braking stages. Ricardo IGNITE does not state what equations it uses in order to determine the inertial forces that dominates the acceleration stages and as such it is difficult to accurately replicate IGNITE's results. The difference in torque calculations are the biggest during the NEDC urban driving cycle when the inertial forces are higher due to acceleration from stand still. It becomes less during the NEDC extra urban driving cycle, where inertial forces are lower, due to the car already moving and only increasing and decreasing in speed. The differences in calculations for the steady state sections of the driving cycle are minor with the maximum difference of about 6 %. The difference can increase to a maximum of around 20 % just after an acceleration event. The difference in these calculated torques can be attributed to the acceleration PID controller in IGNITE (5). The PID controller uses a feed forward model that dynamically estimates a non-dimensional load level required to follow the drive cycle. Just after an acceleration event, the load requirements reduces significantly and it can take the controller some time to adjust accordingly. The settling time can be decreased with a smaller time step to increase its accuracy, at the expense of simulation time. Braking events in the NEDC is also evident with the IGNITE calculated engine torque values being negative. The braking calculations were not in the scope of the validation and these events will be seen as zero demand in torque for the engine in the dynamometer drive cycle program.

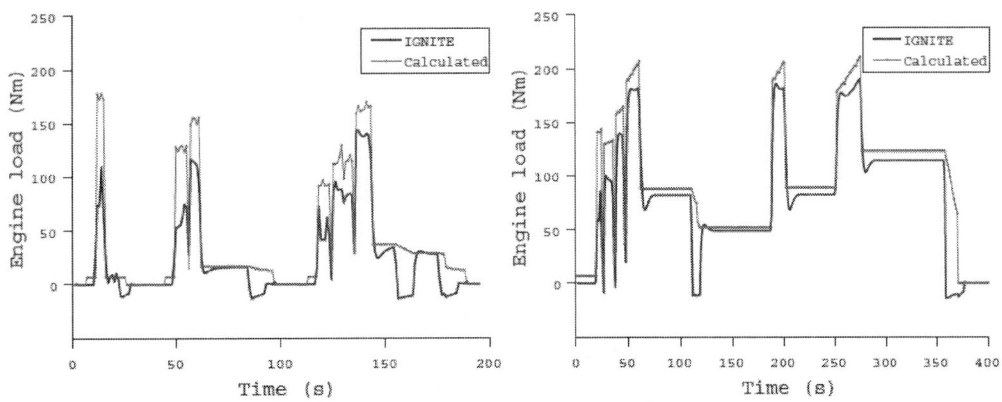

(a) Variation of engine load over the NEDC urban cycle for Ricardo IGNITE simulation and calculated results.

(b) Variation of engine load over the NEDC extra urban cycle for Ricardo IGNITE simulation and calculated results.

Figure 4: Comparisons between Ricardo IGNITE simulation and calculated results for the NEDC speed profile.

Ricardo IGNITE engine load calculations was successfully validated with improvements identified such as PID control optimisation. The same validation procedure for the WLTP drive cycle is unnecessary as IGNITE will use the same vehicle characteristics for calculating the engine load and speed.

3.2 Ricardo WAVE

Steady-state tests were conducted at different engine operating speeds. The engine was run at 2000 rpm, 2500 rpm and 3000 rpm. For each speed set point a 25 %, 50 % and 75 % load were applied on the engine through the dynamometer. The comparisons are shown in Figure 5.

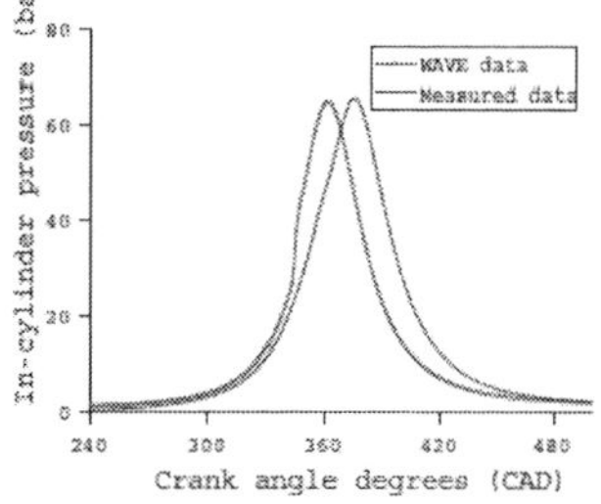

(a) In-cylinder pressure at 2000 rpm and 25 % load.

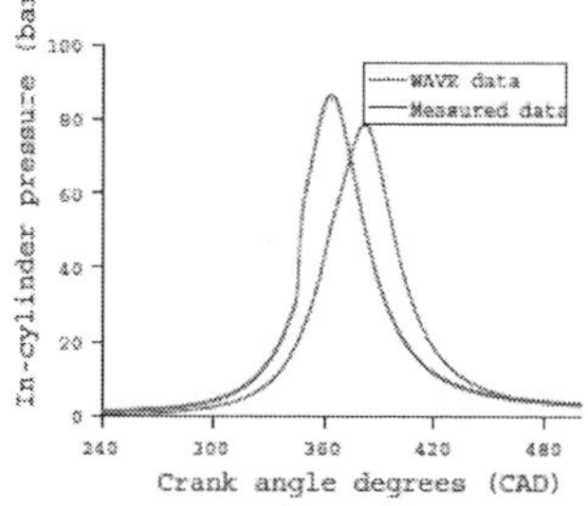

(b) In-cylinder pressure at 2000 rpm and 50 % load.

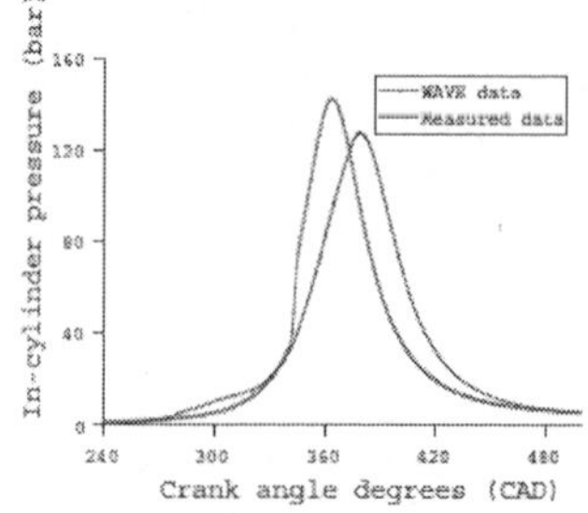

(c) In-cylinder pressure at 2000 rpm and 75 % load.

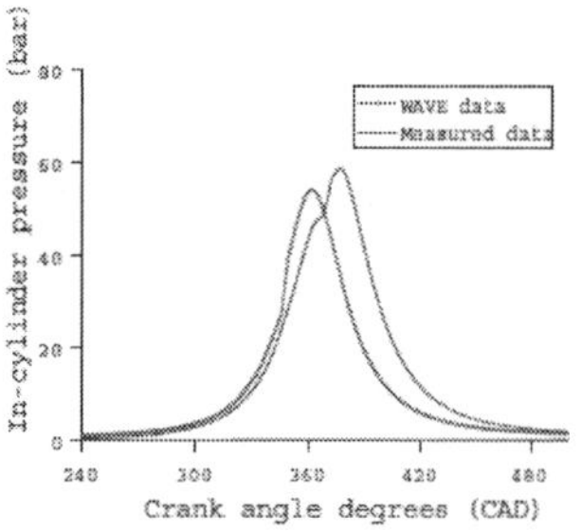

(d) In-cylinder pressure at 2500 rpm and 25 % load.

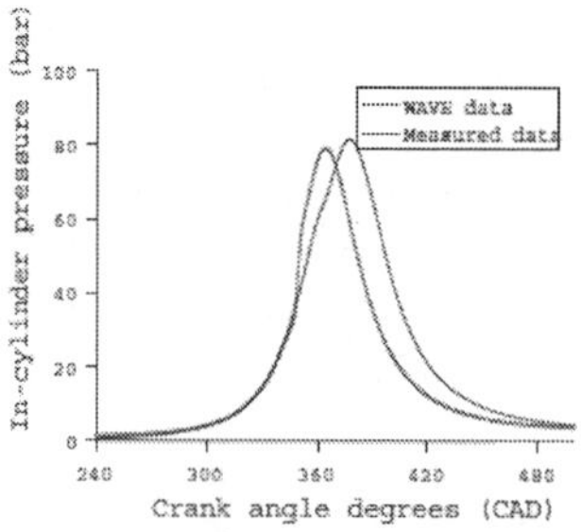

(e) In-cylinder pressure at 2500 rpm and 50 % load.

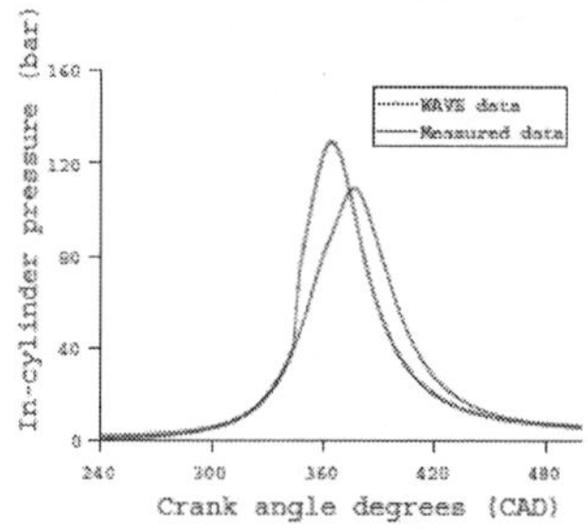

(f) In-cylinder pressure at 2500 rpm and 75 % load.

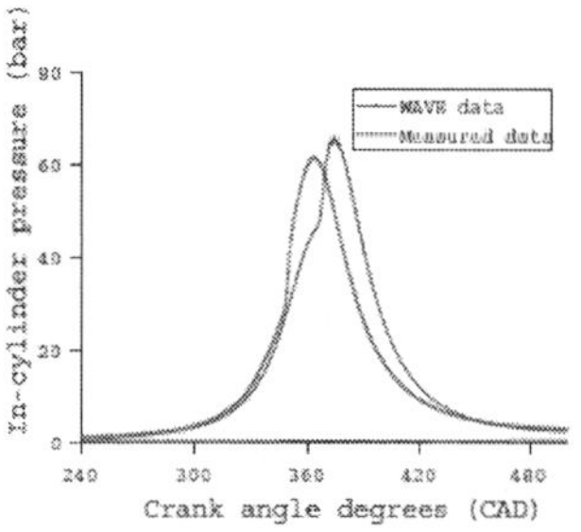

(g) In-cylinder pressure at 3000 rpm and 25 % load.

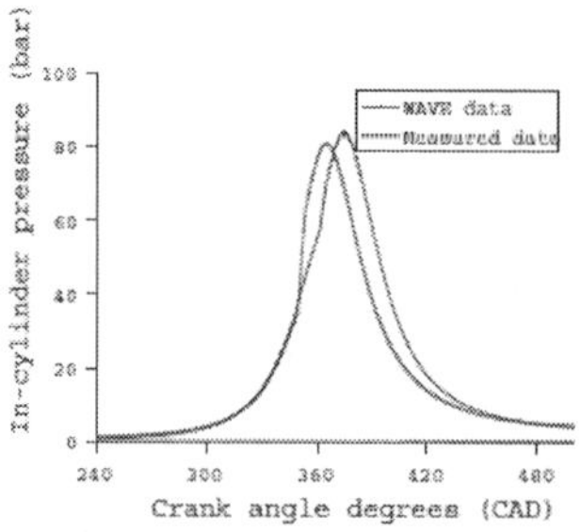

(h) In-cylinder pressure at 3000 rpm and 50 % load.

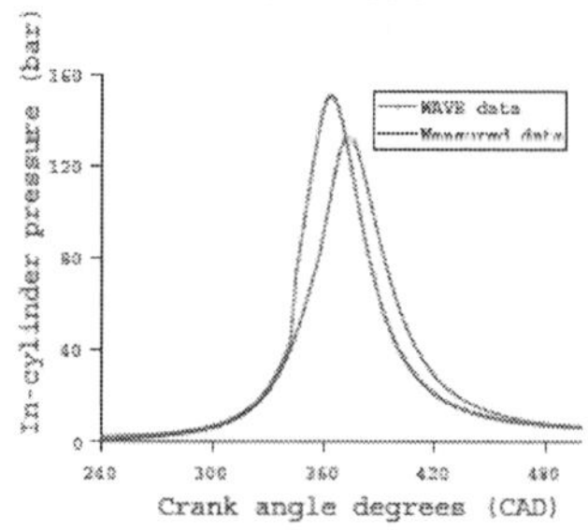

(i) In-cylinder pressure at 3000 rpm and 75 % load.

Figure 5: In-cylinder pressure comparisons between measured data and WAVE simulated data.

3.2.1 In-cylinder pressure

All the simulated in-cylinder pressure profiles are in good agreement with the measured data collected using a pieso-electric pressure transducer mounted in the glow plug. The minimum and maximum pressure deviations in the simulated results when compared to the measured in-cylinder pressures are 1 % for figure 5a and 15 % for figure 5f. It is also clear in all the graphs that the calculated start of combustion is advanced compared to the measured data. WAVE calculates the ignition delay using the procedure described by Watson (11). In this procedure, the ignition delay is calculated using in-cylinder temperature and pressure which are assumed to be homogeneous throughout the cylinder (6). However, in reality there are local temperature and pressure gradients within the vicinity of the injected fuel in the combustion chamber and these would account for a difference in the start of combustion.

3.2.2 Emissions

Comparison data for CO and NO_x emissions can be seen in Figure 6. The carbon monoxide (CO) emissions in figure 6a, figure 6c and figure 6e show significant differences between the measured and calculated data. At 25 % load conditions and 2000 rpm the difference in calculated and measured emissions is approximately 14 %, at 2500 rpm it is 61 % and at 3000 rpm it is approximately 480 %. At 50 % load conditions and 2000 rpm the difference in calculated and measured emissions is approximately 48 %, at 2500 rpm it is 40 % and at 3000 rpm it is approximately 158 %. At 75 % load conditions and 2000 rpm the difference in calculated and measured emissions is approximately 1250 %, at 2500 rpm it is 1220 % and at 3000 rpm it is approximately 1170 %.

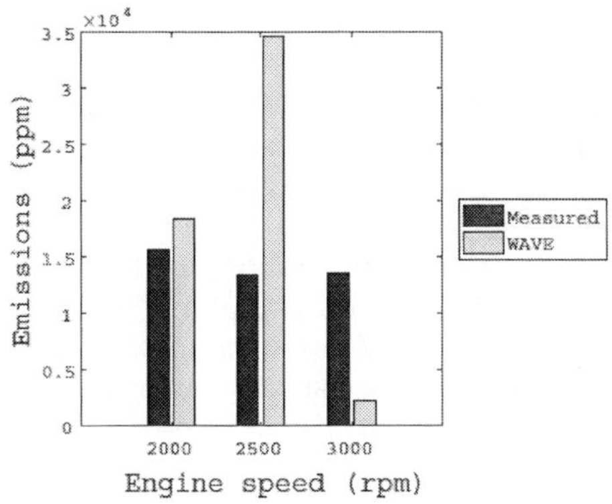

(a) CO engine emissions at 25 % load

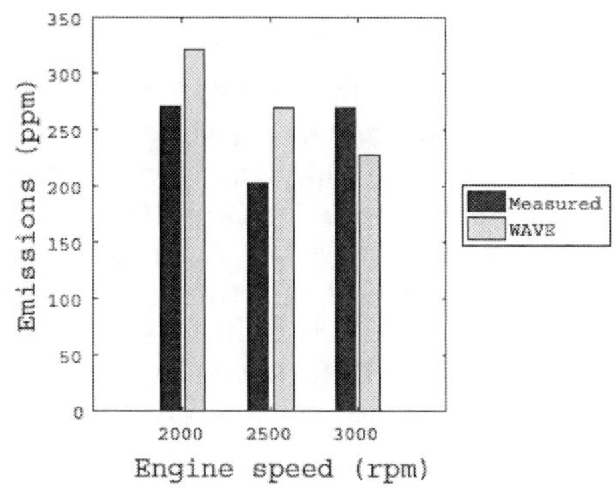

(b) NO engine emissions at 25 % load

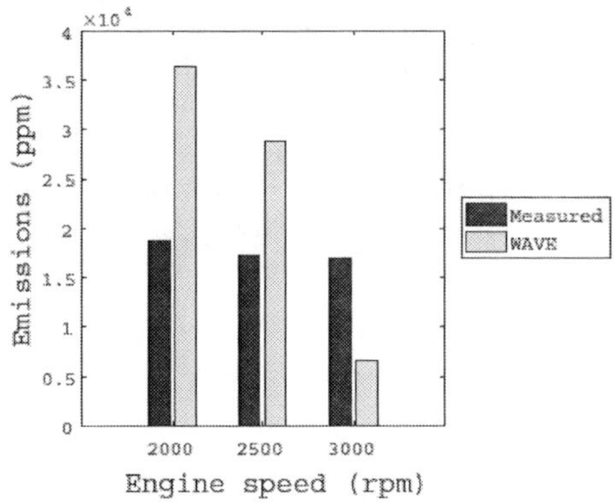

(c) CO engine emissions at 50 % load

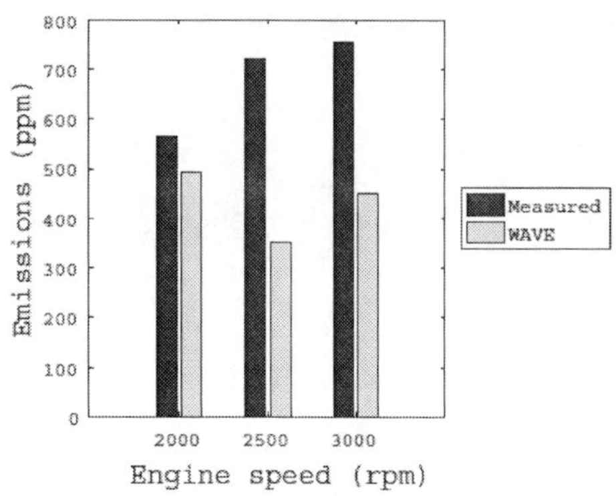

(d) NO engine emissions at 50 % load

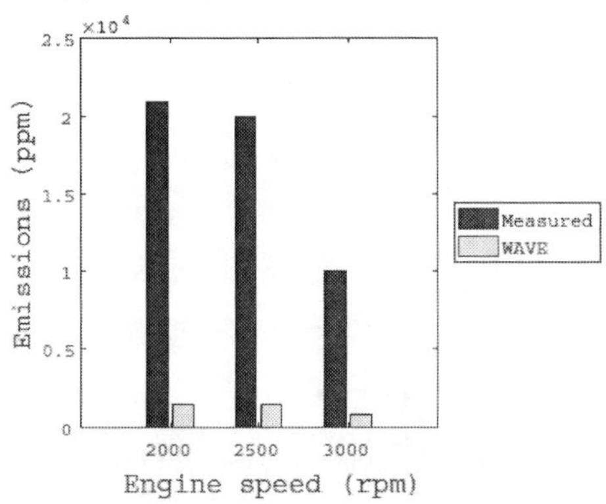

(e) CO engine emissions at 75 % load

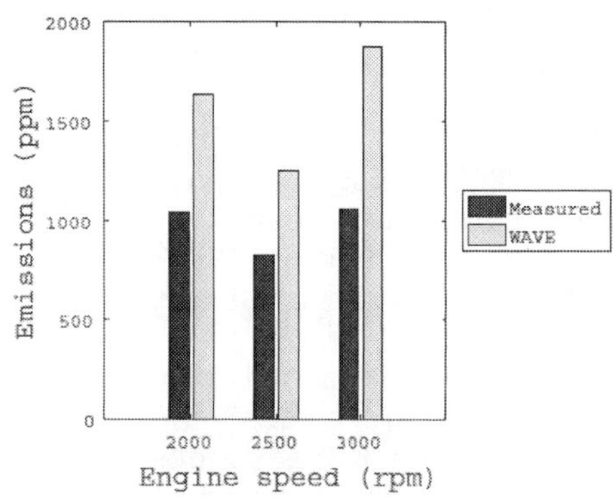

(f) NO engine emissions at 75 % load

Figure 6: Engine emissions comparisons at different engine speeds and loads.

Ricardo WAVE is using a combination of calculations to determine the CO emissions. The software uses a chemical reaction in its gas property calculations that gives good results for scenarios with fuel rich combustion, but poor predictions for lean fuel mixture combustion. A chemical reaction suggested by Newhall (12) gives good results for lean fuel mixture combustion, but its prediction for fuel rich combustion is poor. WAVE combines these two calculations to cover the whole fuel concentration range and adds a weight value based on the O_2 concentration (for fuel rich combustion) and the CO and CO_2 concentrations (for lean fuel combustion). Ricardo compared its calculated results with measured data and the differences were within one order of magnitude (6). The differences in figure 6a and figure 6c are thus within the stated margin of error. The differences in figure 6e are due to WAVE assuming a homogeneous air and fuel mixture which contributes to a more complete combustion.

The oxides of nitrogen (NO_x) emission measurements are compared with the WAVE calculated emissions and shown in figure 6b, figure 6d and figure 6f. At 25 % load conditions and 2000 rpm the difference between the calculated and measured emissions is approximately 16 %, at 2500 rpm it is 25 % and at 3000 rpm it is approximately 19 %. At 50 % load conditions and 2000 rpm the difference between the calculated and measured emissions is approximately 15 %, at 2500 rpm it is 104 % and at 3000 rpm it is approximately 67 %. At 75 % load conditions and 2000 rpm the difference between the calculated and measured emissions is approximately 36 %, at 2500 rpm it is 34 % and at 3000 rpm it is approximately 43 %.

WAVE's NO_x emission sub model accounts for NO formation only. Most of the NO emission values are gained from correlation data that gives the ratio of NO to equilibrium NO as a function of equivalence ratio (6). Constant multipliers in the sub-model can be used to increase or decrease the NO_x results, but the overall inaccuracy can be attributed to using correlation data and the fact that WAVE is using one-dimensional equations to solve for the temperature distribution during combustion, which influences the formation of NO_x gasses. Overall the NO_x emission calculations are in good agreement with the measured emissions.

4 FUTURE FUELS

The suitability of the simulation developed was tentatively tested using a potential future fuel. All the simulation parameters were kept the same except the fuel was changed to Fatty Acid Methyl Ester (FAME). The cumulative emissions were recorded over the NEDC test cycle and are presented in Figure 7. A decrease in CO of about 90 % and HC of about 30 % are evident and an increase in NO emissions of approximately 30 % is also seen. The results presented agree with similar studies in the literature (13, 14).

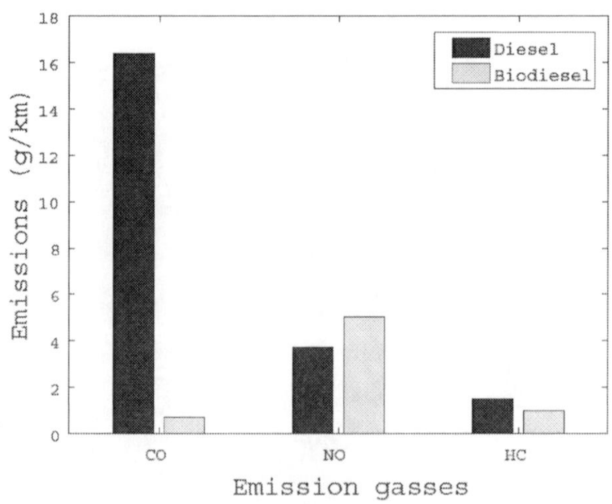

Figure 7: Engine emissions over the NEDC drive cycle for diesel and biodiesel fuel.

5 CONCLUSIONS

This paper presents the development, and validation, of an engine simulation using commercially available software for accessing the impact of utilising future fuels with future drive cycles. The main conclusions from this study are:

- The simulation in-cylinder pressure profiles are broadly in good agreement with the experiments; differences are due to local temperature and pressure gradients.
- The simulated CO and NO_x emissions were validated by exhaust gas measurements and are within the stated margin of error of the sub-model.
- A provisional simulation with a 100 % bio-fuel (FAME) yielded promising results that agreed with similar studies.

This simulation technique is appropriate with new emissions legislation such as the RDE being introduced in 2017. The ability to simulate all aspects of a vehicle and being able to investigate accurate emissions data over different driving cycles makes it a competitive tool in vehicle and engine design and optimisation.

6 FUTURE WORK

The successful validation of the co-simulated simulation makes it possible to investigate different scenarios that can influence engine emissions and contribute to the decreased emission of harmful gasses into the atmosphere. Engine operating characteristics can be altered in Ricardo WAVE and its effect on engine emissions over different drive cycles can be evaluated in Ricardo IGNITE. Future work including the multi-fidelity simulation is summarised below.

6.1 Improve simulation with CFD results

The differences in the start of combustion, mentioned in section 3.2.1, were attributed to local differences in temperature and pressure. It is planned to address this through the use of a higher fidelity simulation. Ricardo VECTIS is a 3D CFD software that can be coupled with WAVE. VECTIS will be utilised to simulate the in-cylinder dynamics and combustion, which will then be fed into WAVE to improve the accuracy of the simulation.

6.2 Renewable fuel

The use of blends of diesel fuel with renewable fuels such as biodiesel and ethanol will be investigated with the use of the multi-fidelity powertrain simulation. The effect on engine emissions by using different binary blends between conventional diesel and biodiesel as well as ternary blends between diesel, biodiesel and ethanol will be investigated. Other areas such as engine torque, engine power and fuel consumption will also be monitored over the available driving cycles.

6.3 Low temperature combustion

The multi-fidelity powertrain simulation will be used to investigate low temperature combustion (LTC) scenarios which include multiple injections and exhaust gas recirculation (EGR) to reduce emissions. The best fuel injection configuration for reducing engine emissions will be determined with the use of a combination of early injections, pilot injections and post injections. The homogeneity of the air fuel mixture will also be investigated when using multiple injections and its effect on promoting low temperature combustion will be evaluated. The optimum percentage of EGR will be determined over the available driving cycles and its effect on engine performance and emissions will be evaluated.

6.4 Validation of multi-fidelity powertrain simulation

The multi-fidelity powertrain simulation will be validated using available engine test facilities. Different driving cycles such as the NEDC and WLTP will be run on the engine test bench and the corresponding engine emissions and performance data will be used to validate the simulation. Engine emissions and performance data from an RDE route will also be used.

REFERENCE LIST

(1) UNECE. *World Harmonized Light vehicles Test Procedure (WLTP)*, 2017. Available from: https://wiki.unece.org/pages/viewpage.action?pageId=2523179 [Accessed 28 June 2017].

(2) UNFCCC. Paris agreement *[online]*. Official Journal L 282, 19.10.2016, p. 4–18, 2015. Available from: http://eur-lex.europa.eu/legal-content/EN/TXT/PDF/?uri=CELEX:22016A1019(01)&from=EN [Accessed: 17 July 2017].

(3) European Union. Directive 2009/28/EC of the European parliament and of the council of 23 April 2009 on the promotion of the use of energy from renewable sources and amending and subsequently repealing directives 2001/77/EC and 2003/30/EC *[online]*. Official Journal L 140, 5.6.2009, p. 16–62, 2009. Available from: http://eur-lex.europa.eu/legal-content/EN/ALL/?uri=CELEX:32009L0028&qid=1498667012776 [Accessed: 28 June 2017].

(4) European Union. Regulation (EC) no 715/2007 of the European parliament and of the council of 20 June 2007 on type approval of motor vehicles with respect to emissions from light passenger and commercial vehicles (euro 5 and euro 6) and on access to vehicle repair and maintenance information *[online]*. Official Journal L 171, 29.6.2007, p. 1–16, 2007. Available from: http://eur-lex.europa.eu/legal-content/EN/ALL/?uri=CELEX:32007R0715 [Accessed: 28 June 2017].

(5) Ricardo. *IGNITE User Manual*, 2016.

(6) Ricardo. *WAVE User Manual*, 2016.

(7) Workshop Manuals. *Ford Workshop Manuals*, 2017. Available from: http://workshop-manuals.com/ford/transit_2006.5_04.2006 [Accessed 28 June 2017].

(8) G Mellios, S Hausberger, M Keller, Z Samaras, and L Ntziachristos. Parameterisation of fuel consumption and CO_2 emissions of passenger cars and light commercial vehicles for modelling purposes. *Publications Office of the European Union*, 2:16–18, 2011.

(9) D Kunninger. *The development of a vehicle and road load simulation for an engine dynamometer intended for emissions testing and powertrain development*. PhD thesis, Faculty of Engineering and the Built Environment, 2013.

(10) European Union. Council directive 91/441/EEC of 26 June 1991 amending directive 70/220/EEC on the approximation of the laws of the member states relating to measures to be taken against air pollution by emissions from motor vehicles *[online]*. Official Journal L 242, 30.08.1991 p. 1-106, 2009. Available from: http://eur-lex.europa.eu/legal-content/EN/TXT/?qid=%201499086220026&uri=CELEX:31991L0441 [Accessed: 28 June 2017].

(11) N Watson, AD Pilley, and Mohamed Marzouk. A combustion correlation for diesel engine simulation. Technical report, SAE Technical Paper, 1980.

(12) H K Newhall. Kinetics of engine-generated nitrogen oxides and carbon monoxide. In *Symposium (International) on Combustion*, volume 12, pages 603–613. Elsevier, 1969.
(13) CD Bannister, JG Hawley, HM Ali, CJ Chuck, P Price, SS Chrysafi, A Brown, and W Pickford. The impact of biodiesel blend ratio on vehicle performance and emissions. *Proceedings of the Institution of Mechanical Engineers, Part D: Journal of Automobile Engineering*, 224(3):405–421, 2010.
(14) M Lapuerta, O Armas, and J Rodriguez-Fernandez. Effect of biodiesel fuels on diesel engine emissions. *Progress in energy and combustion science*, 34(2):198- 223, 2008.

A review of Life Cycle Assessment approaches applied to automotive energy vectors

H J Willmott, J W G Turner, M C McManus
Department of Mechanical Engineering, University of Bath, UK

ABSTRACT

The automotive industry has been working towards EU targets to reduce vehicle tailpipe CO_2 emissions. The use of alternatives to conventional fossil fuels, such as electricity, are currently being implemented to achieve further reductions. It is vital that we understand and monitor the implications of these alternatives in order to minimise ongoing environmental impacts. The potential for use of Life Cycle Assessment (LCA) to determine the true environmental impact of different fuels for use in internal combustion engines is considered in this paper. Six LCA studies are reviewed and their approaches to the goal and scope definition phase discussed.

ABBREVIATIONS

BEV	Battery electric vehicle
CNGV	Compressed natural gas vehicle
EV	Electric vehicle
FCV	Fuel cell vehicle
GHG	Greenhouse gas
GREET	Greenhouse Gases, Regulated Emissions, and Energy Use in Transportation
HICEV	Hydrogen internal combustion engine vehicle
ICE	Internal combustion engine
ICEV	Internal combustion engine vehicle
LCA	Life Cycle Assessment
OEM	Original equipment manufacturer
PEM FCV	Polymer electrolyte membrane fuel cell vehicle
PHEV	Plug-in hybrid electric vehicle
TTW	Tank to wheel
WTT	Well to tank
WTW	Well to wheel

1 INTRODUCTION

In 2013 the transport sector accounted for 27.6% of global energy consumption, with 80% of the total amount used in road transport [1]. This paper will focus on reviewing Life Cycle Assessment (LCA) as a means of evaluating the environmental impact of road transport, specifically that of light duty vehicles.

Growth in the transport sector has been driven by many factors including rising global population and industrial development. Consideration must be made of the balance between human needs, development, and limiting environmental impact. Transport is a major contributor to climate change and environmental degradation [2]. We now have a good understanding of the causes of climate change, and we are able to predict potential future outcomes based on different rates of release of greenhouse gases (GHGs) [3]. A reduction in the emission of GHGs is vital to limit the effects of climate change. It is also important to consider other environmental impacts of road transport in order to protect the quality of the air, land and water.

Over the history of automotive transport many different forms of energy have been utilised to power vehicles. Burning fossil fuels emerged as the predominant energy choice, and 93% of energy for road transport is based on the consumption of fossil fuel products [4]. There is a well-established infrastructure to service this form of energy provision, which has grown symbiotically with the increase in vehicles.

Fossil fuels are a non-renewable resource, and their combustion releases GHGs. As well as the environmental impact of the use of such fuels to power vehicles, there are energy security issues that arise from our reliance on them. There is political instability in key areas of oil supply, which can have an impact on access and availability of the resource [5]. One way for the UK to decarbonise the transport system and improve energy security is to look towards alternative energy sources to diversify automotive fuels.

The automotive industry has been set EU targets to reduce vehicle tailpipe CO_2 emissions with the introduction of mandatory performance standards for new passenger cars [6], and vans [7]. The fleet average to be achieved by all new cars by 2021 is 95 grams of CO_2 per kilometre driven under New European Driving Cycle (NEDC) conditions, which constitutes a 40% reduction compared to 2007 levels. Significant improvements have already been achieved by the vehicle original equipment manufacturers (OEMs), with average tailpipe CO_2 reductions of 33.6% between 2000 and 2016 [8]. However, the rate of progress towards achieving this goal has begun to slow [9], and it is recognised that there is a limit as to what extent technological innovation can improve internal combustion engine (ICE) vehicle efficiency to reduce tailpipe CO_2 emissions [10]. A report by SMMT concluded that for the UK to achieve the EU target there would need to be a significant increase in the rate of progress [8].

The use of alternatives to conventional fuels, such as electricity, is one approach currently being used by OEMs to achieve further reductions in fleet average tailpipe emissions. One way in which OEMs are making progress to meet targets is through the sale of electric vehicles (EVs), which have zero tailpipe emissions. Increased sales of these vehicles could help OEMs meet their overall fleet average tailpipe CO_2 emission targets. The UK government currently offers consumers grants to incentivise the purchase of plug in EV cars and vans. However, any increase in EV sales requires a major infrastructure change, as these vehicles need access to charge points, and although there are no tail pipe emissions there are emissions associated with the electricity generation. The effect of the additional demand on the National Grid also needs to be considered. EVs currently have a significantly limited distance range compared to conventionally fuelled vehicles [11]. This can limit the current practicality of EVs for some users. Hybrid electric vehicles (HEV) and plug in hybrid electric vehicles (PHEV) are also being introduced, however as well as using electricity these vehicles still require conventional fuels. In 2015 HEVs, PHEVs and EVs made up less than 3% of new cars registered in the UK [12]. If EVs emerge as the most environmentally benign future automotive option, and range issues are alleviated, there will still need to be a period of transition during which time the use of low carbon fuels could play a role in lowering automotive transport emissions and meeting climate targets. Research should continue into low carbon fuels that can be used with the current automotive energy infrastructure.

It is essential to understand the full environmental impact of prospective alternative fuels in order to ensure that any developments deliver a true benefit. To gain a better understanding of the likely outcome of different fuel or energy choices for automotive use, tailpipe emission of CO_2 should not be the only consideration. The source of energy, how it was extracted and processed, transported, stored and delivered, are also important and should be explored. This wider approach to measuring impact is often undertaken through the use of Life Cycle Assessment (LCA).

2 LIFE CYCLE ASSESSMENT

LCA is a tool that can be used to give a better understanding of true environmental impact. Rather than focusing on one factor, i.e. tailpipe CO_2 emissions, LCA can be used to explore the entire system that is in place to provide energy for automotive transport. LCA research studies are used to systematically compile environmental impacts of product systems, from acquirement of raw materials to waste end of life [13]. LCA does not include any measure of economic or technological feasibility, however LCA may be carried out in parallel with other decision support tools to incorporate these aspects and help to determine overall sustainability of a product.

LCA was first standardised in 1993 in the Society of Environmental Toxicology and Chemistry code of practice [14], and later amended in the 1997 International Organisation for Standardisation ISO 14040 series. These ISO standards were revised in 2006 [13, 15, 16].

LCA is commonly agreed to be a useful comparative tool, but there are some issues associated with its uptake and comparability. These issues have been fully explored in many other papers, see for example McManus and Taylor [16], McManus *et al.* [17], and Hellweg and Milà i Canals [18].

LCA methodology broadly follows four stages:

- Goal and Scope Definition
- Life Cycle Inventory Analysis
- Life Cycle Impact Assessment
- Interpretation [19]

2.1 Goal and Scope Definition

Goal and scope definition is the first stage of an LCA study. The purpose of this phase is to define goals, functional units, system boundaries, amount and quality of data, assumptions and limits, and environmental issues to be considered. LCA studies can be conducted for many different purposes depending on the needs of the commissioner. The goal and scope definition stage should make clear the specific reasons for carrying out a study. It should set an explicit research question to be answered and state how the results should be used. This should ensure that the number of value choices that remain to be made during subsequent LCA methodology phases are limited [19].

2.1.1 System Boundaries

System boundaries set criteria defining which processes are part of a product system. They are defined during the goal and scope stage of LCA. "Well to wheel" (WTW) is a methodology that is often used as a subset of LCA, focusing on energy consumption and CO_2 emissions of the fuel, from the point of primary energy source extraction to its point of utilisation [20][1]. Vehicle manufacture and vehicle end of life are not included. This approach can be useful to decision makers who are purely looking for simplified results to compare different fuels. Vehicle light weighting is one example of a manufacturing strategy that can improve vehicle fuel efficiency but in some cases could have a higher environmental impact during vehicle production [21]. Where system boundaries are expanded beyond WTW, LCA can highlight impacts outside of the scope of WTW and show at what stage of vehicle life any increased environmental cost of production is balanced out by greater fuel efficiency. This would not be apparent if focusing on WTW alone, as the results would show only the improved fuel efficiency and would not account for any additional environmental impact of manufacturing. WTW is an effective tool to consider vehicle efficiency and the impact that this has on the environment, but it is not a substitute for LCA when the full picture is required.

2.1.2 Functional Unit

The ISO standard defines an LCA functional unit as "the quantified performance of a product system for use as a reference in a life cycle assessment study" [13]. It is important that the functional unit chosen is relevant to the goal and scope of the study, and allows for a fair comparison even though the compared products may not be identical [19].

2.1.3 Impact Categories

Environmental impact categories are selected during goal and scope definition for use in the life cycle impact assessment phase. Impact categories must be selected in line with the goal and scope of the study in order to give results that are suitable for their intended applications. They must cover all of the main environmental issues related to the system, however this can be open to a degree of researcher judgement [22]. This paper examines the use of Life Cycle Assessment (LCA) to analyse different automotive fuel and energy choices.

3 METHODS

A search was conducted using the database Scopus. Figure 1 shows how search terms were applied to identify relevant research on the topic of LCA in automotive energy, with specific reference to internal combustion engines (ICE) [2].

The results are incomplete for 2017 as this study was conducted mid-year. There were 7,200 results for Life Cycle Assessment or LCA within article titles. Of these results 3,753 included the terms fuel, energy or biofuel in the title, abstract or keywords. Papers relating to automotive transport were identified using the search terms automotive, automobile, vehicle or transport, producing 725 results. Finally the term internal combustion engine was included, to identify papers that either focused primarily on ICEVs or used them as a benchmark for comparison with other types of powertrain. This produced 35 search results, of which 16 were discounted on grounds of relevance and availability.

---

[1] Note that the current legislative instrument for assessing the fossil impact of vehicular use, tailpipe CO_2 emissions per kilometre driven on the NEDC, is a tank-to-wheel (TTW) assessment. Well-to-tank (WTT) is the equivalent assessment for fuel supply, and the product of these two values gives the overall WTW result.

[2] These were (TITLE("Life Cycle Assessment" or LCA) AND TITLE-ABS-KEY(fuel or energy or biofuel) AND TITLE-ABS-KEY(automotive or automobile or car or vehicle or transport) AND TITLE-ABS-KEY(internal and combustion and engine)).

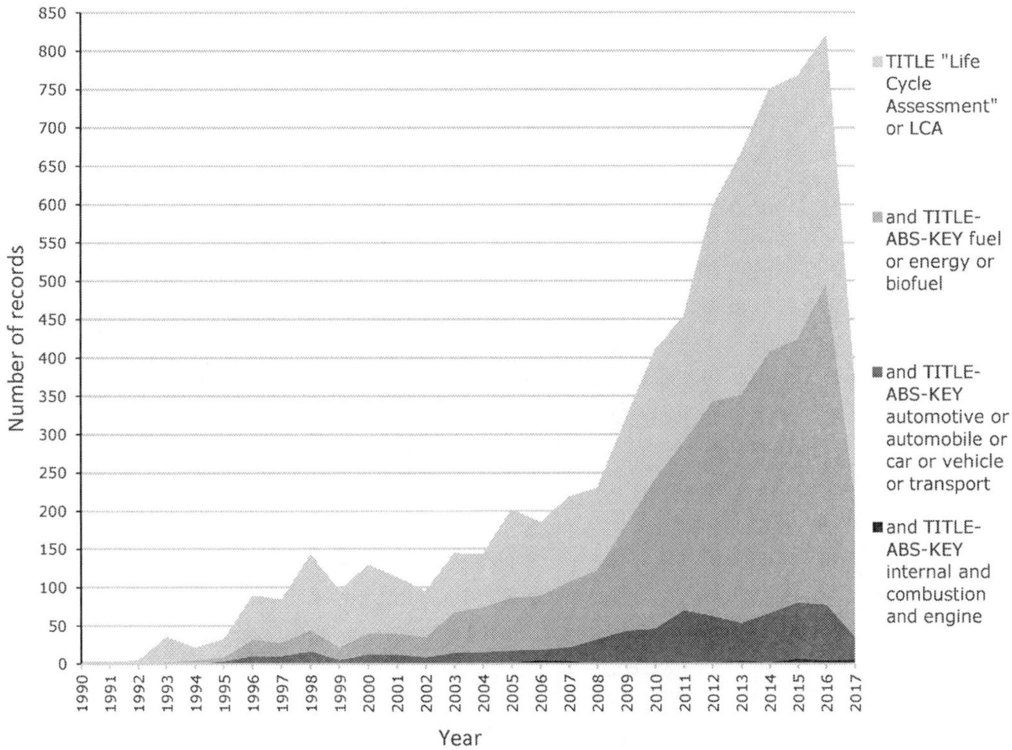

Figure 1. LCA in automotive energy search results from Scopus database (May 2017)

A systematic approach was used to review the data and generate themes, based on researcher judgement of prevalence and potential importance of insight or perspective. It is intended to discuss this process in detail in a later publication. The themes selected for further investigation fall within the goal and scope definition phase, evaluating different approaches made to define system boundaries, functional units, and impact categories. Examination of the literature search results found six LCA case studies that stated adherence to ISO standards. These studies were selected by the researchers for further comparison.

4 RESULTS AND DISCUSSION

4.1 Comparison of LCA study approaches

Common elements of the studies reviewed are shown in Table 1.

Table 1. Comparison of LCA studies

Author and date	Title	Powertrain	Energy/fuel	Functional unit	System boundaries
Bartolozzi *et al.* (2013) [23]	Comparison between hydrogen and electric vehicles by life cycle assessment: A case study in Tuscany, Italy	BEV	Electricity (from renewables, benchmarked against Italian energy mix)	200 km at nominal full load, vehicle used deliver goods within an urban area	Hydrogen production chains, at regional scale. Storage, transportation and distribution of hydrogen. Production, use and end-of-life of vehicle and components. Include all the processes within the life cycle phases: extraction of raw materials, fuel production, transportation, manufacture, use, maintenance and end-of-life of plants used for electricity and hydrogen production and of vehicles. Network for electricity distribution (for electricity requirement for gas compression) excluded.
		FCV	Hydrogen (from renewables and biomass energy sources either by electrolysis or direct separation from biomass gasification syngas.		
		HICEV	Hydrogen		
Bicer and Dincer (2017) [24]	Comparative life cycle assessment of hydrogen, methanol and electric vehicles from well to wheel	BEV	Electricity	Person km	From extraction of raw materials for hydrogen, methanol and electricity to disposal of the vehicles.
		ICEV	Methanol		
		ICEV	Hydrogen		
Dai and Lastoskie (2014) [25]	Life cycle assessment of natural gas-powered personal mobility options	BEV	Electricity from natural gas	120,000 miles evenly delivered over a 10-year period	Starts with natural gas extraction and ends with miles delivered in natural gas propulsion by a CNGV, BEV or FCV. For the 3 natural gas utilization pathways, vehicle manufacturing and assembly, as well as the infrastructure needed to convert natural gas to the vehicle specific fuel, is included as part of the material processing and manufacturing life cycle stage. For BEVs and FCVs, carbon capture at fuel processing facilities is considered when modelling the infrastructure. Recycling of batteries is also considered in the EOL stage. The cut off criterion is set to 1%, i.e. process and material or energy inputs whose contributions to the total environmental impact are less than 1% are not included in the analysis.
		CNGV	Compressed natural gas		
		FCV	Hydrogen from natural gas		
		ICEV	Gasoline		

Table 1. Comparison of LCA studies (continued)

Author and date	Title	Powertrain	Energy/fuel	Functional unit	System boundaries
Girardi *et al.* (2015) [26]	A comparative LCA of an electric vehicle and an internal combustion engine vehicle using the appropriate power mix: the Italian case study	BEV	Electricity	150,000 km driven	Vehicle construction and disposal. Battery construction and disposal. Energy vectors (electricity for EV and gasoline for ICEV) supply chains. Use phase of the vehicles (including vehicle maintenance). Road construction, disposal, and maintenance considering 100 years of road life.
		ICEV	Gasoline		
Notter *et al.* (2015) [27]	Life cycle assessment of PEM FC applications: Electric mobility and μ-CHP	BEV	Electricity	1 average km driven on the European road network	All components of a FC starting with the production of the carbon support material, the catalyst powder with platinum nanoparticles, a membrane, a gas diffusion layer, bipolar flow plates up to the FC stack and the FC unit including end of life treatment.
		PEM FCV	Hydrogen		
		ICEV	Gasoline		
Weinberg and Kaltschmitt (2013) [28]	Life cycle assessment of mobility options using wood based fuels - Comparison of selected environmental effects and costs	BEV	Electricity derived from wood	1 km driven in a mid-range car with 100kW engine power	The whole life cycle of biomass production, conversion and use including the car production and disposal is investigated. All pathways contain the pre-chains of the material and energy supply as well as the infrastructure for: -wood cultivation and transport of the wood to the conversion plant -fuel production -fuel distribution -use of the fuel in the car.
		ICEV	Hydrogen, ethanol, Fischer-Tropsch diesel and methane derived from wood		

4.1.1 Goal and scope definition approach

Bartolozzi *et al.* [23] conducted an LCA to evaluate the environmental sustainability of the use of hydrogen produced from renewable energy sources. The study focused on a fleet of hydrogen vehicles used for urban commercial delivery in Tuscany, Italy. Hydrogen produced using energy from wind and biomass sources was benchmarked against hydrogen produced from the Italian electricity mix. A further benchmarking analysis compared the results with standard EVs using the same renewable energy sources.

Bicer and Dincer [24] conducted a comparative LCA study with the goal of investigating the impacts of methanol, hydrogen and electric vehicles on the environment and human health. The scope was described as a comprehensive environmental impact assessment.

Dai and Lastoskie [25] set out with the goal to assess which form of natural gas powered mobility offers the most environmental benefits, compared with conventional gasoline powered ICEVs.

Girardi *et al.* [29] tailored their LCA study to give a representation of the electricity mix that would actually be used by EVs in the study area, Italy, rather than using average national values. They considered two time frames, a 2013 scenario with a low proportion of EVs in relation to ICEVs, and a 2030 scenario in which the higher number of EVs influences the structure of the electricity provision system. The results were intended to be used by policymakers, stakeholders within the field of EVs, and consumers. The study was also intended to provide information to other researchers. The authors recognise that it is not possible to directly compare an EV with an ICEV as each vehicle offers different user experiences, for example acceleration, range and the typical time taken to refuel. However they suggest that for the purposes of average daily travel distances in Italy, which are lower than commercial EV range, a comparison is valid.

Notter *et al.* [27] analysed the environmental impacts of two types of polymer electrolyte membrane fuel cells (PEM FC) with different carbon substrates in FCEVs using hydrogen as a fuel. These technologies are also benchmarked against ICEVs and EVs. The authors state in the scope of their study that they intend for the results to be used by individuals or society, giving examples such as in consideration of phasing out nuclear power.

Weinberg and Kaltschmitt [28] set out to compare automotive fuels derived from poplar wood (electricity, hydrogen, ethanol, Fischer-Tropsch diesel and methane), analysing both environmental and economic performance. There is no benchmarking conducted against conventionally fuelled vehicles in this study.

Both Bartolozzi *et al.*, [23] and Weinberg and Kaltschmitt [28] included additional decision support tools to run in parallel with the LCA to evaluate the economic and technical feasibility of fuels and vehicles in their studies.

4.1.2 Functional unit selection

All of the six studies reviewed use distance driven in a vehicle as the functional unit for comparison of the fuels assessed (see Table 1). This is because the service provided by these vehicles is distance driven. Other options that could have been selected include comparison of a particular volume of fuel, or amount of energy stored in a fuel. LCA studies typically contain many assumptions at every stage. Giving additional results for a functional unit that describes the amount of fuel or energy stored in fuel before it is used in the vehicle would be useful, as it gives an indication of the impact of the fuel before further assumptions are made during the TTW stage and beyond.

4.1.3 System boundary selection

All of the six studies reviewed describe how system boundaries have been chosen to accurately enable fair comparison between different vehicles (see Table 1). However the system boundaries used in each study differ, therefore it is not possible to directly compare results from different studies. All studies reviewed contain the stages of WTW within their system boundaries, but extend to include certain life cycle factors of the vehicle. Girardi *et al.* [26] chose to also include factors relating to road construction, maintenance and disposal. This is not the case in any of the other studies reviewed. It is unclear how useful this inclusion is, when the goal and scope of the study is to determine the environmental impact of two different powertrain types. As the results of the road infrastructure LCA are applicable to both vehicle technologies it does not necessarily provide additional relevant information.

Weinberg and Kaltschmitt [28] chose not to include CO_2 emissions from biomass or biofuel combustion in their LCA. The reason given by the authors is that these emissions are of biogenic origin. However a recent study by Weiguo *et al.* [30] argues that biogenic CO_2 emissions are a relevant factor and should be included when conducting an LCA.

4.1.4 Impact category selection

The impact categories included in the six LCA case studies are summarised in Table 2.

Table 2. Impact Categories used in LCA studies

	Bartolozzi *et al.* (2013) **[23]**	Bicer and Dincer (2017) **[24]**	Dai and Lastoskie (2014) **[25]**	Girardi *et al.* (2015) **[26]**	Notter *et al.* (2015) **[27]**	Weinberg and Kaltschmitt (2013) **[28]**
Global warming/Climate change	✓	✓	✓	✓	✓	✓
Human toxicity	✓	✓	✓	✓	✓	
Acidification	✓			✓		✓
Ecosystem					✓	
Energy demand of fossil resources						✓
Eutrophication	✓			✓		
Freshwater aquatic toxicity	✓					
Human health					✓	
Marine aquatic ecotoxicity	✓					
Mineral depletion potential			✓			
Natural land transformation			✓			
Ozone layer depletion/ Stratospheric ozone depletion	✓	✓				
Particulate matter			✓	✓	✓	
Photochemical (oxidation)	✓		✓	✓		
Primary energy			✓	✓		
Resource depletion				✓	✓	
Terrestrial acidification			✓			
Terrestrial toxicity	✓		✓			
Water depletion potential			✓			

There is a wide variance in the impact categories selected in each study. This makes comparing LCA studies challenging when trying to assess the full environmental impact of a particular automotive energy vector.

Dai and Lastoskie [25] state that they selected impact categories to give a more holistic overview than that of previous studies in the field using the GREET model [31], which does not model the vehicle cycle in depth or report emissions for non-criteria air pollutants which can have human health and ecosystem quality impacts. Notter *et al.*, [27] also selected environmental impact categories that expand on the work of previous studies in the field, which focused only on energy demands and climate change. While other studies reviewed consider human toxicity, this study also assesses human health as a separate impact category. The authors highlight the issue of LCA not yet including the effects of nanoparticle release on human health, but suggest that any exposure during manufacture, use or end of life phase of PEM FCs is unlikely.

Girardi *et al.* [26] state that they chose environmental impact categories in order to assess almost all main relevant effects of EVs in the vehicle fleet, with the most important being air pollution and global warming, as these are the main factors that led to the introduction of EVs.

Bartolozzi *et al.* [23] consider the greatest number of environmental impact categories, while both Bicer and Dincer [24] and Weinberg and Kaltschmitt [28] are limited to just three. This may be too narrow to give a full overview of potential impact. As expected all studies include either global warming or climate change. Weinberg and Kaltschmitt [28] is the only study reviewed that does not consider any category specifically relating to human health or toxicity. No justification is given for excluding this category. The ISO standard defines human health as a category that should generally be considered [13].

5 DISCUSSION

Studies were focused on different areas due to differing goal and scope. This can make for incomparability. However we can draw some broad conclusions. While decisions can appear subjective, LCA is still a good tool to determine impact. The better papers and studies were ones which were transparent and clear, showed a robust method and system boundary, and gave clear indication of impact categories and functional units. Clearly different studies look at different impacts. Within the automotive industry currently the drivers for improvement are climate change and human health. Climate change is a global issue, human health more local and regional. The two are incomparable to some extent, as clearly we wish to optimise for both. In some circumstances, such as inner cities, one may wish to focus on human health. Traditionally LCA has not been a tool which considers spatial impact. That is to say, emissions are aggregated over time and geography. This is called dynamic spatial modelling. Its use is in its infancy but it is something that will be important to consider in the future.

6 CONCLUSION

The lack of explanation for selection of criteria within the goal and scope definition of some LCA studies can make decisions appear subjective. Without providing robust reasons for choices made it could be argued that results could become skewed in favour of particular technologies. LCA studies that use different system boundaries, functional units and impact categories cannot be compared with each other. In the studies reviewed in this paper this is outside of the goal and scope definition, they aim only to allow comparison of different fuels and energy storage within their own studies. One of the limitations of LCA is conveying its purpose to decision makers. LCA cannot be used to give a definitive answer as to which option is the best in terms of environmental impact since this depends on the interpretation of the results. Robust LCA data also needs to be balanced with operational performance considerations to ensure economic and technological feasibility of future alternative ICE fuels.

For transport to make further advances towards environmental sustainability it is important to consider the wider implications of automotive fuel use, not just those measured in a TTW tailpipe emissions test. Decisions made without this knowledge could be inclined to favour fuels that are more environmentally damaging than others that are notionally less attractive. Policy and legislation that is in place to reduce the environmental burden of the use of automobiles should be guided by the most up to date and relevant information on the true impact of fuel or energy use. LCA can provide important information for decision makers, giving a holistic insight into the entire system.

ACKNOWLEDGEMENTS

This work was supported by an Industrial Cooperative Award in Science & Technology from the Engineering and Physical Sciences Research Council and Jaguar Land Rover.

REFERENCE LIST

1. Organisation For Economic Co-Operation And Development. *World Energy Balances 2016*. 2016; Available from: https://www.iea.org/statistics/relateddatabases/worldenergystatisticsandbalances/.
2. Intergovernmental Panel on Climate Change. *Climate Change 2014: Synthesis Report. Contribution of Working Groups I, II and III to the Fifth Assessment Report of the Intergovernmental Panel on Climate Change*. 2014; Available from: https://www.ipcc.ch/report/ar5/syr/.
3. Tavoni, M., et al., *Post-2020 climate agreements in the major economies assessed in the light of global models.* Nature Climate Change, 2014. **5**(2): p. 119-126.
4. International Energy Agency. *Transport*. 2017 [cited 2017 20/03/2017]; Available from: https://www.iea.org/topics/transport/.
5. Marín-Quemada, J.M., J. García-Verdugo, and G. Escribano, *Energy Security for the EU in the 21st Century: Markets, geopolitics and corridors*. Energy Security for the EU in the 21st Century: Markets, Geopolitics and Corridors. 2012. 1-338.
6. European Parliament, *REGULATION (EC) No 443/2009 OF THE EUROPEAN PARLIAMENT AND OF THE COUNCIL of 23 April 2009. Setting emission performance standards for new passenger cars as part of the Community's integrated approach to reduce CO2 emissions from light-duty vehicles.* Official Journal of the European Union, 2009.
7. European Parliament, *REGULATION (EU) No 510/2011 OF THE EUROPEAN PARLIAMENT AND OF THE COUNCIL of 11 May 2011 setting emission performance standards for new light commercial vehicles as part of the Union's integrated approach to reduce CO 2 emissions from light-duty vehicles.* Official Journal of the European Union 2011.
8. The Society of Motor Manufacturers and Traders. *NEW CAR CO2 REPORT 2017 The 16th Edition*. 2017; Available from:

https://www.smmt.co.uk/wp-content/uploads/sites/2/DEF571-SMMT-Co2-report-2017.pdf.
9. OECD/IEA, *International comparison of light-duty vehicle fuel economy: Ten years of fuel economy benchmarking*. 2017.
10. Johnson, T. and A. Joshi, *Review of Vehicle Engine Efficiency and Emissions.* 2017. **1**.
11. Energy Saving Trust. *Electric cars and vehicles.* 2017 [cited 2017 12.05.2017]; Available from: http://www.energysavingtrust.org.uk/travel/electric-vehicles.
12. The Society of Motor Manufacturers and Traders. *SMMT MOTOR INDUSTRY FACTS 2016*. 2016; Available from: https://www.smmt.co.uk/wp-content/uploads/sites/2/SMMT-Motor-Industry-Facts-2016_v2-1.pdf.
13. ISO 14040, *BS EN ISO 14040:2006 - Environmental management. Life cycle assessment. Principles and framework*, in *Umweltmanagement. Oekobilanz - Grundsaetze und Rahmenbedingungen*. 2006.
14. SETAC, B., *Guidelines for Life-Cycle Assessment: A Code of Practice.* Consoli, Allen, Boustead, Fava, Franklin, Jense de Oude, Parrish, Perriman, Postelthwaite, Quay, Sequin, and Vigon, 1993.
15. ISO, E., *ISO 14044-2006.* Environmental management–life cycle assessment–Requirements and Guidelines, 2006.
16. McManus, M.C. and C.M. Taylor, *The changing nature of life cycle assessment.* Biomass Bioenergy, 2015. **82**: p. 13-26.
17. McManus, M.C., et al., *Challenge clusters facing LCA in environmental decision-making-what we can learn from biofuels.* Int J Life Cycle Assess, 2015. **20**: p. 1399-1414.
18. Hellweg, S. and L. Mila i Canals, *Emerging approaches, challenges and opportunities in life cycle assessment.* Science., 2014. **344**(6188): p. 1109-1113.
19. Baumann, H., *The hitch hiker's guide to LCA : an orientation in life cycle assessment methodology and application*, ed. A.-M. Tillman. 2004, Lund: Lund: Studentlitteratur.
20. Ramachandran, S. and U. Stimming, *Well to wheel analysis of low carbon alternatives for road traffic.* Energy and Environmental Science, 2015. **8**(11): p. 3313-3324.
21. Kim, H.C. and T. Wallington, *Life-Cycle Energy and Greenhouse Gas Emission Benefits of Lightweighting in Automobiles: Review and Harmonization*, in *Environ. Sci. Technol.* 2013. p. 6089-6097.
22. Sustainability, E.C.-J.R.C.-I.f.E.a., *International Reference Life Cycle Data System (ILCD) Handbook - General guide for Life Cycle Assessment - Detailed guidance.* 2010.
23. Bartolozzi, I., F. Rizzi, and M. Frey, *Comparison between hydrogen and electric vehicles by life cycle assessment: A case study in Tuscany, Italy.* Applied Energy, 2013. **101**: p. 103-111.
24. Bicer, Y. and I. Dincer, *Comparative life cycle assessment of hydrogen, methanol and electric vehicles from well to wheel.* International Journal of Hydrogen Energy, 2017. **42**(6): p. 3767-3777.
25. Dai, Q. and C.M. Lastoskie, *Life cycle assessment of natural gas-powered personal mobility options.* Energy and Fuels, 2014. **28**(9): p. 5988-5997.
26. Girardi, P., A. Gargiulo, and P.C. Brambilla, *A comparative LCA of an electric vehicle and an internal combustion engine vehicle using the appropriate power mix: the Italian case study.* The International Journal of Life Cycle Assessment, 2015. **20**(8): p. 1127-1142.
27. Notter, D.A., et al., *Life cycle assessment of PEM FC applications: Electric mobility and μ-CHP.* Energy and Environmental Science, 2015. **8**(7): p. 1969-1985.

28. Weinberg, J. and M. Kaltschmitt, *Life cycle assessment of mobility options using wood based fuels - Comparison of selected environmental effects and costs.* Bioresource Technology, 2013. **150**: p. 420-428.
29. Gupta, S., et al., *Life-cycle analysis of energy and greenhouse gas emissions of automotive fuels in India: Part 1 - Tank-to-Wheel analysis.* Energy, 2016. **96**: p. 684-698.
30. Weiguo, L., et al., *Analysis of the Global Warming Potential of Biogenic CO2 Emission in Life Cycle Assessments.* Scientific Reports, 2017. **7**.
31. Argonne National Laboratory. *The Greenhouse Gases, Regulated Emissions, and Energy Use in Transportation Model*. 2016; Available from: https://greet.es.anl.gov/.

Dynamically downsized gasoline demonstrator vehicle

M Bassett[1], J Hall[1], T Cains[1], S Reader[1], R Wall[2]
[1] MAHLE Powertrain Ltd, UK
[2] Aeristech Ltd, UK

ABSTRACT

Gasoline engine downsizing is already established as a technology for reducing vehicle CO_2 emissions. Further benefits are possible through more aggressive downsizing, however, the trade-off between the CO_2 reduction achieved and vehicle drivability limits the level of engine downsizing currently adopted by vehicle manufacturers.

This paper details an extremely downsized engine, and resulting demonstrator vehicle, featuring eSupercharging in combination with a conventional turbocharger. The original 1.2 litre, 3-cylinder, MAHLE downsizing engine has been re-configured to enable it to achieve a BMEP level of 35 bar and a specific power output in excess of 160 kW/litre. The demonstrator vehicle that also features an advanced 48 V lead-carbon battery pack and a 48 V belt-driven starter generator (BSG). The battery and BSG have been selected to enable the continuous high-output (6 kW) operation of the eSupercharger to support prolonged operation of the engine at low speed and high-torque output.

The fuel consumption of the resulting demonstrator vehicle has been analysed over a number of drive-cycles and the benefits of the downsized engine in conjunction with the complete mild-hybrid system have been assessed.

Keywords: Downsizing, eBoosting, eSupercharging, TGDI, high-specific output.

1. INTRODUCTION

The European Environment Agency (EEA) has recently released the provisional data for the CO_2 emissions of passenger cars registered in the European Union in 2014 [1]. This data shows that all manufacturers have achieved their 2015 targets ahead of time, with average emissions of 123.3 g/km in 2014, a decrease of 3 % compared to 2013 [2]. However, Tietge *et al.* [3] analysed data gathered from real world driving and compared the reported fuel consumption to the type-approval value. This analysis revealed that while the average discrepancy between type approval and on-road CO_2 emissions was around 9 % in 2001, by 2015 it had increased to around 40 %.

The vehicle CO_2 values established via type-approval testing form a basis for consumer information, CO_2 regulation, and CO_2 based vehicle taxation [4]. In order to bridge the gap between current type approval and real-world results, as well as align the various approaches used across the globe, the worldwide harmonized light vehicles test procedure (WLTP) has been developed and will replace the existing test cycle, the new European drive cycle (NEDC), in Europe. The WLTP is applicable to new vehicle types from September 2017 and all vehicles by September 2018 [5]. Additionally, concerns that vehicles are being engineered to perform well in the limited laboratory tests, but producing substantially greater emissions in real world driving conditions, have led to the adoption of real driving emissions (RDE) testing as part of the regulatory approval regime [5].

48 V mild-hybrid architectures, currently receiving increasing attention, pave the way for increased levels of engine downsizing. This is facilitated through the ability to achieve a virtually instantaneous and continuous supply of boost air for the engine, even at low engine speeds, via the addition of an electrically powered supercharger (eSupercharger). The ability to increase the level of engine downsizing, while retaining good fuel efficiency across the majority of the engine operating map, would provide manufacturers with a transferable, and scalable, means for meeting these stringent new CO_2 targets across their fleet.

This paper presents results from a 1.2 litre, 3-cylinder, engine fitted with an eSupercharger in addition to a conventional turbocharger. The results show the potential of eSupercharging as a technology enabler for extreme engine downsizing. Engine structural durability considerations and the potential for further CO_2 reductions via 48 V micro-hybridisation and system control are also discussed.

2. DOWNSIZED GASOLINE ENGINE

Gasoline engine downsizing has long been recognised as an effective technology for engine CO_2 reduction [6-8] and is firmly established as a technology for assisting in achieving fleet CO_2 reduction targets. Gasoline engine downsizing is the process whereby the engine operating load point is shifted to a higher, more efficient region, through the reduction of engine swept volume, whilst maintaining the full load performance of the original engine, via pressure charging. Further improvements in fuel economy have been shown to be possible through increased levels of engine downsizing [6-10]. However, as specific output increases so too do the technical challenges, which include abnormal combustion (pre-ignition and detonation), low speed torque, transient response and engine durability.

2.1 MAHLE Downsizing Demonstrator Engine

In order to conduct research into the requirements for advanced downsizing engines and their components, MAHLE previously developed a demonstrator engine, which was intended to represent 50 % downsizing of a naturally aspirated 2.4 litre engine [11]. The resulting engine was a direct injection, 1.2 litre, 3-cylinder inline, turbocharged, gasoline engine. Equipped with a single stage turbocharger, the resulting engine achieved a peak brake mean effective pressure (BMEP) of 30 bar. The peak power output of the engine was 120 kW (100 kW/litre) from 5000 rev/min. Stoichiometric fuelling (lambda = 1.0) could be maintained over a wide proportion of the engine operating map [11]. A key target during the development of the single-stage turbocharged engine was to ensure that the transient response remained acceptable. The specifications of the single turbocharger derivative are summarised in Table 1.

Table 1: Specifications of the MAHLE Downsizing 3-cylinder (Di3) demonstrator engine.

Parameter	Value
Engine displacement	1.2 litres
No. of cylinders	3 in-line
Bore/stroke	83.0 / 73.9 mm
Compression ratio	9.3:1
Fuel injection	Multi-hole central DI
Turbocharger	Bosch MAHLE Turbo Systems (BMTS)

The engine was installed into a D-segment demonstrator vehicle, which featured stop-start capability. Testing confirmed that the vehicle achieves a fuel consumption figure of 5.8 l/100 km (CO_2 emissions of 135 g/km) over the NEDC. The MAHLE downsizing engine achieves very high BMEP levels and the BMEP curve achieved is shown in Figure 1.

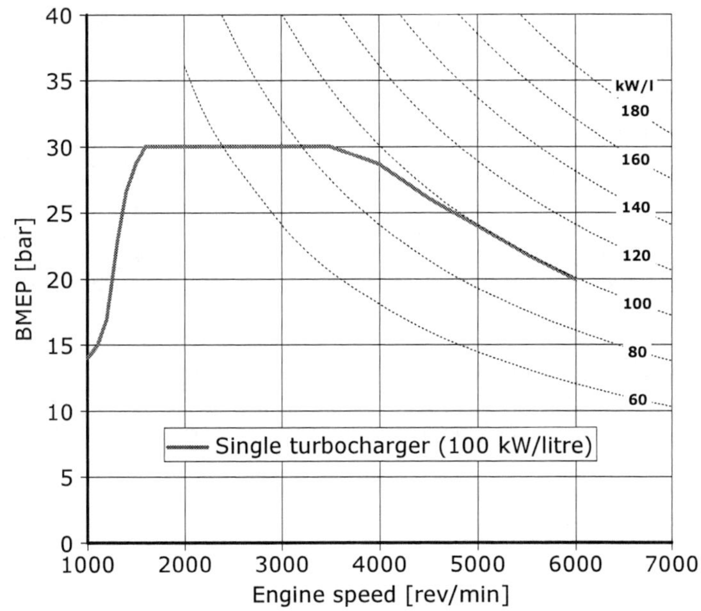

Figure 1: MAHLE Di3 engine performance.

2.2 Increased Engine Downsizing

MAHLE Powertrain has also investigated the possibility of further increasing the level of engine downsizing. With a revised boosting system and charge cooling layout, the low-speed torque characteristics of the engine have been enhanced further. The high specific power output has been achieved through the use of a carefully selected larger turbocharger which, unassisted, has a detrimental effect upon the low-speed torque and transient response of the engine. An electrically powered supercharger (eSupercharger) has been utilised to enable the transient response and low speed torque to be at least recovered.

The eSupercharger, manufactured by Aeristech was found to be well suited to this application, having a compressor which can provide the desired pressure ratio and mass-flow rate, and a motor capable of continuous operation. The specifications of the Aeristech eSupercharger are summarised in Table 2.

Aeristech have developed a compact electric machine, for variable high-speed applications, the power electronics of which are configured for operation at 48 V, in-line with current automotive trends. The unit is capable of running continuously at high boost pressures and can also accelerate from an idle speed to its maximum speed of 120,000 rev/min in under 0.5 seconds [12], allowing it to address the low-speed turbo lag issues associated with heavily downsized engines.

Table 2: Aeristech eSupercharger specification summary.

Parameter	Value
Compressor type	Radial
Motor	High-speed Permanent Magnet
Max operating speed	120,000 rev/min
Max Power	9.0 kW
Continuous Power	5.6 kW
Cooling	Liquid
Input voltage	48 V nominal
Unit mass	4.2 kg

In this application the eSupercharger has been located upstream of the main turbocharger compressor, as shown schematically in Figure 2. In order to provide high-pressure air at low engine speeds, the eSupercharger is optimised for low flow rates. At higher engine air mass-flow rates there is sufficient exhaust energy for the larger main turbocharger to provide high-pressure air. Thus the eSupercharger is not capable of passing the full air flow required by the engine at higher power outputs and must be bypassed. Additionally, there must be sufficient overlap between the eSupercharger operation map and the main turbocharger map to enable a seamless handover between the two devices. This avoids a torque interruption, which would give an undesirable driving feel in the vehicle.

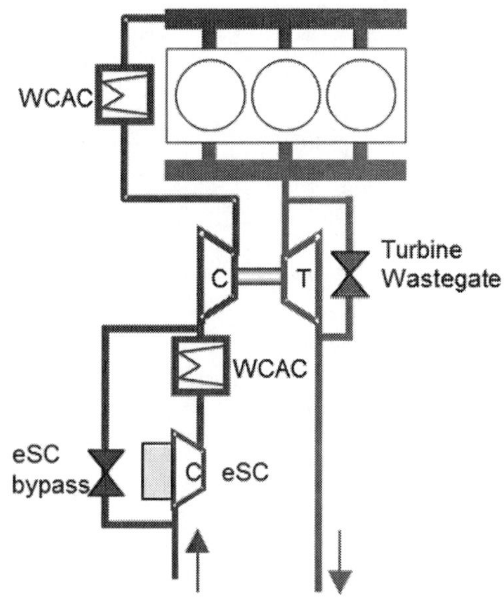

Figure 2: Schematic of the eSupercharged engine arrangement used in this study.

In this application it was also necessary to implement a small inter-stage cooler between the eSupercharger and the main turbocharger compressor, as shown in Figure 2. This was found to be required to maintain operation below the turbocharger compressor maximum outlet temperature limit. This was found to be of most benefit in the high-load operating region close to the hand-over point between the eSupercharger and the turbo-charger, where both the eSupercharger and main compressor are both operating at a high pressure ratio. This inter-stage cooler is located in the inlet system such that it is also by-passed at higher engine operating speeds.

A completely new air intake system was designed, which included an eSupercharger bypass valve that remains closed when the eSupercharger is in operation and provides an alternative low-loss path, once the turbocharger alone is capable of maintaining the desired torque. New components were also designed for the mounting of the eSupercharger, the integration of the new turbocharger, the secondary charge cooler, and a high power BSG, as shown in Figure 3.

Figure 3: MAHLE engine with eSupercharger and revised intake system.

3. ENGINE TEST RESULTS

The eSupercharged version of the engine has been configured with the same 9.3:1 compression ratio (CR) as the baseline 120 kW, single turbocharger, version of the engine. The ability to maintain the same CR as the base engine ensures that the part load fuel consumption map will be unchanged, thus maintaining the drive-cycle fuel consumption benefits already achieved with this engine in the MAHLE downsizing demonstrator vehicle [11].

Figure 4 shows the BMEP and engine power measured during the engine testing compared to the baseline version of the MAHLE downsizing engine. All testing was performed with a fixed charge cooler outlet temperature of 40 °C. From Figure 4 the effects of the revised boosting system can be clearly seen. The peak performance achieved by the engine is summarised in Table 3. These results clearly demonstrate the significant increase in peak power offered by the new turbocharger (from 120 kW to 193 kW).

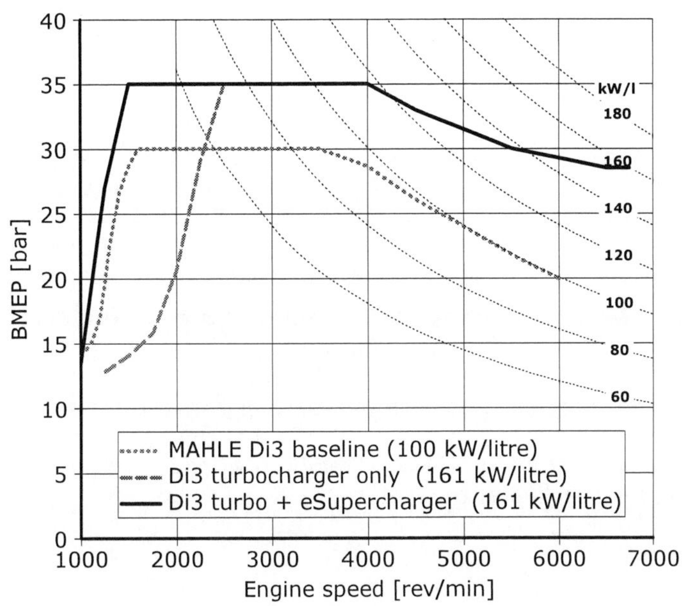

Figure 4: Measured BMEP for the MAHLE downsizing engine with combined eSupercharger and turbocharger.

The penalty that this turbocharger imposes in terms of low speed BMEP output, when operated without the eSupercharger, is also clearly visible in Figure 4. Without the eSupercharger, the new turbocharger has little charging effect until engine speed is above 1500 rev/min, and peak torque is not achieved until 2500 rev/min. If this configuration were to be used in a vehicle, a negative effect on the driveability could be expected, particularly in heavier vehicle applications.

Table 3: eSupercharged Downsizing engine Performance summary.

Parameter	Value
Maximum power	193 kW
Specific power output	161 kW/litre
Maximum BMEP	35 bar (1500 to 1400 rev/min)

When the eSupercharger is activated, the combined system not only restores the performance below 3000 rev/min, it exceeds the previous baseline torque levels at all engine speeds. Of particular note is the increase at 1500 rev/min, where the output of the engine is 35 bar. Furthermore, the BMEP level of the engine below 1500 rev/min has been limited to protect the main and big end bearings and the vehicle driveline.

Clearly, 35 bar BMEP is the gross maximum level achieved by the engine, and if the 48 V battery is depleted, then some torque will be consumed driving the BSG to power the eSupercharger. The eSupercharger electrical power requirement is shown in Figure 5 for steady-state engine operation. During transient tip-in events the eSupercharger is used more extensively to enhance the transient response of the engine. Ideally the 48 V battery will be charged by recuperation, during vehicle deceleration events.

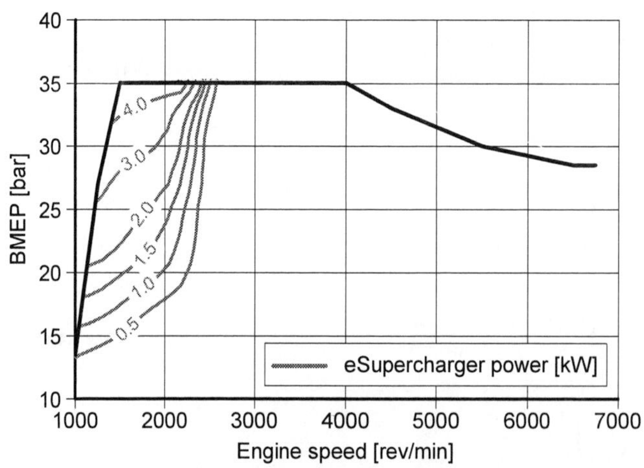

Figure 5: Measured steady-state eSupercharger electrical power requirement.

Figure 6 shows the manifold absolute pressure (MAP) required by the engine to achieve engine torque. Maximum power of 193 kW requires a plenum pressure of 276 kPa. The peak steady-state MAP is required at the "knee point" (1500 min^{-1}, 35 bar BMEP) and is 301 kPa at this point.

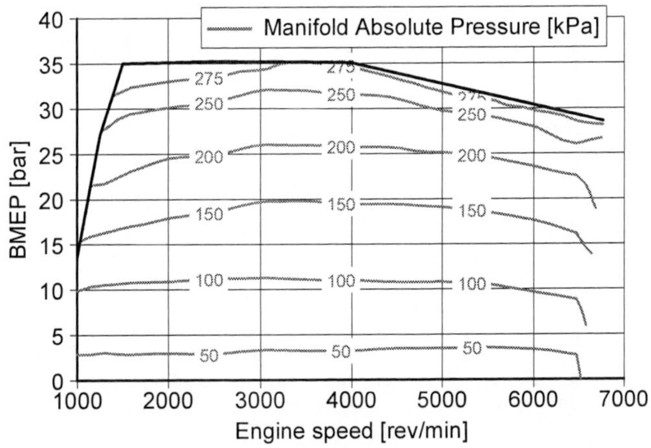

Figure 6: Measured steady-state MAP requirement.

At high specific power levels the engine operates with excess fuel (lambda of less than 1.0) in order to protect the turbine from excessive exhaust gas temperatures. The engine has been calibrated to a turbine inlet temperature limit of 930 °C. The resulting lambda (air/fuel ratio relative to stoichiometric) is shown in Figure 7. From comparison with Figure 4, it can be seen that the lambda 1 boundary sits at a power output of roughly 90 kW/litre and the lambda reduces with increasing power output, up to a minimum value of 0.75 at maximum engine power.

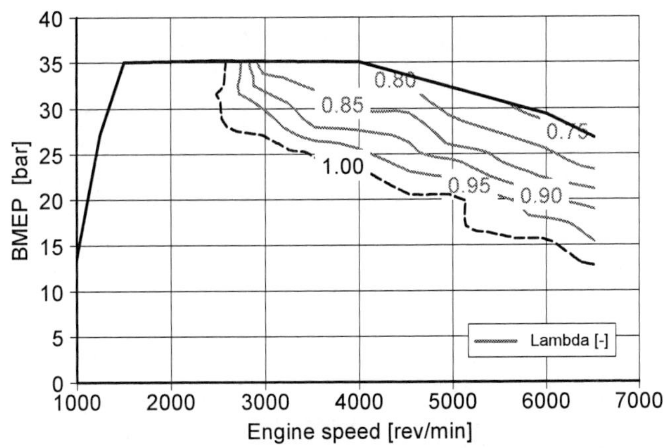

Figure 7: Measured steady-state Lambda requirement.

Typically for gasoline engines peak engine efficiency is achieved with a 50 % mass-fraction burned (MFB) timing of around 8 to 10° crank angle degrees after top dead centre (ATDC) firing. At higher specific loads it is often necessary to retard ignition timing from this optimum, and hence also retard the 50 % MFB timing, to avoid detonation (knock). The 50% MFB timing of the MAHLE downsizing engine is shown in Figure 8. It can be seen that above just of 15 bar BMEP it is necessary to retard the 50% MFB timing to avoid detonation. The amount the 50% MFB is retarded increases almost linearly with increasing BMEP at engine speeds above 3000 rev/min. Below 3000 rev/min the engine is more sensitive to detonation and the 50% MFB is further retarded. The most retarded 50% MFB timing occurs at the knee point, where the 50% MFB timing occurs at 32° ATDC.

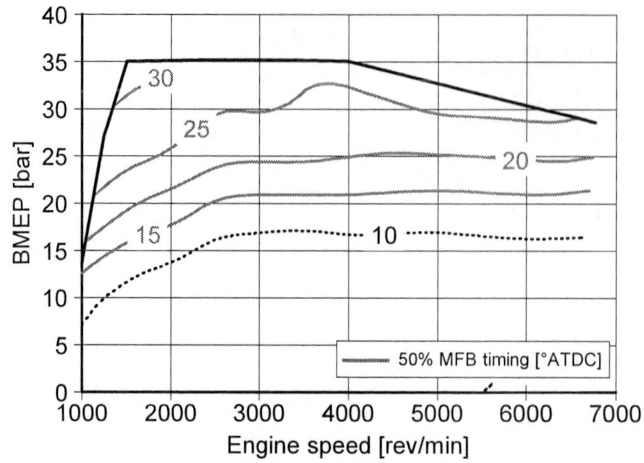

Figure 8: Measured steady-state 50% mass-fraction burned timing.

Figure 9 shows the comparative transient response of the engine to a step increase in torque, at 1500 rev/min, for operation both with and without the eSupercharger, compared to the original 120 kW baseline engine. The transient response of the combined eSupercharged and turbocharged mode has a similar rate of torque rise to that of a naturally aspirated engine, and it can be seen from Figure 9 that the slope of the torque response curve continues at a similar rate beyond the knee point to the pre-knee point portion of the curve.

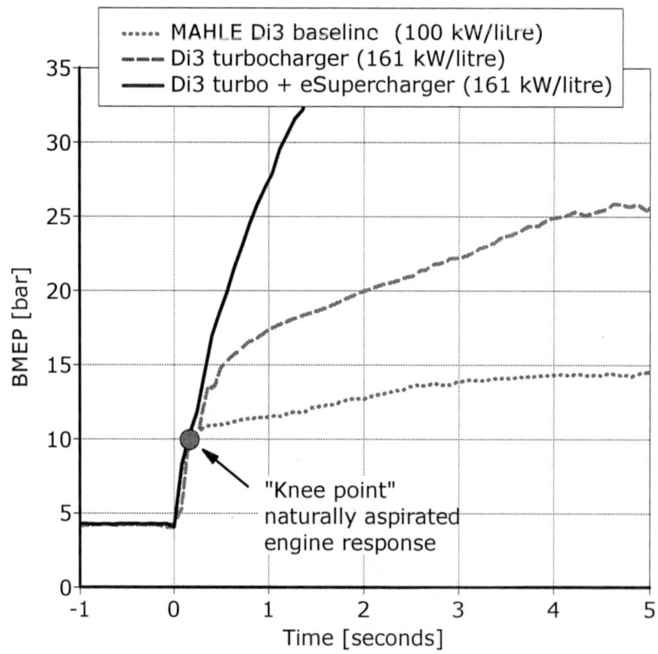

Figure 9: Transient response for a 1500 rev/min load step.

4. 48 V MILD HYBRIDISATION

The degree to which it's possible to mitigate the electrical power consumption of the electric supercharging system is dependent upon the proportion of engine operating time that the eSupercharger is required, how much energy recovery is available during driving, and the storage capacity of the 48 V battery pack. The

48 V platform comprises of an advanced lead acid battery pack, a DC/DC converter to maintain the state of charge of the 12 V battery which supports the existing 12 V systems, the eSupercharger and the speed-start, high power BSG, from Controlled Power Technologies [13].

Analysis of the power requirement at the wheels to propel a typical C-segment vehicle over the world-harmonised light-duty test procedure (WLTP) cycle reveals that the total propulsion energy required, if no energy is recuperated, is 12.5 MJ and the total amount of energy available for recuperation is 2.8 MJ (22.5 % of the total energy required for propulsion). Thus, if all the energy available for recuperation could be captured and deployed without any losses, there is the potential to reduce vehicle fuel consumption by 22.5 % purely through this mechanism. The peak power available for recuperation is 28 kW; if the recuperative capability of the electrical system were less than this then clearly the benefit of hybridisation would be diminished.

Figure 10 presents the results of a simplified analysis into how the CO_2 reduction available due to recuperation is diminished by reduced recuperation power capability for the C-segment vehicle on the WLTP cycle. The "Ideal" line in Figure 10 represents the CO_2 reduction theoretically achievable if the energy available for recuperation could be captured and deployed without any losses.

Also shown in Figure 10 is the likely power limits for both a 12 V and 48 V based electric machine, based on a 350 Amp current limit. Clearly when the eMachine power capability is zero, this idealised CO_2 reduction potential due to recuperation is also zero. Also, the theoretical maximum CO_2 reduction peaks at 22.5 %, and this is achieved when the eMachine power capability reaches 28 kW (the maximum regenerative power requirement of the NEDC), beyond which there is no further CO_2 reduction to be achieved for this drive cycle. It is clear from the trend of the "Ideal" curve that the rate of CO_2 reduction diminishes with increasing recuperation power capability and that there is a 21 % reduction potential at the likely 48 V system power limit of 15 kW and 10 % CO_2 reduction potential with a 12 V system limited to just under 4 kW of recuperative power.

The motor efficiency losses shown in Figure 10 represents the effect of a fixed 85 % electric machine efficiency for the round trip conversion of recuperated power to propulsive power. It can be seen that this simply reduces the CO_2 reduction potential by ~28 % for all recuperative powers relative to the Ideal line. The transmission losses then account for the effect of transmission efficiency on the potential reduction in drive-cycle CO_2, where it can be seen that a further ~10 % reduction occurs.

The engine over-run losses shown in Figure 10 represents an estimation of the drive cycle CO_2 reduction potential of an engine mounted BSG system architecture. In this case it is additionally assumed that the engine must be driven during overrun situations, further reducing the amount energy available for recuperation. In this case the maximum CO_2 reduction potential is diminished to a maximum of 9 %, and is roughly 6 % for a 48 V system with a maximum recuperative power of 10 kW.

Clearly the benefits discussed are theoretical, and further CO_2 reduction is potentially possible through engine load shifting and further downsizing potential through the torque assist capability of these devices.

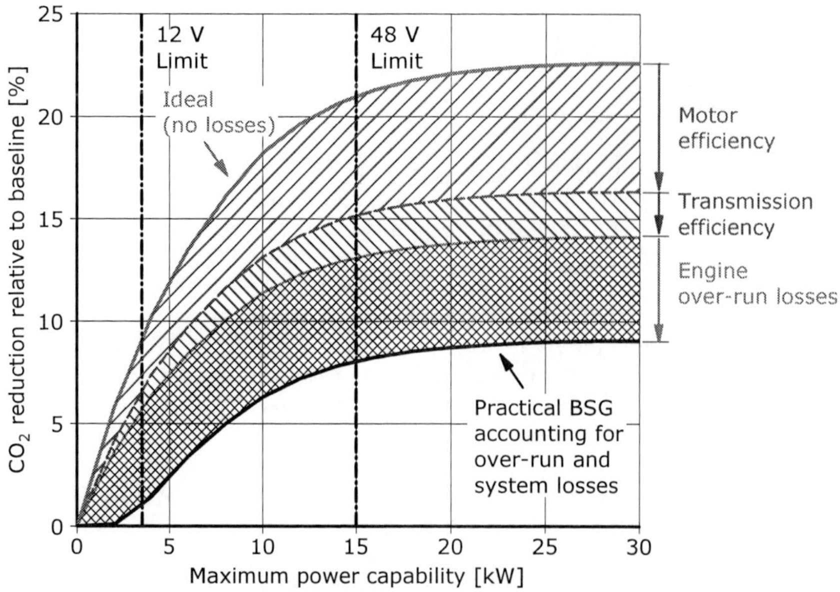

Figure 10: Influence of recuperation power capability and electric machine installation location on CO_2 reduction potential for a C-segment vehicle over the WLTP.

5. DEMONSTRATOR VEHICLE

The 193 kW (262 PS) eSupercharged version of the MAHLE 1.2 litre downsized engine has been installed into a VW Golf GTi, shown in Figure 11, to show case the performance, drivability and fuel economy of the engine. This vehicle was selected because the eSupercharged MAHLE 1.2 litre downsized engine achieves very similar torque and power levels to those of the baseline 2.0 litre turbocharged gasoline direct injection (TGDI) engine fitted as standard to the VW Golf GTi.

Figure 11: eSupercharged heavily downsized demonstrator vehicle.

A comparison of the BSFC improvement offered by the eSupercharged 1.2 litre MAHLE compared to a 2.0 litre TGDI engine, which has a very similar performance specification the baseline 2.0 TGDI engine [14], at an engine speed of 2000 rev/min, is shown in Figure 12. It can be seen from Figure 12 that below 150 Nm the 1.2 litre engine offers significant fuel consumption improvements which are predominantly derived from reduced pumping work due to the engine downsizing.

At higher torque levels the downsized engine exhibits higher fuel consumption than the comparator 2.0 litre TGDI engine, most markedly above 250 Nm (2600 kPa BMEP for the 1.2 litre engine), which is caused by compromised combustion phasing due to the very high specific torque levels, as seen in Figure 8. At the same torque output, the downsized engine operates at a higher specific load than the larger baseline engine. This gives rise to the pumping work benefit at low torque levels, but at higher engine torque outputs the downsized engine will be likely to need to operate with a more retarded ignition timing and hence combustion phasing, giving rise to the poorer fuel efficiency at higher torque levels as observed in Figure 12.

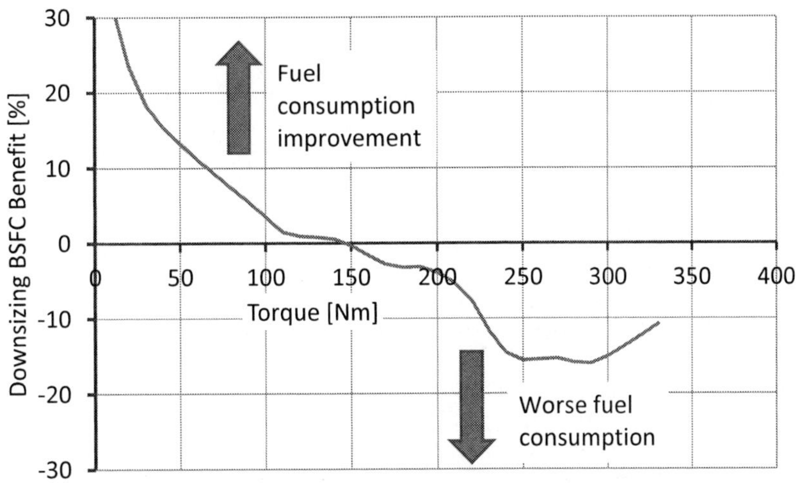

Figure 12: Downsizing fuel consumption benefit at 2000 rev/min, in comparison to the baseline 2.0 litre TGDI engine [14].

Additional technologies could be added to the downsized engine to help further mitigate against the requirement to retard ignition timing at high loads, such as Miller cycle operation [15], exhaust gas recirculation (EGR) [16], water injection [17], or even variable compression ratio [18]. Extreme engine downsizing, combined with such technologies offers a cost effective route to even further CO_2 reduction potential.

The benefit of the downsizing engine, in comparison to the baseline 2.0 litre engine, on the drive-cycle CO_2 (or fuel consumption) performance for the VW Golf vehicle (without any additional mild-hybridisation) has been analysed over a wide range of drive-cycles. The results of this analysis are shown in Figure 13, where it can be seen that a 15 % fuel consumption reduction is achieved on the NEDC and a 12% fuel consumption reduction is achieved on the WLTP.

The fuel consumption benefits calculated from this analysis have been plotted against the cycle average engine power requirement divided by the average speed of the cycle, which represents the energy intensity of the cycle. The higher this value, the higher the energy intensity of the drive-cycle, with drive cycles which have a higher speed or higher acceleration rates having a higher value. Interestingly it can be seen that there is a clear relationship between the fuel consumption benefit achieved through the downsizing and the drive cycle energy intensity, with the lower speed, or least aggressive cycles showing the most benefit from downsizing. This is due to the fact that these cycles have lower engine loading, leading to lower torque operation, where the downsizing engine gives higher fuel consumption benefits. It can also be postulated that if the downsizing engine were fitted to a very heavy vehicle, or predominantly driven in a very aggressive manner, the benefits of downsizing would be significantly reduced.

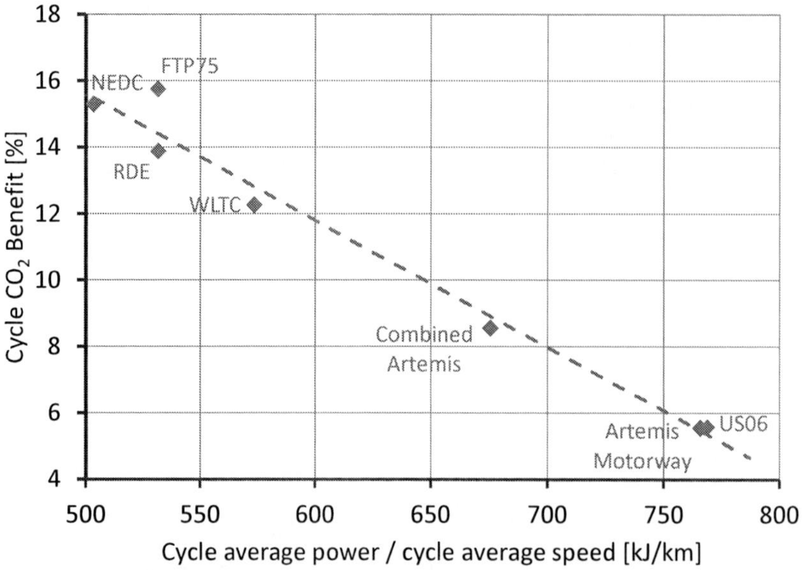

Figure 13: Drive-cycle CO_2 benefit from downsizing from 2.0 litre TGDI engine to 1.2 litre engine in the VW Golf (without mild-hybridisation).

Figure 14 shows the results of analysis to determine the anticipated WLTP CO_2 reduction for the eSupercharged 1.2 litre VW Golf demonstrator vehicle fitted with the 48 V mild-hybrid system, compared to the standard 2.0 litre engine vehicle. The full 6 % benefit from a 48 V hybrid system with a 10 kW power limit has been considered to be available, as the eSupercharger is seldom used during the WLTP cycle. The occasions where it is employed are short, transient, events where the total energy dissipated by the eSupercharger accounts for only a fraction of the recuperated energy available.

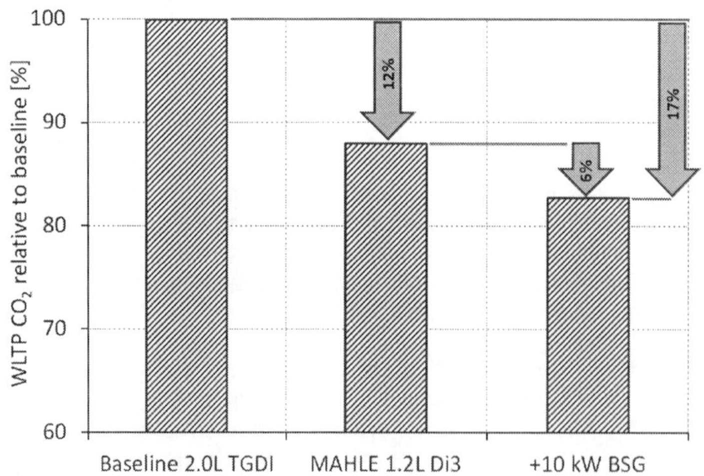

Figure 14: WLTP CO_2 reduction benefit from downsizing and 48 V hybridisation.

Vehicle modelling, using measured fuel consumption data, indicates that swapping the original 2.0 litre engine for the eSupercharged version of the MAHLE 1.2 litre downsized engine will yield a CO_2 reduction of over 12 % on this cycle, as shown in Figure 14. The addition of the 10 kW BSG, based on the analysis presented in Figure 10, will yield a further CO_2 reduction of up to 6 %. Thus, with the combined effect of these measures produces an estimated WLTP benefit of around 17 %, as shown in Figure 14.

6. NEXT STEPS

Recently, with the introduction of real driving emissions (RDE) testing the engine operating region that is under scrutiny during compliance testing has significantly increased. Furthermore, manufacturers may be required to disclose their engine base emissions strategy (BES) and any auxiliary emissions strategy (AES), which may preclude the use of fuel enrichment for component protection in future engines.

At specific power levels much above output levels of 90 kW/litre the engine eSupercharged engine developed during this study operates with excess fuel in order to protect the turbine from excessive exhaust gas temperatures, as shown in Figure 7. A previous study using the MAHLE Di3 engine, investigating both high and low pressure EGR systems, has demonstrated that the introduction of EGR can completely eliminate the requirement for over-fuelling at high engine loads [16].

An analytical study into the boosting system requirements to maintain lambda 1 fuelling across the entire engine operating map is currently underway. Exhaust gas recirculation (EGR) will be utilised, via a mixed-pressure EGR circuit (pre-turbine supply to mid-pressure on the intake), as depicted in Figure 15. Previous testing of using EGR on the MAHLE Di3 engine, at lower specific output levels, indicates that a peak of around 20 % EGR flow will be required to enable the operation of the engine to be maintained at lambda 1 fuelling across the engine operating map [16].

It is likely that the boosting system will need to be revised in order to maintain specific power output levels with high levels of EGR. It is hoped that by locating the eSupercharger downstream of the main compressor will enable a larger turbocharger to be accommodated, whilst marinating good transient response. It is also hoped that the eSupercharger will assist in mitigating any transient response issues associated with gasoline EGR systems.

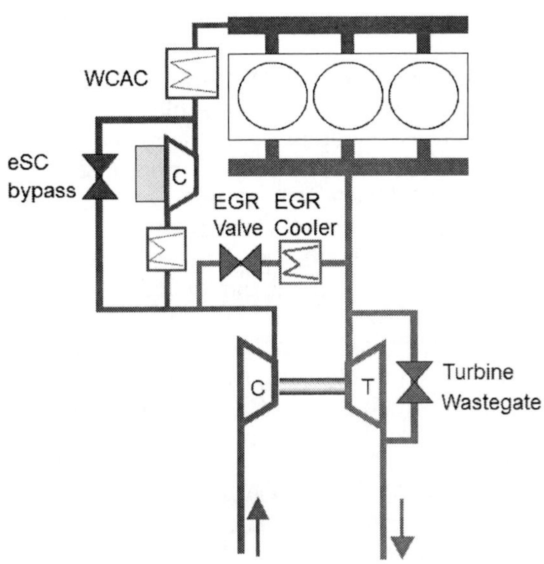

Figure 15: Schematic of the eSupercharged engine arrangement and EGR system under investigation.

7. CONCLUSIONS

MAHLE Powertrain has integrated a 48 V version of the Aeristech eSupercharger into their downsized engine. A maximum power level of 193 kW has been achieved during dynamometer testing, resulting in a specific power output of 161 kW/litre. The concept of combining an eSupercharger and high-flow turbocharger has proved successful and the Aeristech eSupercharger has also been demonstrated to enable continuous, high pressure, operation. The high specific power output and BMEP levels achieved enable fuel economy improvements through high levels of engine downsizing.

In order to enable continuous electrical power to be provided to the eSupercharger, even when the 48 V battery is depleted, the demonstrator vehicle developed by MAHLE Powertrain is also equipped with a 10 kW BSG. The combination of the heavily downsized gasoline engine, together with 48 V hybridisation showcased in this demonstrator vehicle is expected to yield a combined CO_2 reduction of 17 % on the WLTP, of which 12 % is directly attributable to engine downsizing.

An analytical study into the boosting system requirements to maintain lambda 1 fuelling across the entire engine operating map, via the use of EGR, is currently underway.

8. REFERENCES

[1] "Monitoring of CO_2 emissions from passenger cars: Summary data for 2014", European Environmental Agency http://www.eea.europa.eu/data-and-maps/data/co2-cars-emission-8 (accessed 5th May 2017).

[2] Zacharof, N., Tietge, U. and Mock, P., "CO_2 emissions from new passenger cars in the EU: Car manufacturers' performance in 2014", International Council on Clean Transportation, Berlin 2015. http://www.theicct.org/sites/default/files/publications/ICCTbriefing_EU-CO_2_201507.pdf (accessed 5th May 2017).

[3] Tietge, U., Diaz, S., Mock, P., German, J., Bandivadekar, A. and Ligterink, N., "From laboratory to road: A 2016 update of official and 'real-world' fuel consumption and CO_2 values for passenger cars in Europe", International Council on Clean Transportation (ICCT), Berlin 2016. http://www.theicct.org/sites/default/files/publications/ICCT_LaboratoryToRoad_2016.pdf (accessed 8th May 2017).

[4] Report of the Working Party on Pollution and Energy (GRPE) on its 65th session, ECE/TRANS/WP.29/GRPE/65, Geneva 2013. http://www.unece.org/fileadmin/DAM/trans/doc/2013/wp29grpe/ECE-TRANS-WP29-GRPE-65e.pdf (accessed 9th May 2017).

[5] Hayes, J., Department for Transport, Explanatory Memorandum on European Union Legislation, 23rd January 2017: http://europeanmemoranda.cabinetoffice.gov.uk/files/2017/01/170118_-_Real_Driving_Emissions_(1).pdf (accessed 24th May 2017).

[6] Fraser, N., Blaxill, H., Lumsden, G., and Bassett, M., "Challenges for Increased Efficiency through Gasoline Engine Downsizing," SAE Int. J. Engines 2(1):991-1008, 2009.

[7] Bandel, W., Fraidl, G., Kapus, P., Sikinger, H. et al., "The Turbocharged GDI Engine: Boosted Synergies for High Fuel Economy Plus Ultra-low Emission," SAE Paper 2006-01-1266, 2006.

[8] Atkins, M. and Koch, C., "A Well-to-Wheel Comparison of Several Powertrain Technologies," SAE Paper 2003-01-0081, 2003.

[9] Hancock, D., Fraser, N., Jeremy, M., Sykes, R. et al., "A New 3 Cylinder 1.2l Advanced Downsizing Technology Demonstrator Engine," SAE Paper 2008-01-0611, 2008.
[10] Lumsden, G., OudeNijeweme, D., Fraser, N., and Blaxill, H., "Development of a Turbocharged Direct Injection Downsizing Demonstrator Engine," SAE Int. J. Engines 2(1):1420-1432, 2009.
[11] Korte, V., Rueckauf J., Harms K, Miersch J., Brandt M. Muenz S. and Rauscher M., "MAHLE-Bosch Demonstrator Vehicle for Advanced Downsizing", 19th Aachen Colloquium Automobile and Engine Technology, 2010.
[12] Aeristech's 48 V 10 kW eSupercharger datasheet: http://www.aeristech.co.uk/wp-content/uploads/A-Sample-2-English.pdf (accessed 5th May 2017).
[13] Hubert, A. and Bloore, P., "Benefits of a Switched-Reluctance E-Motor for Mild Hybrid 48 V Applications", 24th Aachen Colloquium Automobile and Engine Technology, 2015.
[14] Baker, D. "NAWA Engine Benchmarking Report", MAHLE Powertrain internal report No. NAWA0102252RD. 2014.
[15] Gottschalk, W., Lezius, U., and Mathusall, L., "Investigations on the Potential of a Variable Miller Cycle for SI Knock Control," SAE Paper 2013-01-1122, 2013.
[16] Mahr, B., Taylor, J. and Bassett, M., "Kraftstoffverbrauchsvorteile im realen Fahrbetrieb durch Abgasrückführung bei modernen Downsizing-Motoren", 7. MTZ-Fachtagung, Der Antrieb von morgen, 2012.
[17] Thewes, M. and Lehrheuer, B., "Water Injection – High Power and High Efficiency Combined", 6th Aachen Colloquium China Automobile and Engine Technology, Beijing 2016.
[18] Fraidl, G., Kapus, P., Melde, H., Loesch, S., Schoeffmann, W., Sorger, H., Weissback, J. and Wolkerstorfer, J., "Variable Compression Ratio - in a Technology Competition?", 37th International Vienna Motor Symposium, Vienna 2016.

AUTHOR INDEX